EVOLUTION AND CREATIONISM

EVOLUTION AND CREATIONISM

A Documentary and Reference Guide

Christian C. Young and Mark A. Largent

Greenwood Press

Westport, Connecticut • London

Library of Congress Cataloging-in-Publication Data

Young, Christian C.
Evolution and creationism: a documentary and reference guide/Christian C. Young and Mark A.
Largent.
    p. cm.
    Includes bibliographical references and index.
    ISBN 978–0–313–33953–0 (alk. paper)
    1. Evolution (Biology) 2. Creationism. I. Largent, Mark A. II. Title.
QH366.2.Y68 2007
576.8–dc22   2007010682

British Library Cataloguing in Publication Data is available.

Library of Congress Catalog Card Number: 2007010682
ISBN-13: 978–0–313–33953–0
ISBN-10: 0–313–33953–8

First published in 2007

Greenwood Press, 88 Post Road West, Westport, CT 06881
An imprint of Greenwood Publishing Group, Inc.
www.greenwood.com

Printed in the United States of America

The paper used in this book complies with the
Permanent Paper Standard issued by the National
Information Standards Organization (Z39.48–1984).

10 9 8 7 6 5 4 3 2 1

Copyright Acknowledgments

The authors and the publisher gratefully acknowledge permission for use of the following material:

Excerpts from *Christianity and Evolution*, copyright © 1969 by Editions du Seuil, English translation by
Rene Hague © 1971 by William Collins Sons & Company Limited and Harcourt, Inc. Reprinted by
permission of Harcourt, Inc.

Excerpts from *Life: An Introduction to Biology*, copyright © 1965 by Harcourt, Inc. and renewed in
1993 by William S. Beck, Elizabeth Simpson Wurr, Helen S. Vishniac, and Joan S. Burns, reprinted by
permission of Harcourt, Inc.

Every reasonable effort has been made to trace the owners of copyright materials in this book, but in
some instances this has proven impossible. The authors and publisher will be glad to receive information
leading to more complete acknowledgments in subsequent printings of the book and in the meantime
extend their apologies for any omissions.

To our teacher, John Beatty

CONTENTS

Contents

PREFACE

The American evolution/creation debate raged throughout the twentieth century and shows no sign of letting up in the twenty-first. American attitudes toward evolution typically show sharp divisions between, on the one hand, those who hold that a belief in the existence of God makes it impossible to accept evolution and, on the other hand, those who believe that the fact of evolution undermines religion. Caught somewhere in the middle is the largest segment of the American public, those who somehow accept both evolution and God. Legal battles over the teaching of evolutionary and creationist accounts of biological origins highlight the seemingly unbridgeable gap between beliefs. Such a split suggests that even if evolution receives the unwavering support of scientists, people of a creationist persuasion remain unwilling to adjust their spiritual beliefs far enough to embrace a scientific explanation for human origins. Scientists and their advocates make increasingly confident statements about the reality of evolutionary change in nature. They provide the potential application of modern evolutionary theory to both the natural sciences and the social sciences. At the same time, opponents of both the teaching of evolution in public schools and the claims made by evolution's enthusiasts have found a number of ways to press their case. These battles have a shorter history than the longer debates between evolution and creation.

The lines for today's battles over evolution and creationism were drawn nearly a century ago. Sharp criticism about evolution emerged shortly after the turn of the twentieth century. Encouraged by a deep-seated concern about the state of Western culture, especially after the devastation of World War I, critics developed the antievolution position and used it to attack the very notion of evolution as a natural phenomenon. Antievolution was often couched in terms of religion and took on a newly adopted literal interpretation of *Genesis*. Prior to that time, contrasts between evolution and religion were very rarely founded in biblical literalism. This emerging fundamentalist view of creation placed evolution in more direct conflict with religion than had been the case in the nineteenth century. By 1920, religious antievolutionism had developed into a widespread movement, primarily populated by American Protestants, which characterized evolutionary theory as a corrosive ideology that had wrecked devastation on the cultures that had adopted it. American antievolutionists linked Darwin's thought to everything from social Darwinism to eugenics, imperialism to robber barons, and Friedrich Nietzsche to Margaret Sanger.

While antievolutionism as we know it today is a twentieth-century phenomena, the potential social and theological implications of a Darwinian worldview was evident to Darwin as he wrote *On the Origin of Species*. Several months after he released the first edition of the book, Darwin wrote a letter that exhibited his conception of the relation of the theory of evolution by natural selection to religious and social concerns. The recipient of that letter was Darwin's friend and colleague, Asa Gray. Gray was a botanist at Harvard University, an enthusiastic supporter of Darwin's theory of evolution by natural selection, and an equally ardent supporter of natural theology, the belief that the study of nature reveals the goodness and omnipotence of God. For Gray, evolution and natural selection synthesized nicely. He helped introduce Darwin's work to the United States, and he emphasized how evolution might be a tool by which God worked. He even published an article on the subject, "Natural Selection Not Inconsistent with Natural Theology."

In the letter, dated May 22, 1860, Darwin thanked Gray for collecting royalties for him and discussed both positive and negative reviews of the book. Darwin then launched into a discussion on the "theological view of the question." He asserted at the start that he was "bewildered" by the subject and that he found it "painful." Most significantly, he pointedly stated, "I had no intention to write atheistically." That being said, Darwin found himself caught between two equally unacceptable alternatives. On the one hand, Darwin could not agree with Gray and other natural theologians that a careful study of nature offered "evidence of design & beneficence on all sides of us." Darwin's researches and his personal experiences, most notably the death of his young daughter Annie from a painful stomach ailment, had convinced him that the world contained much more than merely beauty and magnificence. "There seems to me," he wrote to Gray, "too much misery in the world." He could not accept that a "beneficent & omnipotent God" would have created certain types of organisms in the world, types that exhibited behaviors that could only have been intended to create misery and pain. Among them, he described the Ichneumonidae, an insect that laid its eggs in the living bodies of caterpillars so that their larvae could eat their way out. In that vein, Darwin also rejected the idea that cats' compulsions to play with mice before they finally killed and ate them could be part of God's design. Given his unwillingness to lay blame for his daughter's death or the apparent sadistic nature of cats at the feet of God, he was likewise unwilling to credit God for all the good things in the world.

With one alternative dismissed, Darwin explained that he was equally unwilling to accept the atheistic alternative: "On the other hand," he wrote, "I cannot anyhow be contented to view this wonderful universe & especially the nature of man & to conclude that everything is the result of brute force." How, if Darwin rejected both the natural theologians' claim that evolution represented God's continued influence in the world and the atheists' assertion that everything is merely the result of brute force, did he perceive the "theological view of the question"? Darwin explained to Gray that he was "inclined to look at everything as resulting from designed law, with the details, whether good or bad, left to the working out of what we may all chance." This interpretation allowed him to find a specific role for God, that of the original designer of the laws of nature, without making God directly responsible for misery and pain. This compromise was not particularly satisfying to Darwin, however. He confessed, "I feel most deeply that the whole subject is too profound for the human intellect. A dog might as well speculate on the mind of Newton." He concluded the letter by admitting, "The more I think the more bewildered I become; as indeed I have probably shown by this letter."

Then, as now, there were no easy answers to the questions raised by modern evolutionary theory and the criticisms of it offered by antievolutionists. Polls of Americans' views on this subject demonstrate the difficulties that we all confront in the evolution/creation controversy. When asked in recent decades, somewhere between 40 and 45 percent of Americans believe that God created humans as they currently exist. That number dwarfs the total percentage of people who have adopted an atheistic interpretation that evolution occurred without any influence of God. This group of materialistic evolutionists rarely approaches even 15 percent of the American population. Nearly 40 percent of Americans, however, adopt the position that evolution is, in fact, a natural phenomenon, and believe that it is guided by God. Some in this group adopt a position much like Asa Gray's, which we call theistic evolution, while others accept Charles Darwin's view, which is termed deistic evolution. Taken together, the materialist evolutionists and the theistic or deistic evolutionists typically make up over 50 percent of the American public.

It is our hope that the primary source materials found in this volume will help those of you who, like Darwin, struggle with the theological view of the question of evolution. Books that offer strong polemics in support of one side or the other are quite easy to find, and more are published every year. Those of you seeking support for your views, whether they are creationistic, materialistic, deistic, or theistic, will find ample assistance in the many books already available. This one is not a contribution to either side of the debate, but rather it seeks to present, in as objective a manner as possible, the claims made by authors on the subject of evolution and creationism from the last two centuries. Readers will find that the documents are organized into chronological chapters, each of which begins with an overview of the major events and issues addressed by the authors. They will also find short introductions to each of the readings that contextualize the authors' claims and situations. We encourage you to approach these materials with an open mind and an appreciation for each author's concerns and contexts.

Christian C. Young and Mark A. Largent

INTRODUCTION

The ongoing debate over the place of evolution in public schools, which has raged in fits and starts for over eighty years, illuminates a larger struggle between worldviews. It seems so urgent, in part, because the challenges facing humanity from our neighborhoods to the global commonwealth arise from beliefs that are central to both religion and to secular society. Parents, community leaders, politicians, and educators feel extraordinary pressure to ensure that the path of future generations follows a set of beliefs that maintains the dignity and primacy of the human species. The stories that we believe and that we tell our children about the origins of humanity itself serve as a microcosm of the broader discussion, so evolution and creation provide a useful dichotomy. Remarkably, the origin of this dichotomy itself is relatively recent and well documented, so we have the opportunity to explore it in some detail.

Prior to 1800, no one considered the idea that a group of related organisms could change over the course of generations into a fundamentally different sort of organism, an idea we today call evolution. Basic tenets of Western philosophy from Aristotle to the French revolution were inherently focused on stability and everlasting organization of forms. As a result the concept of evolution was so alien to Western thought that it simply never arose. It was philosophy that shaped premodern and early modern views about the nature of species and the origin of the great diversity of life on earth, not ideology. Thus, the absence of evolutionary thought was in no way motivated by the religious or political ideologies that have provoked some twentieth- and twenty-first-century antievolutionists.

Evolution was philosophically untenable prior to the nineteenth century for several reasons, all of which had ancient origins and had been reified by philosophers throughout the middle ages as well as the medieval and early modern periods. First, the emphasis by natural philosophers on empiricism required the direct observation of phenomena or at least the direct observation of the effects of those phenomena. The nature of evolution, with its slow change over very long periods of time, made it unobservable by those standards in the course of the average civilization's lifespan, much less the average human lifespan. While today we have examples of evolution occurring quickly enough to be measured within a relatively short span of time, these examples were unavailable to our predecessors. Second, the dominance of essentialism posited an eternal, unchanging essence that defined all material objects. The very nature of this view required the fixity of species and undermined any notion that one species could evolve into a fundamentally different one. Finally, considerations of the nature of time prior to the modern period were dominated by examples of cyclical time, such as one observes in the passage of days, seasons, and years as well as the action of tides and even of reproduction from one generation to another. Time, it seemed, always returned us to where we started, be it morning, spring, low tide, or infancy. Evolution required that revolutionary thinkers begin to consider the notion of time in a linear fashion. In order to establish the foundations for an evolutionary worldview, philosophers needed to eschew essentialism, alter their notions of empiricism, and adopt a linear view of time. These changes began sporadically and did not meaningfully coalesce until the beginning of the nineteenth century.

Even once the philosophical tenets for an evolutionary worldview were established, it took a full generation before Westerners could begin to imagine that species were mutable and that the earth was incredibly old. Moreover, by the time these realizations emerged, other factors inhibited an evolutionary worldview. For example, naturalists had only recently accepted extinction of species was possible, but they first believed extinction demonstrated the fact that species did not change. In fact, species failed to react to changing environments. Naturalists did not immediately assume the now commonly accepted explanation that some species went extinct, while others emerged in response to environmental change.

Although it is important to point out that it was not religion, but a rather broad philosophical constraints that inhibited the emergence of an evolutionary worldview, it is equally accurate to say that religious influence was quickly felt by early evolutionary thinkers. Charles Darwin himself seriously considered the theological implications of his work as he was writing it; in a May 22, 1860, letter to Asa Gray, Darwin admitted that he had "no intention to write atheistically." But at the same time, he wrote, "I cannot see as plainly as others do, and as I should wish to do, evidence of design and beneficence on all sides of us. There seems to me too much misery in the world." He concluded, "I feel most deeply that the whole subject is too profound for the human intellect. A dog might as well speculate on the mind of Newton. Let each man hope and believe what he can."

We see the influence of religion in many nineteenth-century naturalists' acceptance of the concept of evolution. Gray himself considered evolution by natural selection to be evidence that God was so powerful that He could create natural processes that could carry out his wishes, rather than dirtying his hands creating each species of plant and animal individually. Others, like Pierre Teilhard de Chardin, wrote a century later about how evolution and religion could be synthesized into a powerful, hopeful worldview.

Certainly, religion played some role in the immediate reception of Darwin's *On the Origin of Species*, but in no way did it play as dominant a role in public considerations of evolution that it does today. When did religion, in particular fundamentalist Christianity, come to figure so prominently in the public perception of evolution? It did not happen in Darwin's lifetime, nor in the decades immediately after his death. It was not until shortly after the turn of the twentieth century that fundamentalism itself emerged. The incredible devastation of World War I combined with a growing animosity among many American Protestants toward historical or literary criticism of the Bible. Together, these currents produced a popular uprising against modernism, which was the deliberate departure from tradition and the open critique of subjects previously considered above reproach. The return to the fundamentals of the Christian faith that underlay fundamentalism viewed evolution as part of the attack on traditional values. Several years into the twenty-first century, political and religious animosity toward evolutionary thought is still alive and well.

Throughout the decades when political and religious leaders began to attack evolutionary thought and thinkers, changes took place within evolutionary biology itself. Darwin's theory of evolution by natural selection was indeed a watershed event in the history of evolutionary theory, but it certainly was not the last word on the subject. To the contrary, Darwin's greatest accomplishments were the conversation he started, the new disciplines he initiated, and the structure he provided to scientists interested in thinking about the tremendous diversity of life. The generation that followed Darwin aggressively debated not only his work, but dozens of competing theories that arose between 1880 and 1940. During these years Darwin's work faced its harshest critics from both within the science community and from outside it, and his theory of evolution by natural selection emerged in the mid-twentieth century as a central principle in modern evolutionary theory. The modern evolutionary synthesis, which combined Darwin's work with modern genetics, generated a comprehensive, coherent explanation of evolutionary change that today forms the central organizing principle for the biological sciences. As Theodosius Dobzhansky explained in the title of a 1973 article, "Nothing in Biology Makes Sense Except in the Light of Evolution."

Even as biologists expanded and refined their understanding of the explanatory power of evolutionary thinking, theologians, philosophers, religious leaders, and scientists in other fields struggled with the implications of this new approach to natural explanations. Some embraced evolution in one realm, but kept it compartmentalized, refusing to allow it to interfere with their beliefs and philosophies in another realm. Others attempted to adopt evolution as a means for understanding the way their own beliefs had come into being and changed through generations. Still others looked to science more generally as a means of exploring and understanding their place in the universe, and in that way took evolution and religion alike to be derivative processes of the larger endeavor. On a personal level, it may be that everyone who engaged these questions arrived at a different solution, because these remain the questions that seem most important to understanding one's place in the universe.

What remains to be considered, beyond the individual reactions and directions, are the ways that communities come to share their understanding of evolution and creation, and to demand that others adopt common views. Scientific communities, with rules about evidence and appeals to reason, become downright intolerant when they find certain pieces of evidence and their particular tools of reasons ignored or reinterpreted according to different rules. Religious communities cohere around shared views of salvation, and disruption of those views by accounts that undermine the common symbols and significance of the religion cannot be endured for very long. Finally, as communities, these groups command a political presence within the broader social structure. Seeking a balance in that structure requires enormous effort.

In the United States, the search for balance means constant challenges for science and religion. In most public opinion polls, about a third of respondents profess a belief in a literal interpretation of the Bible, and well over half agree that humans are the product of God's creation in the last 10,000 years, either directly or through some process such as evolution. Such numbers require political leaders to be wary of statements that align them with naturalistic views or evolutionary origins. As a result, various laws and regulations that limit the teaching of evolution are likely to pass in this country, especially in local areas where the political base of fundamentalism is even stronger. And while the judicial system may continue to strike down laws that blur the separation of church and state, the courts will not adjudicate the truthfulness of science or religion.

Introduction

Philosopher of science Larry Lauden hoped, at the close of the 1981 *McLean v. Arkansas* case, that scientists would use the evidence of nature to unravel the arguments offered by creation scientists. He suggested that legal maneuvers only provided creationists with more leverage to argue their beliefs against scientific beliefs. In the decades since, his suggestion seems to have served creationists as sound advice, as they have distanced their arguments further from biblical creationism and attached themselves more firmly to questions of science that have not been answered by scientific reasoning or physical evidence. Intelligent design, as the movement is now called, argues for the insertion of supernatural explanations wherever natural explanations fall short. Given the open-ended process of science and the constant appearance of new and unanswered questions, intelligent design appears to have a strong philosophical foothold in the United States.

So at the end of two centuries of investigation, after the publication and wide acceptance of an enormous volume of biological, geological, chemical, and physical evidence, not to mention contributions from the fields of psychology, anthropology, and astronomy, scientists have managed to shift the beliefs of only a small portion of the American public. Philosophically and religiously, many remain unmoved by the scientific evidence, even where philosophers, theologians, and religious leaders advise them to acknowledge the implications of science for the physical universe and to seek for answers to supernatural questions beyond the natural realm.

The situation in the United States is somewhat unique. In most of the European and Asian cultures that share similar technological advantages and scientific enlightenment with the United States, questioning the reality of evolution strikes an overwhelming majority of people as odd. How, they wonder, can Americans embrace their modern lifestyle in every other respect, and yet cling to the hope of divine creation. The answer to this question lies not in the history of science, but in the broader cultural history that recalls the experience of Protestant denominations, isolationism and eventual leadership in global conflicts, adherence to a state without an official religion, and a two-party political system that plays up individual issues in every election cycle.

Americans' experiences and convictions are woven through the documents in this guide and are described in the accompanying introductions. As the debate continues, each chapter may require us to rethink the connections between the documents and the emerging contexts and claims. Certainly more chapters would need to be added; nonetheless, the editors' hopes are pinned on the possibility that each person engaged in the debate or considering his or her place within it will find here a wealth of perspectives. It is our sincere wish that readers use this volume to learn how others have articulated beliefs about the history of humanity and how those beliefs may help us serve one another with generosity and respect.

Christian C. Young and Mark A. Largent

1

EVOLUTION BEFORE DARWIN'S *ON THE ORIGIN OF SPECIES*

Most Americans today assume that in the years before the publication of Darwin's *On the Origin of Species* mostly everyone rejected the notion of evolution for the very same reasons that some people oppose it now. Encouraged by today's arguments in which naturalistic evolutionary explanations are considered in opposition to literal biblical accounts of creation, many twenty-first-century citizens believe that people living two hundred years ago shared their notions about evolution. That is, most people today presume that people who lived before 1859 were antievolutionists because, like most vocal antievolutionists now, they then took for granted a literal, six-day creation. In fact, this is an incorrect assumption; there was very little religious motivation in the pre-Darwinian rejection of an evolutionary worldview.

In the years before Darwin published his theory of evolution by natural selection, the notion of evolution was rejected by most natural scientists, including those scientists who had no religious motivations. Even some of the most atheistic natural scientists rejected the concept of linear change through time because it defied common sense and ran counter to a number of other increasingly popular explanations of natural phenomena. For example, as natural scientists came to believe that some species of animals did in fact go extinct, they simultaneously grew increasingly confident that species did not evolve. Extinction, they believed, was caused by changes in the environment. Since they assumed that animals were intentionally created for specific environments, as the environment changed, natural scientists believed that extinct animals had been simply incapable of adapting to new conditions. If it had been possible

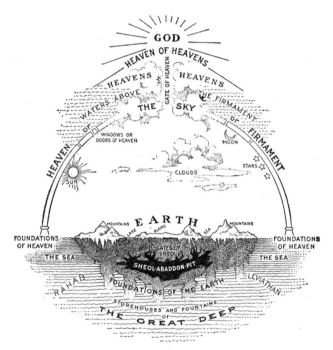

THE ANCIENT HEBREW CONCEPTION OF THE UNIVERSE
TO ILLUSTRATE THE ACCOUNT OF CREATION AND THE FLOOD

Accounts of Earth history including the biblical flood generally continued to rely on an ancient Hebrew conception of the universe. In the early twentieth century, it was assumed the ancient Hebrews had assumed the Earth was flat and could be flooded as illustrated here. Frontispiece from G.L. Robinson, Leaders of Israel. New York: Association Press. [Author's collection.]

for those animals to evolve and become better adapted to their new environments, natural scientists concluded, they surely would have and in doing so would have avoided extinction. Increasing acceptance of the fact of extinction, therefore, served as evidence against evolution.

Early nineteenth-century, antievolutionary worldviews were also motivated by conceptions about appropriate scientific methodology and even the very definition of science. As the notion of science emerged as a separate activity from philosophy generally, practitioners of science more explicitly defined the nature of their work. By the 1830s, just as the word *scientist* was coined by William Whewell, practitioners of natural science emphasized the role of empiricism and the direct observation of natural phenomenon. At the same time, they increasingly differentiated the activity of science from the work of philosophers, which were often described as speculative in comparison with the certain facts of science. Science, in the hands of most early nineteenth-century scientists, involved the classification of confirmable facts about nature. Evolution was too philosophical, too abstract, and too speculative to be included among the emerging sciences. In addition, it was not directly observable within the context of one's lifetime.

Despite the obstacles, beginning shortly after the turn of the nineteenth century, some natural scientists offered explanations about how some species of animals might transmutate into fundamentally different species. Among the first to allude to the groups of organisms linearly changing from one species to an essentially different one was Erasmus Darwin, Charles Darwin's grandfather. Shortly thereafter, the French natural scientist, Lamarck, offered an evolutionary explanation. Both works met with little interest among working naturalists. Instead, cutting-edge early nineteenth-century biological research focused on the classification of living things and the increasingly popular system of comparative anatomy, which was founded on Cuvier's work. Both activities presupposed the immutability of species, thus supporting common sense notions that organisms did not evolve. What little evolutionary thought existed in the generations before Charles Darwin's 1859 *On the Origin of Species* was generally rejected by working natural scientists. The fixity of the species was well established by early nineteenth-century scientists without allusion to literal biblical accounts of creation.

ERASMUS DARWIN, ZOONOMIA (1794–1796)

Erasmus Darwin (1731–1802) was one of the leading English intellectuals of the eighteenth century and the grandfather of the more famous Charles Darwin. He was trained as a physician and published several works on medicine and on botany as well as being a poet. His writings on the subject of galvanism inspired Mary Shelley to write *Frankenstein*, and his poetry was widely admired by significant nineteenth-century poets, including Wordsworth. *Zoonomia* was by far his most important scientific work. Published throughout 1794, 1795, and 1796, the book offered an evolutionary narrative about the present existence of the wide range of living organisms. It differed from his grandson's work by telling a story in which "certain habits of action" might influence organisms and be transmitted from one generation to the next. It did not offer an explanation of the mechanism for the inheritance of these modified traits.

Section XXXIX
Of Generation

The ingenious Dr. Hartley in his work on man, and some other philosophers, have been of the opinion, that our immortal part acquires during this life certain habits of action or of sentiment, which become for ever indissoluble, continuing after death in a future state of existence; and add, that if these habits are of the malevolent kind, they must render the possessor miserable even in heaven. I would apply this ingenious idea to the generation or production of the embryo, or new animal, which partakes so much of the form and propensities of the parent.

Owing to the imperfection of language the offspring is termed and *new* animals, but is in truth a branch or elongation of the parent; since a part of the embryo-animal is, or was, a part of the parent; and therefore in strict language it cannot be said to be entirely *new* at the time of its production; and therefore it may retain some of the habits of the parent-system.

At the earliest period of its existence the embryo, as secreted from the blood of the male, would seem to consist of a living filament with certain capabilities of irritation, sensation, volition, and association; and also with some acquired habits or propensities peculiar to the parent: the former of these are in common with other animals; the latter seem to distinguish or produce the kind of animal, whether man or quadruped, with the similarity of feature or form to the parent. It is difficult to be conceived, that a living entity can be separated or produced from the blood by the action of a gland; and which shall afterwards become an animal similar to that in whose vessels it is formed; even though we should suppose with some modern theorists, that the blood is alive; yet every other hypothesis concerning generation rests on principles still more difficult to our comprehension.

At the time of procreation this speck of entity is received into an appropriated nidus, in which it must acquire two circumstances necessary to its life and growth; one of these is food or sustenance, which is to be received by the absorbent mouths of its vessels; and the other is that part of atmospherical air, or of water, which by the new chemistry is termed oxygen, and which affects the blood by passing through the coats of the vessels which contain it. The fluid surrounding the embryo in its new habitation, which is called liquor amnii, supplies it with nourishment; and as some air cannot but be introduced into the uterus along with the new embryo, it would seem that this same fluid would for a short time, suppose for a few hours, supply likewise a sufficient quantity of the oxygen for its immediate existence.

On this account the vegetable impregnation of aquatic plants is preformed in the air; and it is probable that the honey-cup or nectary of vegetables requires to be open to the air, that the anthers and stigmas of the flower may have food of a more oxygenated kind than the common vegetable sap-juice.

On the introduction of this primordium of entity into the uterus of the irritation of the liquor amnii, which surrounds it, excites the absorbent mouths of the new vessels into action; they drink up a part of it, and a pleasurable sensation accompanies this new action; at the same time the chemical affinity of the oxygen acts through the vessels of the rubescent blood; and a previous want, or disagreeable sensations, is relieved by this process.

As the want of this oxygenation of the blood is perpetual (as appears from the incessant necessity of breathing by lungs or gills), the vessels become extended by the efforts of pain or desire to seek this necessary object of oxygenation, and to remove the disagreeable sensation, which want occasions. At the same time new particles of matter are absorbed, or applied to these extended vessels, and they become permanently elongated, as the fluid in contact with them soon looses the oxygenous part, which it at first possessed, which was owing to the introduction of air along with the embryo. These new blood-vessels approach the sides of the uterus, and penetrate with their fine terminations into the vessels of the mother; or adhere to them, acquiring oxygen through their coats from the passing currents of the arterial blood of the mother.

This attachment of the placental vessels to the internal side of the uterus by their own proper efforts appears further illustrated by the many instances of extra-uterine fetuses, which have thus attached or inserted their vessels into the peritoneum; or on the viscera, exactly in the same manner as they naturally insert or attach them to the uterus.

The absorbent vessels of the embryo continue to drink up nourishment from the fluid in which they swim, or liquor amnii; and which at first needs no previous digestive preparation; but which, when the whole apparatus of digestion becomes complete, is swallowed by the mouth into the stomach, and being mixed with saliva, gastric juice, bile, pancreatic juice, and mucus of the intestines, becomes digested, and leaves a excrement, which produces the first feces of the infant, called meconium.

The liquor amnii is secreted into the uterus, as the fetus requires it, and may probably be produced by the irritation of the fetus as an extraneous body; since a similar fluid is acquired from the peritoneum in cases of extra-uterine gestation. The young caterpillars of the gadfly paced in the skins of cows, and the young of the ichneumon fly placed in the back of the caterpillars on cabbages, seem to produce their nourishment by their irritating the sides of their nidus. A vegetable secretion and concretion is thus produced on oak-leaves by the gall insect, and by the cynips in the bedeguar of the rose; and by the young grasshopper on many plants, by which the animal surrounds itself with froth. But in no circumstance is extra-uterine gestation so exactly resembled as by the eggs of a fly, which are deposited in the frontal sinus of sheep and calves. These eggs float in some ounces of fluid collected in a thin pellicle or hydatide. This bag of fluid compresses the optic nerve on one side, by which the vision being less distinct in that eye, the animal turns in perpetual circles towards the side affected, in order to get a more accurate view of objects; for the same reason as in squinting the affected eye is turned away from the object contemplated. Sheep in the warm months keep their noses close to the group to prevent this fly from readily getting into their nostrils.

The liquor amnii is secreted into the womb as it is required, not only in respect to quantity, but, as the digestive powers of the fetus become formed, this fluid becomes of a different consistency and quality, till it is exchanged for milk after nativity. (Haller, Physiol. V. I.). In the egg of the white part, which is analogous to the liquor amnii of quadrupeds, consists of two distinct parts; one of which is more viseid, and probably more difficult of digestion, and more nutritive than the other; and this latter is used in the last week of incubation. The yolk of the egg is a still stronger or more nutritive fluid, which is drawn up into the bowels of the chick just at its exclusion from the shell, and serves it for nourishment for a day or two, till it is able to digest, and has learnt to chose the harder feeds or grains, which are to afford it sustenance. Nothing analogous to this yolk is found in the fetus of lactiferous animals, as the milk is another nutritive fluid ready prepared for the young progeny.

The yolk therefore is not necessary to the spawn of fish, the eggs of insects, or the feeds of vegetables; as their embryos have probably their food presented to them as soon as they are excluded from their shells, or have extended their roots. Whence it happens that some insects produce a living progeny in the spring and summer, and eggs in the autumn; and some vegetables having living roots or buds produced in the place of feeds, as the polygonum vivparum, and magical onions. . . .

There seems however to be a reservoir of nutriment prepared for some seeds besides their cotyledons or feed-leaves, which may be supposed in some measure analogous to the yolk of the egg. Such are the saccarine juices of apples, grapes and other fruits, which supply nutrition to the feeds after they fall on the ground. And such is the milky juice in the centre of a coconut, and part of the kernel of it; the same I suppose of all other monocotyledon feeds, as of the palms, grasses, and lilies.

The process of generation is still involved in impenetrable obscurity, conjectures may nevertheless be formed concerning some of its circumstances. First, the egges of fish and frogs are impregnated, after they leave the body of the female; because they are deposited in a fluid, and are not therefore covered with a hard shell. It is however remarkable, that neither frogs nor fish will part with the spawn without the presence of the male; on which account female carp and gold-fish in small ponds, where there are no males, frequently die from the distention of their growing spawn. 2. The eggs of fowls, which are laid without being impregnated, are seen to contain only the yolk and white, which are evidently the food or sustenance for the future chick. 3. As the cicatricula of these egges is given by the cock, and is evidently the rudiment of the new animal; we may conclude, that the embryo is produced by the male, and the proper food and nidus by the female. For if the female be supposed to form an equal part of the embryo, why should the form of the whole of the apparatus for nutriment and for oxygenation? The male in many animals is larger, stronger, and digests more food than the female, and therefore should contribute as much or more towards the reproduction of the species; but if he contributes only half the embryo, and none of the apparatus for sustenance and oxygenation, the division is unequal; the strength of the male, and his consumption of food are too great for the effect, compared with that of the female, which is contrary to the usual course of nature.

In objection to this theory of generation it may be said, if the animalcula in femine, as seen by the microscope, be all of them rudiments of homunculi, when but one of them can find a nidus, what a waste nature has made of her productions? I do not assert that these moving particles, visible by the microscope, are homunciones; perhaps no creatures at all; but if they are supposed to be rudiments of homunculi, or embryos, such a profusion of them corresponds with the general efforts of nature to provide for the continuance of her species of animals. Every individual tree produces innumerable feeds, and every individual fish innumerable spawn, in such inconceivable abundances as would in a short space of time crowd the earth and ocean with inhabitants; and these are much more perfect animals than the animalcula in femine can be supposed to be, and perish in uncounted millions. This argument only shows, that the productions of nature are governed by general laws; and that by a wise superfluity of provision she has ensured their continuance.

That the embryo is secreted or produced by the male, and not by the conjunction of fluids from both male and female, appears from the analogy of vegetable feeds. In the large flowers, as the tulip, there is no similarity of apparatus between the anthers and the stimga: the feed is produced according to the observations of Spallanzani long before the flowers open, and in consequence long before it can be impregnated, like the egg in the pullet. And after the prolific dust is shed on the stigma, the feed becomes coagulated in one point first, like the cicatricula of the impregnated egg . . . . Now in these simple products of nature, if the female contributed to produce the embryo equally with the male, there would probably have been some visible similarity of parts for this purpose, besides those necessary for the nidus and sustenance of the new progeny. Besides in many flowers the males are more numerous than the females, or than the separate uterine cells in their germs, which would show, that the office of the male was at least as important as that of the female; whereas if the female, besides producing the egg or feed, was to produce an equal part of the embryo, the office of reproduction would be unequally divided between them.

SOURCE: Darwin, Erasmus. *Zoonomia.* London, 1794–1796.

JEAN-BAPTISTE LAMARCK, ZOOLOGICAL PHILOSOPHY (1809)

Jean-Baptiste Lamarck (1744–1829) was the youngest of eleven children, and he followed his family's tradition of joining the military when he was seventeen. After seven years of service an injury forced his retirement, and he turned to the study of medicine and botany. In 1778 he published a book on botany for which he was rewarded with the low-paying position of assistant botanist at the royal botanical garden. There he made his reputation for his study of invertebrate zoology and paleontology, publishing a series of books on the subjects. One of these, *Zoological Philosophy*, offered his theory of evolution, which came to be known as the theory of acquired characteristics. Lamarck's theory of evolution never became popular during his lifetime, and he never commanded the respect accorded to colleagues like Buffon or Cuvier. He spent most of his life in poverty and his last years completely blind. When he died, his children had to sell his scientific collections and library to pay his debts. Decades later, after Darwin published *On the Origin of Species*, some natural scientists reexamined Lamarck's work and accorded him a place among the first evolutionists. By the end of the nineteenth century, some considered Lamarck's theory of acquired characteristics an alternative to Darwin's theory of evolution by natural selection.

. . . [T]he infinitely diversified but slowly changing environment in which the animals of each race have successively been placed, has involved each of them in new needs and corresponding alterations in their habits. This is a truth which, once recognized, cannot be disputed. Now we shall easily discern how the new needs may have been satisfied, and the new habits acquired, if we pay attention to the two following laws of nature, which are always verified by observation.

First Law

In every animal which has not passed the limit of its development, a more frequent and continuous use of any organ gradually strengthens, develops and enlarges that organ, and gives it a power proportional to the length of time it has been so used; while the permanent disuse of any organ imperceptibly weakens and deteriorates it, and progressively diminishes its functional capacity, until it finally disappears.

Second Law

All the acquisitions or losses wrought by nature on individuals, through the influence of the environment in which their race has long been placed, and hence through the influence of the predominant use or permanent disuse of any organ; all these are preserved by reproduction to the new individuals which arise, provided that the acquired modifications are common to both sexes, or at least to the individuals which produce the young.

Here we have two permanent truths, which can only be doubted by those who have never observed or followed the operations of nature, or by those who have allowed themselves to be drawn into the error which I shall now proceed to combat.

Naturalists have remarked that the structure of animals is always in perfect adaptation to their functions, and have inferred that the shape and condition of their parts have determined the use of them. Now this is a mistake for it may be easily proved by observation that it is on the contrary the needs and uses of the parts which have caused the development of these same parts, which have even given birth to them when they did not exist, and which consequently have given rise to the condition that we find in each animal.

If this were not so, nature would have had to create as many different kinds of structure in animals, as there are different kinds of environment in which they have to live; and neither structure nor environment would ever have varied.

This is indeed far from the true order of things. If things were really so, we should not have race-horses shaped like those in England; we should not have big draught-horses so heavy and so different from the former, for none such are produced in nature; in the same way we should not have basset-hounds with crooked legs, nor grey-hounds so fleet of foot, nor water-spaniels, etc.; we should not have fowls without tails, fantail pigeons, etc.; finally, we should be able to cultivate wild plants as long as we liked in the rich and fertile soil of our gardens, without the fear of seeing them change under long cultivation.

A feeling of the truth in this respect has long existed; since the following maxim has passed into a proverb and is known by all, *Habits form a second nature.*

Assuredly if the habits and nature of each animal could never vary, the proverb would have been false and would not have come into existence, nor been preserved in the event of any one suggesting it.

If we seriously reflect upon all that I have just set forth, it will be seen that I was entirely justified when in my work entitled *Recherches sur lee corps vivants* (p. 50), I established the following proposition:

> "It is not the organs, that is to say, the nature and shape of the parts of an animal's body, that have given rise to its special habits and faculties; but it is, on the contrary, its habits, mode of life and environment that have in course of time controlled the shape of its body, the number and state of its organs and, lastly, the faculties which it possesses."

If this proposition is carefully weighed and compared with all the observations that nature and circumstances are incessantly throwing in our way, we shall see that its importance and accuracy are substantiated in the highest degree.

Time and a favorable environment are as I have already said nature's two chief methods of bringing all her productions into existence: for her, time has no limits and can be drawn upon to any extent.

As to the various factors which she has required and still constantly uses for introducing variations in everything that she produces, they may be described as practically inexhaustible.

The principal factors consist in the influence of climate, of the varying temperatures of the atmosphere and the whole environment, of the variety of localities and their situation, of habits, the commonest movements, the most frequent activities, and, lastly, of the means of self-preservation, the mode of life and the methods of defense and multiplication.

Now as a result of these various influences, the faculties become extended and strengthened by use, and diversified by new habits that are long kept up. The conformation, consistency and, in short, the character and state of the parts, as well as of the organs, are imperceptibly affected by these influences and are preserved and propagated by reproduction.

These truths, which are merely effects of the two natural laws stated above, receive in every instance striking confirmation from facts; for the facts afford a clear indication of nature's procedure in the diversity of her productions.

But instead of being contented with generalities which might be considered hypothetical, let us investigate the facts directly, and consider the effects in animals of the use or disuse of their organs on these same organs, in accordance with the habits that each race has been forced to contract.

Now I am going to prove that the permanent disuse of any organ first decreases its functional capacity, and then gradually reduces the organ and causes it to disappear or even become extinct, if this disuse lasts for a very long period throughout successive generations of animals of the same race.

I shall then show that the habit of using any organ, on the contrary, in any animal which has not reached the limit of the decline of its functions, not only perfects and increases the functions of that organ, but causes it in addition to take on a size and development which imperceptibly alter it; so that in course of time it becomes very different from the same organ in some other animal which uses it far less.

The permanent disuse of an organ, arising from a change of habits, causes a gradual shrinkage and ultimately the disappearance and even extinction of that organ.

Since such a proposition could only be accepted on proof, and not on mere authority, let us endeavor to make it clear by citing the chief known facts which substantiate it.

The vertebrates, whose plan of organization is almost the same throughout, though with much variety in their parts, have their jaws armed with teeth; some of them, however, whose environment has induced the habit of swallowing the objects they feed on without any preliminary mastication, are so affected that their teeth do not develop. The teeth then remain hidden in the bony framework of the jaws, without being able to appear outside; or indeed they actually become extinct down to their last rudiments.

In the right-whale, which was supposed to be completely destitute of teeth, M. Geoffroy has nevertheless discovered teeth concealed in the jaws of the foetus of this animal. The professor has moreover discovered in birds the groove in which the teeth should be placed, though they are no longer to be found there.

Even in the class of mammals, comprising the most perfect animals, where the vertebrate plan of organization is carried to its highest completion, not only is the right-whale devoid of teeth, but the ant-eater (Myrmecophaga) is also found to be in the same condition, since it has acquired a habit of carrying out no mastication, and has long preserved this habit in its race.

Eyes in the head are characteristic of a great number of different animals, and essentially constitute a part of the plan of organization of the vertebrates.

Yet the mole, whose habits require a very small use of sight, has only minute and hardly visible eyes, because it uses that organ so little.

Olivier's Spalax (*Voyage en Egypte et en Perse*), which lives underground like the mole, and is apparently exposed to daylight even less than the mole, has altogether lost the use of sight: so that it shows nothing more than vestiges of this "organ." Even these vestiges are entirely hidden under the skin and other parts, which cover them up and do not leave the slightest access to light.

The Proteus, an aquatic reptile allied to the salamanders, and living in deep dark caves under the water, has, like the Spalax, only vestiges of the organ of sight, vestiges which are covered up and hidden in the same way.

The following consideration is decisive on the question which I am now discussing: Light does not penetrate everywhere; consequently animals which habitually live in places where it does not penetrate, have no opportunity of exercising their organ of sight, if nature has endowed them with one. Now animals belonging to a plan of organization of which eyes were a necessary part, must have originally had them. Since, however, there are found among them some which have lost the use of this organ and which show nothing more than hidden and covered up vestiges of them, it becomes clear that the shrinkage and even disappearance of the organ in question are the results of a permanent disuse of that organ.

This is proved by the fact that the organ of hearing is never in this condition, but is always found in animals whose organization is of the kind that includes it: and for the following reason.

The substance of sound, that namely which, when set in motion by the shock or the vibration of bodies, transmits to the organ of hearing the impression received, penetrates everywhere and passes through any medium, including even the densest bodies: it follows that every animal, belonging to a plan of organization of which hearing is an essential part, always has some opportunity for the exercise of this organ wherever it may live. Hence among the vertebrates we do not find any that are destitute of the organ of hearing; and after them, when this same organ has come to an end, it does not subsequently recur in any animal of the posterior classes.

It is not so with the organ of sight; for this organ is found to disappear, re-appear and disappear again according to the use that the animal makes of it.

In the acephalic molluscs, the great development of the mantle would make their eyes and even their head altogether useless. The permanent disuse of these organs has thus brought about their disappearance and extinction, although molluscs belong to a plan of organization which should comprise them.

Lastly, it was part of the plan of organization of the reptiles, as of other vertebrates, to have four legs in dependence on their skeleton. Snakes ought consequently to have four legs, especially since they are by no means the last order of the reptiles and are farther from the fishes than are the batrachians (frogs, salamanders, etc.).

Snakes, however, have adopted the habit of crawling on the ground and hiding in the grass; so that their body, as a result of continually repeated efforts at elongation for the purpose of passing through narrow spaces, has acquired a considerable length, quite out of proportion to its size. Now, legs would have been quite useless to these animals and consequently unused. Long legs would have interfered with their need of crawling, and very short legs would have been incapable of moving their body, since they could only have had four. The disuse of these parts thus became permanent in the various races of these animals, and resulted in the complete disappearance of these same parts, although legs really belong to the plan of organization of the animals of this class.

Many insects, which should have wings according to the natural characteristics of their order and even of their genus, are more or less completely devoid of them through disuse. Instances are furnished by many Coleoptera, Orthoptera, Hymenoptera and Hemiptera, etc., where the habits of these animals never involve them in the necessity of using their wings.

But it is not enough to give an explanation of the cause which has brought about the present condition of the organs of the various animals, condition that is always found to be the same in animals of the same species; we have in addition to cite instances of changes wrought in the organs of a single individual during its life, as the exclusive result of a great mutation in the habits of the individuals of its species. The following very remarkable fact will complete the proof of the influence of habits on the condition of the organs, and of the way in which permanent changes in the habits of an individual lead to others in the condition of the organs, which come into action during the exercise of these habits.

M. Tenon, a member of the Institute, has notified to the class of sciences, that he had examined the intestinal canal of several men who had been great drinkers for a large part of their lives, and in every case he had found it shortened to an extraordinary degree, as compared with the same organ in all those who had not adopted the like habit.

It is known that great drinkers, or those who are addicted to drunkenness, take very little solid food, and eat hardly anything; since the drink which they consume so copiously and frequently is sufficient to feed them.

Now since fluid foods, especially spirits, do not long remain either in the stomach or intestine, the stomach and the rest of the intestinal canal lose among drinkers the habit of being distended, just as among sedentary persons, who are continually engaged on mental work and are accustomed to take very little food; for in their case also the stomach slowly shrinks and the intestine shortens.

This has nothing to do with any shrinkage or shortening due to a binding of the parts which would permit of the ordinary extension, if instead of remaining empty these viscera were again filled; we have to do with a real shrinkage and shortening of considerable extent, and such that these organs would burst rather than yield at once to any demand for the ordinary extension.

Compare two men of equal ages, one of whom has contracted the habit of eating very little, since his habitual studies and mental work have made digestion difficult, while the other habitually takes much exercise, is often out-of-doors, and eats well; the stomach of the first will have very little capacity left and will be filled up by a very small quantity of food, while that of the second will have preserved and even increased its capacity.

Here then is an organ which undergoes profound modification in size and capacity, purely on account of a change of habits during the life of the individual.

The frequent use of any organ, when confirmed by habit, increases the functions of that organ, leads to its development and endows it with a size and power that it does not possess in animals which exercise it less.

We have seen that the disuse of any organ modifies, reduces and finally extinguishes it. I shall now prove that the constant use of any organ, accompanied by efforts to get the most out of it, strengthens and enlarges that organ, or creates new ones to carry on functions that have become necessary.

The bird which is drawn to the water by its need of finding there the prey on which it lives, separates the digits of its feet in trying to strike the water and move about on the surface. The skin which unites these digits at their base acquires the habit of being stretched by these continually repeated separations of the digits; thus in course of time there are formed large webs which unite the digits of ducks, geese, etc., as we actually find them. In the same way efforts to swim, that is to push against the water so as to move about in it, have stretched the membranes between the digits of frogs, sea-tortoises, the otter, beaver, etc.

On the other hand, a bird which is accustomed to perch on trees and which springs from individuals all of whom had acquired this habit, necessarily has longer digits on its feet and differently shaped from those of the aquatic animals that I have just named. Its claws in time become lengthened, sharpened and curved into hooks, to clasp the branches on which the animal so often rests.

We find in the same way that the bird of the water-side which does not like swimming and yet is in need of going to the water's edge to secure its prey, is continually liable to sink in the mud. Now this bird tries to act in such a way that its body should not be immersed in the liquid, and hence makes its best efforts to stretch and lengthen its legs. The long-established habit acquired by this bird and all its race of continually stretching and lengthening its legs, results in the individuals of this race becoming raised as though on stilts, and gradually obtaining long, bare legs, denuded of feathers up to the thighs and often higher still (Système des Animauxsans vertèbres, p. 14).

We note again that this same bird wants to fish without wetting its body, and is thus obliged to make continual efforts to lengthen its neck. Now these habitual efforts in this individual and its race must have resulted in course of time in a remarkable lengthening, as indeed we actually find in the long necks of all water-side birds.

If some swimming birds like the swan and goose have short legs and yet a very long neck, the reason is that these birds while moving about on the water acquire the habit of plunging their head as deeply as they can into it in order to get the aquatic larvae and various animals on which they feed; whereas they make no effort to lengthen their legs.

If an animal, for the satisfaction of its needs, makes repeated efforts to lengthen, its tongue, it will acquire a considerable length (ant-eater, green-woodpecker); if it requires to seize anything with this same organ, its tongue will then divide and become forked. Proofs of my statement are found in the humming-birds which use their tongues for grasping things, and in lizards and snakes which use theirs to palpate and identify objects in front of them.

Needs which are always brought about by the environment, and the subsequent continued efforts to satisfy them, are not limited in their results to a mere modification, that is to say, an increase or decrease of the size and capacity of organs; but they may even go so far as to extinguish organs, when any of these needs make such a course necessary.

Fishes, which habitually swim in large masses of water, have need of lateral vision; and, as a matter of fact, their eyes are placed on the sides of their head. Their body, which is more or less flattened according to the species, has its edges perpendicular to the plane of the water; and their eyes are placed so that there is one on each flattened side. But such fishes as are forced by their habits to be constantly approaching the shore, and especially slightly inclined or gently sloping beaches, have been compelled to swim on their flattened surfaces in order to make a close approach to the water's edge. In this position, they receive more light from above than below and stand in special need of paying constant attention to what is passing above them; this requirement has forced one of their eyes to undergo a sort of displacement, and to assume the very remarkable position found in the soles, turbots, dabs, etc. (Pleuronectes and Achirus). The position of these eyes is not symmetrical, because it results from an incomplete mutation. Now this mutation is entirely completed in the skates, in which the transverse flattening of the body is altogether horizontal, like the head. Accordingly the eyes of skates are both situated on the upper surface and have become symmetrical.

Snakes, which crawl on the surface of the earth, chiefly need to see objects that are raised or above them. This need must have had its effect on the position of the organ of sight in these animals, and accordingly their eyes are situated in the lateral and upper parts of their head, so as easily to perceive what is above them or at their sides; but they scarcely see at all at a very short distance in front of them. They are, however, compelled to make good the deficiency of sight as regards objects in front of them which might injure them as they move forward. For this purpose they can only use their tongue, which they are obliged to thrust out with all their might. This habit has not only contributed to making their tongue slender and very long and contractile, but it has even forced it to undergo division in the greater number of species, so as to feel several objects at the same time; it has even permitted of the formation of an aperture at the extremity of their snout, to allow the tongue to pass without having to separate the jaws.

Nothing is more remarkable than the effects of habit in herbivorous mammals.

A quadruped, whose environment and consequent needs have for long past inculcated the habit of browsing on grass, does nothing but walk about on the ground; and for the greater part of its life is obliged to stand on its four feet, generally making only few or moderate movements. The large portion of each day that this kind of animal has to pass in filling itself with the only kind of food that it cares for, has the result that it moves but little and only uses its feet for support in walking or running on the ground, and never for holding on, or climbing trees.

From this habit of continually consuming large quantities of food-material, which distend the organs receiving it, and from the habit of making only moderate movements, it has come about that the body of these animals has greatly thickened, become heavy and massive and acquired a very great size: as is seen in elephants, rhinoceroses, oxen, buffaloes, horses, etc.

The habit of standing on their four feet during the greater part of the day, for the purpose of browsing, has brought into existence a thick horn which invests the extremity of their digits; and since these digits have no exercise and are never moved and serve no other purpose than that of support like the rest of the foot, most of them have become shortened, dwindled and, finally, even disappeared.

Thus in the pachyderms, some have five digits on their feet invested in horn, and their hoof is consequently divided into five parts; others have only four, and others again not more than three; but in the ruminants, which are apparently the oldest of the mammals that are permanently confined to the ground, there are not more than two digits on the feet and indeed, in the solipeds, there is only one (horse, donkey).

Nevertheless some of these herbivorous animals, especially the ruminants, are incessantly exposed to the attacks of carnivorous animals in the desert countries that they inhabit, and they can only find safety in headlong flight. Necessity has in these cases forced them to exert themselves in swift running, and from this habit their body has become more slender and their legs much finer; instances are furnished by the antelopes, gazelles, etc.

In our own climates, there are other dangers, such as those constituted by man, with his continual pursuit of red deer, roe deer and fallow deer; this has reduced them to the same necessity, has impelled them into similar habits, and had corresponding effects.

Since ruminants can only use their feet for support, and have little strength in their jaws, which only obtain exercise by cutting and browsing on the grass, they can only fight by blows with their heads, attacking one another with their crowns.

In the frequent fits of anger to which the males especially are subject, the efforts of their inner feeling cause the fluids to flow more strongly towards that part of their head; in some there is hence deposited a secretion of horny matter, and in others of bony matter mixed with horny matter, which gives rise to solid protuberances: thus we have the origin of horns and antlers, with which the head of most of these animals is armed.

It is interesting to observe the result of habit in the peculiar shape and size of the giraffe (Camelo-pardalis): this animal, the largest of the mammals, is known to live in the interior of Africa in places where the soil is nearly always arid and barren, so that it is obliged to browse on the leaves of trees and to make constant efforts to reach them. From this habit long maintained in all its race, it has resulted that the animal's fore-legs have become longer than its hind legs, and that its neck is lengthened to such a degree that the giraffe, without standing up on its hind legs, attains a height of six metres (nearly 20 feet).

Among birds, ostriches, which have no power of flight and are raised on very long legs, probably owe their singular shape to analogous circumstances.

The effect of habit is quite as remarkable in the carnivorous mammals as in the herbivores; but it exhibits results of a different kind.

Those carnivores, for instance, which have become accustomed to climbing, or to scratching the ground for digging holes, or to tearing their prey, have been under the necessity of using the digits of their feet: now this habit has promoted the separation of their digits, and given rise to the formation of the claws with which they are armed.

But some of the carnivores are obliged to have recourse to pursuit in order to catch their prey: now some of these animals were compelled by their needs to contract the habit of tearing with their claws, which they are constantly burying deep in the body of another animal in order to lay hold of it, and then make efforts to tear out the part seized. These repeated efforts must have resulted in its claws reaching a size and curvature which would have greatly impeded them in walking or running on stony ground: in such cases the animal has been compelled to make further efforts to draw back its claws, which are so projecting and hooked as to get in its way. From this there has gradually resulted the formation of those peculiar sheaths, into which cats, tigers, lions, etc. withdraw their claws when they are not using them.

Hence we see that efforts in a given direction, when they are long sustained or habitually made by certain parts of a living body, for the satisfaction of needs established by nature or environment, cause an enlargement of these parts and the acquisition of a size and shape that they would never have obtained, if these efforts had not become the normal activities of the animals exerting them. Instances are everywhere furnished by observations on all known animals.

Can there be any more striking instance than that which we find in the kangaroo? This animal, which carries its young in a pouch under the abdomen, has acquired the habit of standing upright, so as to rest only on its hind legs and tail; and of moving only by means of a succession of leaps, during which it maintains its erect attitude in order not to disturb its young. And the following is the result

1. Its fore legs, which it uses very little and on which it only supports itself for a moment on abandoning its erect attitude, have never acquired a development proportional to that of the other parts, and have remained meagre, very short and with very little strength.

2. The hind legs, on the contrary, which are almost continually in action either for supporting the whole body or for making leaps, have acquired a great development and become very large and strong.

3. Lastly, the tail, which is in this case much used for supporting the animal and carrying out its chief movements, has acquired an extremely remarkable thickness and strength at its base.

These well-known facts are surely quite sufficient to establish the results of habitual use on an organ or any other part of animals. If on observing in an animal any organ particularly well-developed, strong, and powerful, it is alleged that its habitual use has nothing to do with it, that its continued disuse involves it in no loss, and finally, that this organ has always been the same since the creation of the species to which the animal belongs, then I ask, Why can our domestic ducks no longer fly like wild ducks? I can, in short, cite a multitude of instances among ourselves, which bear witness to the differences that accrue to us from the use or disuse of any of our organs, although these differences are not preserved in the new individuals which arise by reproduction: for if they were their effects would be far greater.

I shall show in Part II, that when the will guides an animal to any action, the organs which have to carry out that action are immediately stimulated to it by the influx of subtle fluids (the nervous fluid), which become the determining factor of the movements required. This fact is verified by many observations, and cannot now be called in question.

Hence it follows that numerous repetitions of these organized activities strengthen, stretch, develop and even create the organs necessary to them. We have only to watch attentively what is happening all around us, to be convinced that this is the true cause of organic development and changes.

Now every change that is wrought in an organ through a habit of frequently using it, is subsequently preserved by reproduction, if it is common to the individuals who unite together in fertilization for the propagation of their species. Such a change is thus handed on to all succeeding individuals in the same environment, without their having to acquire it in the same way that it was actually created.

Furthermore, in reproductive unions, the crossing of individuals who have different qualities or structures is necessarily opposed to the permanent propagation of these qualities and structures. Hence it is that in man, who is exposed to so great a diversity of environment, the accidental qualities or defects which he acquires are not preserved and propagated by reproduction. If, when certain peculiarities of shape or certain defects have been acquired, two individuals who are both affected were always to unite together, they would hand on the same peculiarities; and if successive generations were limited to such unions, a special and distinct race would then be formed. But perpetual crossings between individuals, who have not the same peculiarities of shape, cause the disappearance of all peculiarities acquired by special action of the environment. Hence, we may be sure that if men were not kept apart by the distances of their habitations, the crossing in reproduction would soon bring about the disappearance of the general characteristics distinguishing different nations.

If I intended here to pass in review all the classes, orders, genera and species of existing animals, I should be able to show that the, conformation and structure of individuals, their organs, faculties, etc., etc., are everywhere a pure result of the environment to which each species is exposed by its nature, and by the habits that the individuals composing it have been compelled to acquire; I should be able to show that they are not the result of a shape which existed from the beginning, and has driven animals into the habits they are known to possess.

It is known that the animal called theai or sloth (Bradypustridactylus) is permanently in a state of such extreme weakness that it only executes very slow and limited movements, and walks on the ground with difficulty. So slow are its movements that it is alleged that it can only take fifty steps in a day. It is known, moreover, that the organization of this animal is entirely in harmony with its state of feebleness and incapacity for walking; and that if it wished to make other movements than those which it actually does make it could not do so.

Hence on the supposition that this animal had received its organization from nature, it has been asserted that this organization forced it into the habits and miserable state in which it exists.

This is very far from being my opinion; for I am convinced that the habits which theai was originally forced to contract must necessarily have brought its organization to its present condition.

If continual dangers in former times have led the individuals of this species to take refuge in trees, to live there habitually and feed on their leaves, it is clear that they must have given up a great number of movements which animals living on the ground are in a position to perform. All the needs of theai will then be reduced to clinging to branches and crawling and dragging themselves among them, in order to reach the leaves, and then to remaining on the tree in a state of inactivity in order to avoid falling off. This kind of inactivity, moreover, must have been continually induced by the heat of the climate; for among warm-blooded animals, heat is more conducive to rest than to movement.

Now the individuals of the race of theai have long maintained this habit of remaining in the trees, and of performing only those slow and little varied movements which suffice for their needs. Hence their organization will gradually have come into accordance with their new habits; and from this it must follow

1. That the arms of these animals, which are making continual efforts to clasp the branches of trees, will be lengthened;
2. That the claws of their digits will have acquired a great length and a hooked shape, through the continued efforts of the animal to hold on;
3. That their digits, which are never used in making independent movements, will have entirely lost their mobility, become united and have preserved only the faculty of flexion or extension all together;
4. That their thighs, which are continually clasping either the trunk or large branches of trees, will have contracted a habit of always being separated, so as to lead to an enlargement of the pelvis and a backward direction of the cotyloid cavities;
5. Lastly, that a great many of their bones will be welded together, and that parts of their skeleton will consequently have assumed an arrangement and form adapted to the habits of these animals, and different from those which they would require for other habits.

This is a fact that can never be disputed; since nature shows us in innumerable other instances the power of environment over habit and that of habit over the shape, arrangement and proportions of the parts of animals.

Since there is no necessity to cite any further examples, we may now turn to the main point elaborated in this discussion.

It is a fact that all animals have special habits corresponding to their genus and species, and always possess an organization that is completely in harmony with those habits.

It seems from the study of this fact that we may adopt one or other of the two following conclusions, and that neither of them can be verified.

Conclusion adopted hitherto: Nature (or her Author) in creating animals, foresaw all the possible kinds of environment in which they would have to live, and endowed each species with a fixed organization and with a definite and invariable shape, which compel each species to live in the places and climates where we actually find them, and there to maintain the habits which we know in them.

My individual conclusion: Nature has produced all the species of animals in succession, beginning with the most imperfect or simplest, and ending her work with the most perfect, so as to create a gradually increasing complexity in their organization; these animals have spread at large throughout all the habitable regions of the globe, and every species has derived from its environment the habits that we find in it and the structural modifications which observation shows us.

The former of these two conclusions is that which has been drawn hitherto, at least by nearly everyone: it attributes to every animal a fixed organization and structure which never have varied and never do vary; it assumes, moreover, that none of the localities inhabited by animals ever vary; for if they were to vary, the same animals could no longer survive, and the possibility of finding other localities and transporting themselves thither would not be open to them.

The second conclusion is my own: it assumes that by the influence of environment on habit, and thereafter by that of habit on the state of the parts and even on organization, the structure and organization of any animal may undergo modifications, possibly very great, and capable of accounting for the actual condition in which all animals are found.

In order to show that this second conclusion is baseless, it must first be proved that no point on the surface of the earth ever undergoes variation as to its nature, exposure, high or low situation, climate, etc., etc.; it must then be proved that no part of animals undergoes even after long periods of time any modification due to a change of environment or to the necessity which forces them into a different kind of life and activity from what has been customary to them.

Now if a single case is sufficient to prove that an animal which has long been in do-mestication differs from the wild species whence it sprang, and if in any such domesticated species, great differences of conformation are found between the individuals exposed to such a habit and those which are forced into different habits, it will then be certain that the first conclusion is not consistent with the laws of nature, while the second, on the contrary, is entirely in accordance with them.

Everything then combines to prove my statement, namely: that it is not the shape either of the body or its parts which gives rise to the habits of animals and their mode of life; but that it is, on the contrary, the habits, mode of life and all the other influences of the environment which have in course of time built up the shape of the body and of the parts of animals. With new shapes, new faculties have been acquired, and little by little nature has succeeded in fashioning animals such as we actually see them.

Can there be any more important conclusion in the range of natural history, or any to which more attention should be paid than that which I have just set forth?

SOURCE: Lamarck, Jean-Baptiste. *Philosophie zoologique*. Paris, 1809. Trans. as *Zoological Philosophy* by H. Elliot. London: Macmillan, 1914.

BARON GEORGES CUVIER, *DISCOURSE ON THE REVOLUTIONARY UPHEAVALS ON THE SURFACE OF THE GLOBE AND ON THE CHANGES WHICH THEY HAVE PRODUCED IN THE ANIMAL KINGDOM* (1825)

Baron Georges Léopold Chrétien Frédéric Dagobert Cuvier (1769–1832) was a French zoologist and among the most respected naturalists of the eighteenth and nineteenth centuries. In 1799 he became professor of natural history at the College of France and a year later published his classic work *Lessons of Comparative Anatomy*, which presented his system of comparative anatomy. Formulated a generation earlier, comparative anatomy consisted of the study of the structure of organisms in relation to one another. Previous authors had portrayed a single, linear series from simplest to most complex. Cuvier arranged all animals into four large groups—vertebrates, articulates, mollusks, and radiates—and asserted that members within each of the four groups shared similar body plans. Cuvier's comparative anatomy was inherently antievolutionary, as he believed that organisms were complex entities that were specifically designed for their particular environments. Any change in environment or any attempt by them to adapt would result in the species collapsing and going extinct. His 1825 *Discourse on the Revolutionary Upheavals on the Surface of the Globe* described the fundamental changes in organisms' environments that sometimes lead to their demise.

Introduction

In my work on Fossil Bones, I set myself the task of identifying the animals whose fossilized remains fill the surface strata of the earth. This project meant I had to travel along a path where we had so far taken only a few tentative steps. As a new sort of antiquarian, I had to learn to restore these memorials to past upheavals and, at the same time, to decipher their meaning. I had to collect and put together in their original order the fragments which made up these animals, to reconstruct the ancient creatures to which these fragments belonged, to recreate their proportions and characteristics, and finally to compare them to those alive today on the surface of the earth. This was an almost unknown art, which assumed a science hardly touched upon up until now, that of the laws which govern the formal coexistence of the various parts in organic beings. Thus, I had to prepare myself for these studies through a much longer research into animals which presently exist. Only an almost universal review of present creation could provide some proof for my results concerning created life long ago. But at the same time such a study had to provide me with a large collection of equally demonstrable rules and interconnections. In the course of this exploration into a small part of the theory of the earth, I would have to be able to subject the entire animal kingdom in some way to new laws.

I was sustained in this double task by the constant interest which it promised to have and by service to the universal science of anatomy, the essential basis of all those sciences dealing with organic entities, and to the physical history of the earth, the foundation of mineralogy, geography, and,

Baron Georges Cuvier [Library of Congress Prints and Photographs Department, LC-USZ62-134030].

we can say, even of human history and everything really important for human beings to know about themselves.

If one finds it interesting to follow in the infancy of our species the almost eradicated traces of so many extinct nations, how could one not also find it interesting to search in the shadows of the earth's infancy for the traces of revolutionary upheavals which have preceded the existence of all nations? We admire the force with which the human spirit has measured the movements of planets which nature seemed to have concealed for ever from our view; human genius and science have stepped beyond the limits of space; some observations developed by reasoning have unveiled the mechanical workings of the world. Would there not also be some glory for human beings to know how to step beyond the limits of time and to recover, through some observations, the history of this earth and a succession of events which have preceded the birth of the human genus? No doubt the astronomers have proceeded more rapidly than the naturalists. The theory of the earth at the present time is rather like the one in which some philosophers believed that the sky was made of freestone [fine-grained sandstone or limestone] and the moon was as big as the Peloponnese. But, following Anaxagoras, Copernicus and Kepler opened up the road to Newton. And why one day should natural history not also have its own Newton?

Exposition

In this discourse I propose above all to present the plan and result of my work on fossil bones. I will try also to sketch a rapid picture of the attempts made so far to reconstruct the history of the earth's upheavals. No doubt, the facts which I have discovered form only a really small part of those which must make up this ancient history; but several of these lead to significant consequences, and the rigorous way in which I have proceeded in determining them encourages me to believe that people will look on them as points definitely settled, things which will constitute a special age in science. Finally, I hope that their newness will excuse the fact that I focus the major attention of my readers on them.

My object will be, first, to show by what connections the history of the fossil bones of land animals is linked to the theory of the earth and why they have a particular importance in this respect. Then I will develop the principles on which rests the art of sorting out these bones, or, in other words, of recognizing a genus and distinguishing a species by a single bone fragment, an art on whose reliability depends the reliability of all my work. I will give a quick indication of new species, of genera previously unknown, which the application of these principles has led me to discover, as well as of the various sorts of formations which contain them. And since the difference between these species and those today does not exceed certain limits, I will show that these limits are considerably greater than those which today distinguish the varieties of a common species. I will thus reveal just where these varieties could go, whether by the influence of time, or climate, or finally domestication.

In this way, I will proceed to the conclusion (and I shall invite my readers to conclude with me), that there must have been great events to bring about the much greater differences which I have recognized. I will develop then the particular revisions which my research must introduce into the opinions accepted up to the present time about the earth's revolutions. Finally I will examine up to what point the civil and religious history of people agrees with the results of the observations dealing with the physical history of the earth and the probabilities which these observations set concerning the time when human societies could have established permanent homes and arable fields and when, consequently, societies could have taken on a lasting form.

The Geological Record of Ancient Upheavals

The First Appearance of the Earth

When the traveler goes through fertile plains where tranquil waters nourish with their regular flow an abundant vegetation and where the ground, trodden by numerous people and decorated with flourishing villages, rich cities, and superb monuments, is never troubled except by ravages of war or by the oppression of men in power, he is not tempted to believe that nature has also had its internal wars and that the surface of the earth has been overthrown by revolutions and catastrophes. But his ideas change as soon as he seeks to dig through this soil, today so calm, or when he takes himself up into the hills which border the plain; his ideas expand, so to speak, with what he is looking at. They begin to embrace the extent and the grandeur of the ancient events as soon as he climbs up the higher mountains of which these are the foothills, or when he follows the stream beds which descend from these mountains and moves into their interior.

The First Proofs of Upheavals

The lowest and most level land areas show us, especially when we dig there to very great depths, nothing but horizontal layers of material more or less varied, which almost all contain innumerable products of the sea. Similar layers, with similar products, form the hills up to quite high elevations. Sometimes the shells are so numerous that they make up the entire mass of soil by themselves. They occur at elevations higher than the level of all seas, where no sea could be carried today by present causes. Not only are these shells encased in loose sand, but the hardest rocks often encrust them and are penetrated by them throughout. All the parts of the world, both hemispheres, all continents, and all islands of any size provide evidence of the same phenomenon. The time is past when ignorance could continue to maintain that these remains of organic bodies were simple games of nature, products conceived in the bosom of the earth by its creative forces, and the renewed efforts of certain metaphysicians will probably not be enough to make these old opinions acceptable. A scrupulous comparison of the shapes of these deposits, of their make up and often even their chemical composition shows not the slightest difference between these fossil shells and those which the sea nourishes. Their preservation is no less perfect. Very often one observes there neither shattering nor fractures, nothing which signifies a violent movement. The smallest of them keep their most delicate parts, their most subtle crests, their slenderest features. Thus, not only have they lived in the sea, but they have been deposited by the sea, which has left them in the places where we find them. Moreover, this sea has remained in these locations, with a sufficient calm and duration to form deposits so regular, so thick, so extensive, and in places so solid, that they are full of the remains of marine animals. The sea basin therefore has provided evidence of at least one change, whether in extent or location. See what results already from the first inspections and the most superficial observation.

The traces of upheavals become more impressive when one moves a little higher, when one gets even closer to the foot of the great mountain ranges. There are still plenty of shell layers. We notice them, even thicker and more solid ones. The shells there are just as numerous and just as well preserved. But they are no longer the same species. Also, the strata which contain them are no longer generally horizontal. They lie obliquely, sometimes almost vertically. In contrast to the plains and the low hills, where it was necessary to dig deep to recognize the succession of layers, here we see them on the mountain flank, as we follow the valleys produced by their tearing apart. At the foot of the escarpments, immense masses of debris form rounded hillocks, whose height is increased by each thawing and each storm.

And those upright layers which form the crests of secondary mountains do not rest on the horizontal layers of hills which serve as their lower stages. By contrast, they sink under these hills, which rest on the slopes of these oblique strata. When we bore into the horizontal strata near mountains with oblique layers, we find these oblique layers deep down. Sometimes when the oblique layers are not very high, their summits are even crowned with horizontal strata. The oblique layers are therefore older than the horizontal layers. Since it is impossible, at least for most of them, not to have been formed horizontally, evidently they have been lifted up again and were in existence before the others which rest on top of them.

Thus, before forming these horizontal layers, the sea had formed other strata. These were for some reason or other broken, raised up, and overturned in thousands of ways. As several of these oblique layers which the sea formed in a previous age rise higher than the horizontal layers which succeeded them and which surrounded them, the causes which gave these layers their oblique orientation also made them protrude above the level of the sea and turned them into islands or at least reefs and uneven structures, whether they were raised again by an extreme condition or whether a contrasting subsidence made the waters sink. The second result is no less clear or less proven than the first for anyone who will take the trouble to study the monuments which provide evidence for these results.

Proofs That These Revolutions Have Been Numerous

But the revolutions and changes which are responsible for the present state of the earth are not limited to the upsetting of the ancient strata and to the ebbing of the sea after the formations of new layers. When we compare together in greater detail the various layers and the products of life which they conceal, we soon realize that this ancient sea did not continuously deposit the same type of stones nor the remains of animals of the same species, and that each of its deposits did not extend over all the surface which the sea covered. Successive variations took place, of which only the first ones were almost universal; the others appear to have been considerably less. The older the layers, the more each of them is uniform over a great extent; the newer the layers, the more they are limited and subject to variation within small distances. Thus, the changes in the strata were accompanied and followed by changes in the nature of the liquid and of the materials which it held in solution. When certain layers, appearing above the sea, split the surface with islands and protruding ranges, different changes could have taken place in several particular ocean basins.

We know that in the midst of such variations in the nature of the liquid, the animals which it nourished could not have stayed the same. Their species, even their genera, changed with the layers; and although there are some returns of species within small distances, it is true to state, in general, that the shells of the ancient layers have forms unique to them, that they disappear gradually and do not show up in the recent layers, even less in the present sea, where we never discover species analogous to them. Even several of their genera are not found there. The shells of recent layers, by contrast, are generically similar to those which live in our seas. In the most recent and least solid of these layers and in certain recent and limited deposits there are some species which the most practiced eye would not be able to distinguish from those which the neighbouring coasts nourish.

Thus in animal nature a succession of variations has taken place, brought about by changes in the liquid where the animals lived or at least by variations which corresponded to those changes. And these variations brought by degrees the classes of aquatic animals to their present condition. At last, when the sea left our continents for the last time, its inhabitants did not differ much from those which the sea still feeds today.

Finally, we say that if we examine with even greater care the remains of these organic creatures, we come to discover in the middle of the marine strata, even the most ancient ones, layers full of animal or vegetable products from land and fresh water. In the most recent layers (i.e., the ones closest to the surface) there are some where land animals are buried under masses of marine creatures. Thus, not only did the different catastrophes which moved the layers gradually make the various parts of our continent rise up from the bottom of the sea and reduce the size of the sea basin; but this basin has been moved in several directions. Often the regions converted into dry land have been covered again by the seas, whether they have sunk or the waters have been carried above them. As for the particular matter of the soil which the sea uncovered in its last retreat, the part which human beings and terrestrial animals live on right now, it had already been dry land once and had nourished at that time quadrupeds, birds, plants, and land forms of all sorts. Thus, the sea which left that land had previously invaded. The changes in the heights of the oceans did not therefore consist only in one withdrawal more or less gradual, more or less universal. It was a matter of a succession of various eruptions and retreats. The result of these has definitely been, however, a general lowering of the sea level.

Proofs That These Revolutions Have Been Sudden

But it is also really important to note that these eruptions and repeated retreats were not at all slow and did not all take place gradually. On the contrary, most of the catastrophes which brought them on have been sudden. That is especially easy to demonstrate for the last of these catastrophes, which by a double movement inundated and later left dry our present continents or at least a great part of the land which forms them today. That catastrophe also left in the northern countries the cadavers of great quadrupeds locked in the ice, preserved right up to our time with their skin, hair, and flesh. If they had not been frozen as soon as they were killed, decay would have caused them to decompose. On the other hand, this permanent freezing was not a factor previously in the places where these animals were trapped. For they would not have been able to live in such a temperature. Hence the same instant which killed the animals froze the country where they lived. This event was sudden, instantaneous, without any gradual development. What is so clearly demonstrated for this most recent catastrophe is hardly less so for the earlier ones. The rending, rearranging, and overturning of more ancient layers leave no doubt that sudden and violent causes placed them in the state in which we see them. The very force of the movements which the bodies of water experienced is still attested to by the mountain of remains and rounded pebbles interposed in many places between the solid layers. Thus, life on this earth has often been disturbed by dreadful events. Innumerable living creatures have been victims of these catastrophes. Some inhabitants of dry land have seen themselves swallowed up by floods; others living in the ocean depths when the bottom of the sea was lifted up suddenly were placed on dry land. Their very races were extinguished for ever, leaving behind nothing in the world but some hardly recognizable debris for the natural scientist.

Such are the conclusions to which we are necessarily led by the objects which we meet at every step and which we can verify at every instant in almost every country. These huge and terrible events are clearly printed everywhere for the eye which knows how to read the story in their monuments.

But what is even more astonishing and what is no less certain is that life has not always existed on the earth and that it is easy for the observer to recognize the point where life began to deposit her productions.

SOURCE: Cuvier, Georges. *Discourse on the Revolutionary Upheavals on the Surface of the Globe and on the Changes Which They Have Produced in the Animal Kingdom.* Paris: G. Dufour et Ed. D'Ocagne, 1825.

WILLIAM PALEY, *NATURAL THEOLOGY* (1802)

William Paley (1743–1805) showed an early interest and talent in mathematics. In college he had a choice between mathematics and theology, choosing the latter but maintaining a life-long interest in the former. He excelled as a writer of textbooks and was able to explain complex arguments in simple, concise terms. *A View of the Evidences of Christianity* (1794) was followed by *Natural Theology* (1802), and both offered an apologetic explanation of the relationship between the study of nature and of theology. Ultimately, *Natural Theology*, was among the most influential popular works in nineteenth-century theology; in his autobiography, Darwin claimed that Paley's works were among his favorites, and their influence on his worldview is apparent.

Chapter I: State of the Argument

William Paley [Library of Congress Prints and Photographs Department, LC-USZ62-91494].

In crossing a heath, suppose I pitched my foot against a stone, and were asked how the stone came to be there, I might possibly answer, that, for any thing I knew to the contrary, it had lain there for ever: nor would it perhaps be very easy to show the absurdity of this answer. But suppose I had found a watch upon the ground, and it should be inquired how the watch happened to be in that place; I should hardly think of the answer which I had before given, that, for any thing I knew, the watch might have always been there. Yet why should not this answer serve for the watch as well as for the stone? why is it not as admissible in the second case, as in the first? For this reason, and for no other, viz. that, when we come to inspect the watch, we perceive (what we could not discover in the stone) that its several parts are framed and put together for a purpose, *e.g.* that they are so formed and adjusted as to produce motion, and that motion so regulated as to point out the hour of the day; that, if the different parts had been differently shaped from what they are, of a different size from what they are, or placed after any other manner, or in any other order, than that in which they are placed, either no motion at all would have been carried on in the machine, or none which would have answered the use that is now served by it. To reckon up a few of the plainest of these parts, and of their offices, all tending to one result;—We see a cylindrical box containing a coiled elastic spring, which, by its endeavour to relax itself, turns round the box. We next observe a flexible chain (artificially wrought for the sake of flexure), communicating the action of the spring from the box to the fusee. We then find a series of wheels, the teeth of which catch in, and apply to, each other, conducting the motion from the fusee to the balance, and from the balance to the pointer; and at the same time, by the size and shape of those wheels, so regulating that motion, as to terminate in causing an index, by an equable and measured progression, to pass over a given space in a given time. We take notice that the wheels are made of brass in order to keep them from rust; the springs of steel, no other metal being so elastic; that over the face of the watch there is placed a glass, a material employed in no other part of the work, but in the room of which, if there had been any other than a transparent substance, the hour could not be seen without opening the case. This mechanism being observed (it requires indeed an examination of the instrument, and perhaps some previous knowledge of the subject, to perceive and understand it; but being once, as we have said, observed and understood), the inference, we think, is inevitable, that the watch must have had a maker: that there must have existed, at some time, and at some place or other, an artificer or artificers who formed it for the purpose which we find it actually to answer; who comprehended its construction, and designed its use.

Nor would it, I apprehend, weaken the conclusion, that we had never seen a watch made; that we had never known an artist capable of making one; that we were altogether incapable of executing such a piece of workmanship ourselves, or of understanding in what manner it was performed; all this being no more than what is true of some exquisite remains of ancient art, of some lost arts, and, to the generality of mankind, of the more curious productions of modern manufacture. Does one man in a million know how oval frames are turned? Ignorance of this kind exalts our opinion of the unseen and unknown artist's skill, if he be unseen and unknown, but raises no doubt in our minds of the existence and agency of such an artist, at some former time, and in some place or other. Nor can I perceive that it varies at all the inference, whether the question arise concerning a human agent, or concerning an agent of a different species, or an agent possessing, in some respects, a different nature.

Neither, secondly, would it invalidate our conclusion, that the watch sometimes went wrong, or that it seldom went exactly right. The purpose of the machinery, the design, and the designer, might be evident, and in the case supposed would be evident, in whatever way we accounted for the irregularity of the movement, or whether we could account for it or not. It is not necessary that a machine be perfect, in order to show with what design it was made: still less necessary, where the only question is, whether it were made with any design at all.

Nor, thirdly, would it bring any uncertainty into the argument, if there were a few parts of the watch, concerning which we could not discover, or had not yet discovered, in what manner they conduced to the general effect; or even some parts, concerning which we could not ascertain, whether they conduced to that effect in any manner whatever. For, as to the first branch of the case; if by the loss, or disorder, or decay of the parts in question, the movement of the watch were found in fact to be stopped, or disturbed, or retarded, no doubt would remain in our minds as to the utility or intention of these parts, although we should be unable to investigate the manner according to which, or the connection by which, the ultimate effect depended upon their action or assistance; and the more complex is the machine, the more likely is this obscurity to arise. Then, as to the second thing supposed, namely, that there were parts which might be spared, without prejudice to the movement of the watch, and that we had proved this by experiment,—these superfluous parts, even if we were completely assured that they were such, would not vacate the reasoning which we had instituted concerning other parts. The indication of contrivance remained, with respect to them, nearly as it was before.

Nor, fourthly, would any man in his senses think the existence of the watch, with its various machinery, accounted for, by being told that it was one out of possible combinations of material forms; that whatever he had found in the place where he found the watch, must have contained some internal configuration or other; and that this configuration might be the structure now exhibited, viz. of the works of a watch, as well as a different structure.

Nor, fifthly, would it yield his inquiry more satisfaction to be answered, that there existed in things a principle of order, which had disposed the parts of the watch into their present form and situation. He never knew a watch made by the principle of order; nor can he even form to himself an idea of what is meant by a principle of order, distinct from the intelligence of the watch-maker.

Sixthly, he would be surprised to hear that the mechanism of the watch was no proof of contrivance, only a motive to induce the mind to think so:

And not less surprised to be informed, that the watch in his hand was nothing more than the result of the laws of *metallic* nature. It is a perversion of language to assign any law, as the efficient, operative cause of any thing. A law presupposes an agent; for it is only the mode, according to which an agent proceeds: it implies a power; for it is the order, according to which that power acts. Without this agent, without this power, which are both distinct from itself, the *law* does nothing; is nothing. The expression, "the law of metallic nature," may sound strange and harsh to a philosophic ear; but it seems quite as justifiable as some others which are more familiar to him, such as "the law of vegetable nature," "the law of animal nature," or indeed as "the law of nature" in general, when assigned as the cause of phenomena, in exclusion of agency and power; or when it is substituted into the place of these.

Neither, lastly, would our observer be driven out of his conclusion, or from his confidence in its truth, by being told that he knew nothing at all about the matter. He knows enough for his argument: he knows the utility of the end: he knows the subserviency and adaptation of the means to the end. These points being known, his ignorance of other points, his doubts concerning other points, affect not the certainty of his reasoning. The consciousness of knowing little, need not beget a distrust of that which he does know.

Chapter II: State of the Argument (Continued)

Suppose, in the next place, that the person who found the watch, should, after some time, discover that, in addition to all the properties which he had hitherto observed in it, it possessed the unexpected property of producing, in the course of its movement, another watch like itself (the thing is conceivable); that it contained within it a mechanism, a system of parts, a mould for instance, or a complex adjustment of lathes, files, and other tools, evidently and separately calculated for this purpose; let us inquire, what effect ought such a discovery to have upon his former conclusion.

The first effect would be to increase his admiration of the contrivance, and his conviction of the consummate skill of the contriver. Whether he regarded the object of the contrivance, the distinct apparatus, the intricate, yet in many parts intelligible mechanism, by which it was carried on, he would perceive, in this new observation, nothing but an additional reason for doing what he had already done,—for referring the construction of the watch to design, and to supreme art. If that construction *without* this property, or which is the same thing, before this property had been noticed, proved intention and art to have been employed about it; still more strong would the proof appear, when he came to the knowledge of this further property, the crown and perfection of all the rest.

He would reflect that though the watch before him were, *in some sense*, the maker of the watch, which was fabricated in the course of its movements, yet it was in a very different sense from that, in which a carpenter, for instance, is the maker of a chair; the author of its contrivance, the cause of the relation of its parts to their use. With respect to these, the first watch was no cause at all to the second: in no such sense as this was it the author of the constitution and order, either of the parts which the new watch contained, or of the parts by the aid and instrumentality of which it was produced. We might possibly say, but with great latitude of expression, that a stream of water ground corn: but no latitude of expression would allow us to say, no stretch of conjecture could lead us to think, that the stream of water built the mill, though it were too ancient for us to know who the builder was. What the stream of water does in the affair, is neither more nor less than this; by the application of an unintelligent impulse to a mechanism previously arranged, arranged independently of it, and arranged by intelligence, an effect is produced, viz. the corn is ground. But the effect results from the arrangement. The force of the stream cannot be said to be the cause or author of the effect, still less of the arrangement. Understanding and plan in the formation of the mill were not the less necessary, for any share which the water has in grinding the corn: yet is this share the same, as that which the watch would have contributed to the production of the new watch, upon the supposition assumed in the last section. Therefore, though it be now no longer probable, that the individual watch, which our observer had found, was made immediately by the hand of an artificer, yet doth not this alteration in anywise affect the inference, that an artificer had been originally employed and concerned in the production. The argument from design remains as it was. Marks of design and contrivance are no more accounted for now, than they were before. In the same thing, we may ask for the cause of different properties. We may ask for the cause of the colour of a body, of its hardness, of its head; and these causes may be all different. We are now asking for the cause of that subserviency to a use, that relation to an end, which we have remarked in the watch before us. No answer is given to this question, by telling us that a preceding watch produced it. There cannot be design without a designer; contrivance without a contriver; order without choice; arrangement, without any thing capable of arranging; subserviency and relation to a purpose, without that which could intend a purpose; means suitable to an end, and executing their office, in accomplishing that end, without the end ever having been contemplated, or the means accommodated to it. Arrangement, disposition of parts, subserviency of means to an end, relation of instruments to a use, imply the presence of intelligence and mind. No one, therefore, can rationally believe, that the insensible, inanimate watch, from which the watch before us issued, was the proper cause of the mechanism we so much admire in it;–could be truly said to have constructed the instrument, disposed its parts, assigned their office, determined their order, action, and mutual dependency, combined their several motions into one result, and that also a result connected with the utilities of other beings. All these properties, therefore, are as much unaccounted for, as they were before.

Nor is any thing gained by running the difficulty farther back, *i.e.* by supposing the watch before us to have been produced from another watch, that from a former, and so on indefinitely. Our going back ever so far, brings us no nearer to the least degree of satisfaction upon the subject. Contrivance is still unaccounted for. We still want a contriver. A designing mind is neither supplied by this supposition, nor dispensed with. If the difficulty were diminished the further we went back, by going back indefinitely we might exhaust it. And this is the only case to which this sort of reasoning applies. Where there is a tendency, or, as we increase the number of terms, a continual approach towards a limit, *there*, by supposing the number of terms to be what is called infinite, we may conceive the limit to be attained: but where there is no such tendency, or approach, nothing is effected by lengthening the series. There is no difference as to the point in question (whatever there may be as to many points), between one series and another; between a series which is finite, and a series which is infinite. A chain, composed of an infinite number of links, can no more support itself, than a chain composed of a finite number of links. And of this we are assured (though we never *can* have tried the experiment), because, by increasing the number of links, from ten for instance to a hundred, from a hundred to a thousand, &c. we make not the smallest approach, we observe not the smallest tendency, towards self-support. There is no difference in this respect (yet there may be a great difference in several respects) between a chain of a greater or less length, between one chain and another, between one that is finite and one that is infinite. This very much resembles the case before us. The machine which we are inspecting, demonstrates, by its construction, contrivance and design. Contrivance must have had a contriver; design, a designer; whether the machine immediately proceeded from another machine or not. That circumstance alters not the case. That other machine may, in like manner, have proceeded from a former machine: nor does that alter the case; contrivance must have had a contriver. That former one from one preceding it: no alteration still; a contriver is still necessary. No tendency is perceived, no approach towards a diminution of this necessity. It is the same with any and every succession of these machines; a succession of ten, of a hundred, of a thousand; with one series, as with another; a series which is finite, as with a series which is infinite. In whatever other respects they may differ, in this they do not. In all equally, contrivance and design are unaccounted for.

The question is not simply, How came the first watch into existence? which question, it may be pretended, is done away by supposing the series of watches thus produced from one another to have been infinite, and consequently to have had no-such *first*, for which it was necessary to provide a cause. This, perhaps, would have been nearly the state of the question, if no thing had been before us but an unorganized, unmechanized substance, without mark or indication of contrivance. It might be difficult to show that such substance could not have existed from eternity, either in succession (if it were possible, which I think it is not, for unorganized bodies to spring from one another), or by individual perpetuity. But that is not the question now. To suppose it to be so, is to suppose that it made no difference whether we had found a watch or a stone. As it is, the metaphysics of that question have no place; for, in the watch which we are examining, are seen contrivance, design; an end, a purpose; means for the end, adaptation to the purpose. And the question which irresistibly presses upon our thoughts, is, whence this contrivance and design? The thing required is the intending mind, the adapting hand, the intelligence by which that hand was directed. This question, this demand, is not shaken off, by increasing a number or succession of substances, destitute of these properties; nor the more, by increasing that number to infinity. If it be said, that, upon the supposition of one watch being produced from another in the course of that other's movements, and by means of the mechanism within it, we have a cause for the watch in my hand, viz. the watch from which it proceeded. I deny, that for the design, the contrivance, the suitableness of means to an end, the adaptation of instruments to a use (all which we discover in the watch), we have any cause whatever. It is in vain, therefore, to assign a series of such causes, or to allege that a series may be carried back to infinity; for I do not admit that we have yet any cause at all of the phenomena, still less any series of causes either finite or infinite. Here is contrivance, but no contriver; proofs of design, but no designer.

Our observer would further also reflect, that the maker of the watch before him, was, in truth and reality, the maker of every watch produced from it; there being no difference (except that the latter manifests a more exquisite skill) between the making of another watch with his own hands, by the mediation of files, lathes, chisels, &c. and the disposing, fixing, and inserting of these instruments, or of others equivalent to them, in the body of the watch already made in such a manner, as to form a new watch in the course of the movements which he had given to the old one. It is only working by one set of tools, instead of another.

The conclusion of which the *first* examination of the watch, of its works, construction, and movement, suggested, was, that it must have had, for the cause and author of that construction, an artificer, who understood its mechanism, and designed its use. This conclusion is invincible. A *second* examination presents us with a new discovery. The watch is found, in the course of its movement, to produce another watch, similar to itself; and not only so, but we perceive in it a system or organization, separately calculated for that purpose. What effect would this discovery have, or ought it to have, upon our former inference? What, as hath already been said, but to increase, beyond measure, our admiration of the skill, which had been employed in the formation of such a machine? Or shall it, instead of this, all at once turn us round to an opposite conclusion, viz. that no art or skill whatever has been concerned in the business, although all other evidences of art and skill remain as they were, and this last and supreme piece of art be now added to the rest? Can this be maintained without absurdity? Yet this is atheism.

SOURCE: Paley, William. *Natural Theology, or, Evidences of the Existence and Attributes of the Deity.* Philadelphia: John Morgan (printed by H. Maxwell), 1802.

ROBERT CHAMBERS, *VESTIGES OF THE NATURAL HISTORY OF CREATION* (1844)

Robert Chambers (1802–1871) was a journalist and well-known literary figure in mid-nineteenth-century Britain. He published *Vestiges of the Natural History of Creation* anonymously, which caused considerable speculation about the author and his or her intentions. The book was a sensation and was widely read and discussed by professional naturalists and the general public. While it was on the surface an evolutionary account of the diversity of life, it was laced with a political argument against political conservatives generally and especially conservative Presbyterianism. Published just as Darwin was thinking and writing about his theory of evolution by natural selection, the public controversy over the book led Darwin to shelve his plans to write on evolution for another fifteen years.

It has been already intimated, as a general fact, that there is an obvious gradation amongst the families of both the vegetable and animal kingdoms, from the simple lichen and animalcule respectively up to the highest order of dicotyledonous trees and the mammalia. Confining our attention, in the meantime, to the animal kingdom—it does not appear that this gradation passes along one line, on which every form of animal life can be, as it were, strung; there may be branching or double lines at some places; or the whole may be in a circle composed of minor circles, as has been recently suggested. But still it is incontestable that there are general appearances of a scale beginning with the simple and advancing to the complicated. The animal kingdom was divided by Cuvier into four sub-kingdoms, or divisions, and these exhibit an unequivocal gradation in the order in which they are here enumerated:—Radiata, (polypes, &c.;) mollusca, (pulpy animals;) articulate, (jointed animals;) vertebrate, (animals with internal skeleton.) The gradation can, in like manner, be clearly traced in the classes into which the sub-kingdoms are Subdivided, as, for instance, when we take those of the vertebrate in this order—reptiles, fishes, birds, mammals.

While the external forms of all these various animals are so different, it is very remarkable that the whole are, after all, variations of a fundamental plan, which can be traced as a basis throughout the whole, the variations being merely modifications of that plan to suit the particular conditions in which each particular animal has been designed to live. Starting from the primeval germ, which, as we have seen, is the *representative* of a particular order of full-grown animals, we find all others to be merely advances from that type, with the extension of endowments and modification of forms which are required in each particular case; each form, also, retaining a strong affinity to that which precedes it, and tending to impress its own features on that which succeeds. This unity of structure, as it is called, becomes the more remarkable, when we observe that the organs, while preserving a resemblance, are often put to different uses. For example: the ribs become, in the serpent, organs of locomotion, and the snout is extended, in the elephant, into a prehensile instrument.

It is equally remarkable that analogous purposes are served in different animals by organs essentially different. Thus, the mammalia breathe by lungs; the fishes, by gills. These are not modifications of one organ, but distinct organs. In mammifers, the gills exist and act at an early stage of the foetal state, but afterwards go back and appear no more; while the lungs are developed. In fishes, again, the gills only are fully developed; while the lung structure either makes no advance at all, or only appears in the rudimentary form of an air-bladder. So, also, the baleen of the whale and the teeth of the land mammalia are different organs. The whale, in embryo, shows the rudiments of teeth; but these, not being wanted are not developed, and the baleen is brought forward instead. The land animals, we may also be sure, have the rudiments of baleen in their organization. In many instances, a particular structure is found advanced to a certain point in a particular set of animals, (for instance, feet in the serpent tribe,) although it is not there required in any degree; but the peculiarity, being carried a little farther forward, is perhaps useful in the next set of animals in the scale. Such are called rudimentary organs. With this class of phenomena are to be ranked the useless mammae of the male human being, and the unrequired process of bone in the male opossum, which is needed in the female for supporting her pouch. Such curious features are most conspicuous in animals which form links between various classes . . . .

These facts clearly show how all the various organic forms of our world are bound up in one—how a fundamental unity pervades and embraces them all, collecting them, from the humblest lichen up to the highest mammifer, in one system, the whole creation of which must have depended upon one law or decree of the Almighty, though it did not all come forth at one time. After what we have seen, the idea of a separate exertion for each must appear totally inadmissible. The single fact of abortive or rudimentary organs condemns it; for these, on such a supposition, could be regarded in no other light than as blemishes or blunders—the thing of all others most irreconcilable with that idea of Almighty Perfection which a general view of nature so irresistibly conveys. On the other hand, when the organic creation is admitted to have been effected by a general law, we see nothing in these abortive parts but harmless peculiarities of development, and interesting evidences of the manner in which the Divine Author has been pleased to work.

We have yet to advert to the most interesting class of facts connected with the laws of organic development. It is only in recent times that physiologists have observed that each animal passes, in the course of its germinal history, through a series of changes resembling the *permanent forms* of the various orders of animals inferior to it in the scale. Thus, for instance, an insect, standing at the head of the articulated animals, is, in the larva state, a true annelid, or worm, the annelida being the lowest in the same class. The embryo of a crab resembles the perfect animal of the inferior order myriapoda, and passes through all the forms of transition which characterize the intermediate tribes of crustacea. The frog, for some time after its birth, is a fish with external gills, and other organs fitting it for an aquatic life, all of which are changed as it advances to maturity, and becomes a land animal. The mammifer only passes through still more stages, according to its higher place in the scale. Nor is man himself exempt from this law. His first form is that which is permanent in the animalcule. His organization gradually passes through conditions generally resembling a fish, a reptile, a bird, and the lower mammalia, before it attains its specific maturity. At one of the last stages of his foetal career, he exhibits an intermaxillary bone, which is characteristic of the perfect ape; this is suppressed, and he may then be said to take leave of the simial type, and become a true human creature. Even, as we shall see, the varieties of his race are represented in the progressive development of an individual of the highest, before we see the adult Caucasian, the highest point yet attained in the animal scale.

To come to particular points of the organization. The brain of man, which exceeds that of all other animals in complexity of organization and fullness of development, is, at one early period, only "a simple fold of nervous matter, with difficulty distinguishable into three parts, while a little tail-like prolongation towards the hinder parts, and which had been the first to appear, is the only representation of a spinal marrow. Now, in this state it perfectly resembles the brain of an adult fish, thus assuming *in transitu* the form that in the fish is permanent. In a short time, however, the structure is become more complex, the parts more distinct, the spinal marrow better marked; it is now the brain of a reptile. The change continues; by a singular motion, certain parts (*corpora quadragemina*)

CRUST OF THE EARTH AS RELATED TO ZOÖLOGY.

Cuvier had inspired Agassiz and others to see the different forms of animals as fixed within separate branches of creation. The organizing principles of that system of nature were shaped by commitment to a Creator. [Author's collection.]

which had hitherto appeared on the upper surface, now pass towards the lower; the former is their permanent situation in fishes and reptiles, the latter in birds and mammalia. This is another advance in the scale, but more remains yet to be done. The complication of the organ increases; cavities termed *ventricles* are formed, which do not exist in fishes, reptiles, or birds; curiously organized parts, such as the corpora striata, are added; it is now the brain of the mammalia. Its last and final change alone seems wanting, that which shall render it the brain of MAN." And this change in time takes place.

So also with the heart. This organ, in the mammalia, consists of four cavities, but in the reptiles of only three, and in fishes of two only, while in the articulated animals it is merely a prolonged tube. Now in the mammal foetus, at a certain early stage, the organ has the form of a prolonged tube; and a human being may be said to have then the heart of an insect. Subsequently it is shortened and widened, and becomes divided by a contraction into two parts, a ventricle and an auricle; it is now the heart of a fish. A subdivision of the auricle afterwards makes a triple-chambered form, as in the heart of the reptile tribes; lastly, the ventricle being also subdivided, it becomes a full mammal heart . . . .

The tendency of all these illustrations is to make us look to *development* as the principle which has been immediately concerned in the peopling of this globe, a process extending over a vast space of time, but which is nevertheless connected in character with the briefer process by which an individual being is evoked from a simple germ. What mystery is there here—and how shall I proceed to enunciate the conception which I have ventured to form of what may prove to be its proper solution! It is an idea by no means calculated to impress by its greatness, or to puzzle by its profoundness. It is an idea more marked by simplicity than perhaps any other of those which have explained the great secrets of nature. But in this lies, perhaps, one of its strongest claims to the faith of mankind.

The whole train of animated beings, from the simplest and oldest up to the highest and most recent, are, then, to be regarded as a series of *advances of the principle of development*, which have depended upon external physical circumstances, to which the resulting animals are appropriate. I contemplate the whole phenomena as having been in the first place arranged in the counsels of Divine Wisdom, to take place, not only upon this sphere, but upon all the others in space, under necessary modifications, and as being carried on, from first to last, here and elsewhere, under immediate favour of the creative will or energy. The nucleated vesicle, the fundamental form of all organization, we must regard as the meeting-point between the inorganic and the organic—the end of the mineral and beginning of the vegetable and animal kingdoms, which thence start in different directions, but in perfect parallelism and analogy. We have already seen that this nucleated vesicle is itself a type of mature and independent being in the infusory animalcules, as well as the starting point of the foetal progress of every higher individual in creation, both animal and vegetable. We have seen that it is a form of being which electric agency will produce—though not perhaps usher into full life-in albumen, one of those compound elements of animal bodies, of which another (urea) has been made by artificial means. Remembering these things, we are drawn on to the supposition, that the first step in the creation of life upon this planet was *a chemico-electric operation, by which simple germinal vesicles were produced*. This is so much, but what were the next steps? Let a common vegetable infusion help us to an answer. There, as we have seen, simple forms are produced at first, but afterwards they become more complicated, until at length the life-producing powers of the infusion are exhausted. Are we to presume that, in this case, the simple engender the complicated? Undoubtedly, this would not be more wonderful as a natural process than one which we never think of wondering at, because familiar to us—namely, that in the gestation of the mammals, the animalcule-like ovum of a few days is the parent, in a sense, of the chick-like form of a few weeks, and that in all the subsequent stages-fish, reptile, &c.—the one may, with scarcely a metaphor, be said to be the progenitor of the other. I suggest, then, as an hypothesis already countenanced by much that is ascertained, and likely to be further sanctioned by much that remains to be known, that the first step was *an advance under favour of peculiar conditions, from the simplest forms of being, to the next more complicated, and this through the medium of the ordinary process of generation* . . . .

It has been seen that, in the reproduction of the higher animals, the new being passes through stages in which it is successively fish-like and reptile-like. But the resemblance is not to the adult fish or the adult reptile, but to the fish and reptile at a certain point in their foetal progress; this holds true with regard to the vascular, nervous, and other systems alike. It may be illustrated by a simple diagram. The foetus of all the four classes may be supposed to advance in an identical condition to the point A. The fish there diverges and passes along a line apart, and peculiar to itself, to its mature state at F. The reptile, bird, and mammal, go on together to C, where the reptile diverges in like manner, A and advances by itself to R. The bird diverges at D, and goes on to B. The mammal then goes forward in a straight line to the highest point of organization at M. This diagram shows only the main ramifications; but the reader must suppose minor ones, representing the subordinate differences of orders, tribes, families, genera, &c., if he wishes to extend his views to the whole varieties of being in the animal kingdom. Limiting ourselves at present to the outline afforded by this diagram, it is apparent that the only thing required for an advance from one type to another in the generative process is that, for example, the fish embryo should not diverge at A, but go on to C before it diverges, in which case the progeny will be, not a fish, but a reptile. To protract the *straightforward part of the gestation over a small space*—and from species to species the space would be small indeed—is all that is necessary.

This might be done by the force of certain external conditions operating upon the parturient system. The nature of these conditions we can only conjecture, for their operation, which in the geological eras was so powerful, has in its main strength been long interrupted, and is now perhaps only allowed to work in some of the lowest departments of the organic world, or under extraordinary casualties in some of the higher, and to these points the attention of science has as yet been little directed. But though this knowledge were never to be clearly attained, it need not much affect the present argument, provided it be satisfactorily shown that there must be some such influence within the range of natural things.

To this conclusion it must be greatly conducive that the law of organic development is still daily seen at work to certain effects, only somewhat short of a transition from species to species. Sex we have seen to be a matter of development. There is an instance, in a humble department of the animal world, of arrangements being made by the animals themselves for adjusting this law to the production of a particular sex. Amongst bees, as amongst several other insect tribes, there is in each community but one true female, the queen bee, the workers being false females or neuters; that is to say, sex is carried on in them to a point where it is attended by sterility. The preparatory states of the queen bee occupy sixteen days; those of the neuters, twenty; and those of males, twenty-four. Now it is a fact, settled by innumerable observations and experiments, that the bees can so modify a worker in the larva state, that, when it emerges from the pupa, it is found to be a queen or true female. For this purpose they enlarge its cell, make a pyramidal hollow to allow of its assuming a vertical instead of a horizontal position, keep it warmer than other larvæ are kept, and feed it with a peculiar kind of food. From these simple circumstances, leading to a shortening of the embryotic condition, results a creature different in form, and also in dispositions, from what would have otherwise been produced. Some of the organs possessed by the worker are here altogether wanting. We have a creature "destined to enjoy love, to burn with jealousy and anger, to be incited to vengeance, and to pass her time without labour," instead of one "zealous for the good of the community, a defender of the public rights, enjoying an immunity from the stimulus of sexual appetite and the pains of parturition; laborious, industrious, patient, ingenious, skilful; incessantly engaged in the nurture of the young, in collecting honey and pollen, in elaborating wax, in constructing cells and the like!-paying the most respectful and assiduous attention to objects which, had its ovaries been developed, it would have hated and pursued with the most vindictive fury till it had destroyed them!" All these changes may be produced by a mere modification of the embryotic progress, which it is within the power of the adult animals to effect. But it is important to observe that this modification is different from working a direct change upon the embryo. It is not the different food which effects a metamorphosis. All that is done is merely to accelerate the period of the insect's perfection. By the arrangements made and the food given, the embryo becomes sooner fit for being ushered forth in its image or perfect state. Development may be said to be thus arrested at a particular stage—that early one at which the female sex is complete. In the other circumstances, it is allowed to go on four days longer, and a stage is then reached between the two sexes, which in this species is designed to be the perfect condition of a large portion of the community. Four days more make it a perfect male. It is at the same time to be observed that there is, from the period of oviposition, a destined distinction between the sexes of the young bees. The queen lays the whole of the eggs which are designed to become workers, before she begins to lay those which become males. But probably the condition of her reproductive system governs the matter of sex, for it is remarked that when her impregnation is delayed beyond the twenty-eighth day of her entire existence, she lays only eggs which become males.

We have here, it will be admitted, a most remarkable illustration of the principle of development, although in an operation limited to the production of sex only. Let it not be said that the phenomena concerned in the generation of bees may be very different from those concerned in the reproduction of the higher animals. There is a unity throughout nature which makes the one case an instructive reflection of the other.

We shall now see an instance of development operating within the production of what approaches to the character of variety of species. It is fully established that a human family, tribe, or nation, is liable, in the course of generations, to be either advanced from a mean form to a higher one, or degraded from a higher to a lower, by the influence of the physical conditions in which it lives. The coarse features, and other structural peculiarities of the negro race only continue while these people live amidst the circumstances usually associated with barbarism. In a more temperate clime, and higher racial state, the face and figure become greatly refined. The few African nations which possess any civilization also exhibit forms approaching the European; and when the same people in the United States of America have enjoyed a within-door life for several generations, they assimilate to the whites amongst whom they live. On the other hand, there are authentic instances of a people originally well-formed and good-looking, being brought, by imperfect diet and a variety of physical hardships, to a meaner form. It is remarkable that prominence of the jaws, a recession and diminution of the cranium, and an elongation and attenuation of the limbs, are peculiarities always produced by these miserable conditions, for they indicate an unequivocal retrogression towards the type of the lower animals. Thus we see nature alike willing to go back and to go forward. Both effects are simply the result of the operation of the law of development in the generative system. Give good conditions, it advances; bad ones, it recedes. Now, perhaps, it is only because there is no longer a possibility, in the higher types of being, of giving sufficiently favourable conditions to carry on species to species, that we see the operation of the law so far limited.

Let us trace this law also in the production of certain classes of monstrosities. A human foetus is often left with one of the most important parts of its frame imperfectly developed: the heart, for instance, goes no farther than the three-chambered form, so that it is the heart of a reptile. There are even instances of this organ being left in the two-chambered or fish form. Such defects are the result of nothing more than a failure of the power of development in the system of the mother, occasioned by weak health or misery. Here we have apparently a realization of the converse of those conditions which carry on species to species, so far, at least, as one organ is concerned. Seeing a complete specific retrogression in this one point, how easy it is to imagine an access of favourable conditions sufficient to reverse the phenomenon, and make a fish mother develop a reptile heart, or a reptile mother develop a mammal one. It is no great boldness to surmise that a super-adequacy in the measure of this under-adequacy (and the one thing seems as natural an occurrence as the other) would suffice in a goose to give its progeny the body of a rat, and produce the ornithorynchus, or might give the progeny of an ornithorynchus the mouth and feet of a true rodent, and thus complete at two stages the passage from the aves to the mammalia.

Perhaps even the transition from species to species does still take place in some of the obscurer fields of creation, or under extraordinary casualties, though science professes to have no such facts on record. It is here to be remarked, that such facts might often happen, and yet no record be taken of them, for so strong is the prepossession for the doctrine of invariable like-production, that such circumstances, on occurring, would be almost sure to be explained away on some other supposition, or, if presented, would be disbelieved and neglected. Science, therefore, has no such facts, for the very same reason that some small sects are said to have no discreditable members—namely, that they do not receive such persons, and extrude all who begin to verge upon the character There are, nevertheless, some facts which have chanced to be reported without any reference to this hypothesis, and which it seems extremely difficult to explain satisfactorily upon any other. One of these has already been mentioned—a progression in the forms of the animalcules in a vegetable infusion from the simpler to the more complicated, a sort of microcosm, representing the whole history of the progress of animal creation as displayed by geology. Another is given in the history of the Acarus Crossii which may be only the ultimate stage of a series of similar transformations effected by electric agency in the solution subjected to it. There is, however, one direct case of a translation of species, which has been presented with a respectable amount of authority. It appears that, whenever oats sown at the usual time are kept cropped down during summer and autumn, and allowed to remain over the winter, a thin crop of rye is the harvest presented at the close of the ensuing summer. This experiment has been tried repeatedly, with but one result; invariably the *secale cereale* is the crop reaped where the *avena sativa*, a recognised different species, was sown. Now it will not satisfy a strict inquirer to be told that the seeds of the rye were latent in the ground and only superseded the dead product of the oats; for if any such fact were in the case, why should the usurping grain be always rye? Perhaps those curious facts which have been stated with regard to forests of one kind of trees, when burnt down, being succeeded (without planting) by other kinds, may yet be found most explicable, as this is, upon the hypothesis of a progression of species which takes place under certain favouring conditions, now apparently of comparatively rare occurrence. The case of the oats is the more valuable, as bearing upon the suggestion as to a protraction of the gestation at a particular part of its course. Here, the generative process is, by the simple mode of cropping down, kept up for a whole year beyond its usual term. The type is thus allowed to advance, and what was oats becomes rye.

The idea, then, which I form of the progress of organic life upon the globe—and the hypothesis is applicable to all similar theatres of vital being—is, *that the simplest and most primitive type, under a law to which that of like-production is subordinate, gave birth to the type next above it, that this again produced the next higher, and so on to the very highest*, the stages of advance being in all cases very small—namely, from one species only to another; so that the phenomenon has always been of a simple and modest character. Whether the whole of any species was at once translated forward, or only a few parents were employed to give birth to the new type, must remain undetermined; but, supposing that the former was the case, we must presume that the moves along the line or lines were simultaneous, so that the place vacated by one species was immediately taken by the next in succession, and so on back to the first, for the supply of which the foundation of a new germinal vesicle out of inorganic matter was alone necessary. Thus, the production of new forms, as shown in the pages of the geological record, has never been anything more than a new stage of progress in gestation, an event as simply natural, and attended as little by any circumstances of a wonderful or startling kind, as the silent advance of an ordinary mother from one week to another of her pregnancy. Yet, be it remembered, the whole phenomena are, in another point of view, wonders of the highest kind, for in each of them we have to trace the effect of an Almighty Will which had arranged the whole in such harmony with external physical circumstances, that both were developed in parallel steps—and probably this development upon our planet is but a sample of what has taken place, through the same cause, in all the other countless theatres of being which are suspended in space.

SOURCE: Chambers, Robert. *Vestiges of the Natural History of Creation*. London: J. Churchill, 1844.

TERMS

Argument from Design—an argument in favor of the existence of a supernatural deity or deities based on the recognition that nature is complex, self-regulating, and appears to operate according to fixed, universal laws. The origin of the argument is ancient; in *Timaeus* Plato posited the existence of a demiurge, who could create something from nothing, and Aristotle described the Prime Mover, who set the universe in motion. In the middle ages and early modern period, both the scholastics and the empiricists often employed the argument from design. At the beginning of the nineteenth century William Paley published his *Natural Theology*, which was an especially influential version of the argument from design. More recently, the argument from design has been advanced by advocates of intelligent design, who see in the regularity and complexity of nature evidence for the existence of a supernatural designer.

Deism—the belief that a God or gods created the universe and everything in it, as well as the laws by which it operates. Deists believe that they can come to understand the nature and intentions of the Creator or creators through a careful, logical analysis of nature and sometimes scripture. However, unlike theists, deists reject the use of revelation as a useful tool in studying the supernatural or the natural. Moreover, deists generally reject the notion that the original Creator or creators continue to influence the universe today.

Revelation—the communication of knowledge directly from God to humans. It is often contrasted with reason in that humans acquire revealed knowledge in whole, rather than generating it themselves. There are two types of revelation: natural and special. Natural revelation comes through the analysis of creation and reveals God's power, appreciation of order and regularity, and His concern for human welfare. Natural revelation alone cannot bring about salvation according to Christian teachings; salvation comes only through special revelation, which is the provided to humans through the Prophets and through scripture.

Theism—a term that emerged in the seventeenth century as a contrast to atheism. Theists generally believe in the existence of a God or gods who created the universe as well as everything in it including the laws by which it operates. Most theists believe that this deity or deities are still actively engaged in the world and can be understood indirectly through the study of nature or scripture and directly through revelation.

2

DARWIN'S THEORY OF
NATURAL SELECTION

Charles Darwin, a young man when he left England aboard the *H.M.S. Beagle* in 1831, joined Captain Robert Fitzroy as a companion for the long voyage to South America. Darwin's lack of experience as a naturalist made him a distant choice for that job, but he quickly proved himself superior to the man who had originally been assigned the task of collecting zoological and botanical specimens. Officially taking on the role of ship's naturalist, Darwin observed carefully, read widely, and collected incessantly as the *Beagle* made its way around South America.

Textbook accounts of Darwin's work often dramatize the voyage. The story typically goes: He awoke to the diversity of life and "discovered" evolution on remote islands; he gathered fossils that showed the stepwise change in species from primitive to complex, connecting the unthinkable depths of time to the enlightened present. In fact, the evidence of history provides a far more interesting tale. Darwin wrestled with his discoveries, hoped against certain conclusions, and withheld judgment on countless species. He even failed, in some respects, to grasp the full meaning of the observations he made. In the decades that followed, he relied on the notes of Captain Fitzroy and the analysis of leading naturalists and philosophers in England, Europe, and around the world to provide a compelling case that he eventually revealed somewhat reluctantly.

Darwin worked for decades to accumulate evidence for his theory of natural selection, knowing how other naturalists would scrutinize his ideas. He wanted to be sure, so he planned a massive and encyclopedic account of change in populations. When Alfred Russel Wallace's own ideas threatened to eclipse the priority of Darwin's theory, Darwin had to quickly write a concise volume that summarized the mechanism of natural selection with key examples. He chose those examples carefully, attending especially to the analogy with domestic breeding, where the process of selection appeared explicitly.

Selection also related, in Darwin's view, to the well-established observations of the eighteenth-century naturalist Carolus Linnaeus. Features such as camouflage provided some protection from predators, and yet every species was subject to destruction by others. For Linnaeus, the Creator had established a self-preserving and well-ordered nature. Darwin's nature could operate without divine guidance, and he described in detail how the interactions among species produced change with new and beneficial variations. Natural selection examined the survival potential of individuals in the wild just as surely as breeders examined the traits of dogs, cattle, sheep, and pigeons. He pointed out that survival traits could include physical characters as well as behavioral features of animals. Readers who have missed the significance of this analogy have wondered why a book about dogs and pigeons could have caused such a stir.

Like Darwin, Alfred Russel Wallace traveled extensively and collected species around the world. He began to turn his attention to the distribution of organisms in search of a mechanism that might explain the unity and diversity of living things. He relied on Linnaeus and others for the frameworks that suggested a connection between geology and geography that would explain biological change. Again, like Darwin, he acknowledged that Linnaeus had relied heavily on the concept of a creator. Unlike Darwin, Wallace had few connections to the elite of British or European science. He had financial difficulties that Darwin did not face, but perhaps more significantly, he struggled to link his observations and collections with the broader community of naturalists who might help him make sense of what he found.

Ultimately, Darwin recognized Wallace's contributions. They gave a joint presentation of their conclusions regarding questions of the unity and diversity of living things distributed around the globe. For the general community of naturalists who had considered these questions, the two men had clearly arrived at a nearly identical theory at approximately the same time. Given that Darwin had begun his work earlier, documented it more consistently (much of Wallace's work was lost in a shipwreck), and amassed a much greater quantity and quality of evidence, Wallace could not command the same priority for the theory. Darwin generously shared recognition for their accomplishment with Wallace at the time, although most accounts have focused on Darwin's role.

Soon after Darwin published *On the Origin of Species*, his friend Thomas Huxley took up the role as "Darwin's Bulldog," a tireless advocate for the ideas of his somewhat reluctant colleague. Famously, Huxley debated opponents of natural selection, including Bishop Samuel Wilberforce, in open forums and public letters. He made unique and important contributions in defense of Darwin, and also opened new avenues of argument.

In his earlier publications, Darwin so carefully avoided discussion of the place of humans in the broader scheme of evolution that he could scarcely escape the topic without a full volume on the subject. Coming near the end of his life, *The Descent of Man* represented as full an exploration of the human condition as it related to biological change as anyone at the time could have achieved. Darwin's own considered reflections provided an enduring and compelling set of questions about human origins and behavior that remain at the heart of behavioral biology and psychology.

In the end, it was his fairest hope that awareness of human descent would inspire individuals and societies to work toward the furtherance of a trend toward improvement that had begun at the dawn of human ancestry and could continue beyond the current distinction.

CHARLES DARWIN, *THE VOYAGE OF THE BEAGLE* (1839)

The collections that Charles Darwin (1809–1882) took from the Galapagos Islands famously record the diversity of species from that unique archipelago. In this excerpt from his travelogue account of the journey, the young naturalist revealed his careful attention to geological details, his fascination with the giant tortoises, his awareness of the changes brought by human agriculture in recent years, and the unique characteristics of bird species on various islands. Perhaps most significantly, Darwin recognized that a geological timeframe suggested features had more in common across space and time. He began to suggest that the distances between islands, and from the mainland of South America might correlate with the appearance of a few ancestral forms that changed in distinctive ways to suit their new environs. Such changes, he noted in understated fashion, could happen within geological timeframes to which most readers at the time would be unaccustomed. As a naturalist and reader of Charles Lyell's *Principles of Geology*, however, Darwin did not seek explanations that required the compression of ancestry into a few thousand years' time. Four species of birds he collected there formed the core of a grouping now referred to as Darwin's finches. At the time, the young naturalist misidentified all four. Back in England, ornithologist John Gould corrected Darwin. Gould relied on the additional collections of Fitzroy, noting that the finches formed a cohesive group. From this, Darwin could comment on the "perfect gradation in the size of the beaks. . .." This conclusion, the result of months of additional investigation and collaboration belies the typical textbook history that Darwin realized his radical theory by brilliant insight alone.

Chapter 17

September 15th—This archipelago consists of ten principal islands, of which five exceed the others in size. They are situated under the Equator, and between five and six hundred miles westward of the coast of America. They are all formed of volcanic rocks; a few fragments of granite curiously glazed and altered by the heat, can hardly be considered as an exception. Some of the craters, surmounting the larger islands, are of immense size, and they rise to a height of between three and four thousand feet. Their flanks are studded by innumerable smaller orifices. I scarcely hesitate to affirm, that there must be in the whole archipelago at least two thousand craters. These consist either of lava or scoriae, or of finely-stratified, sandstone-like tuff. Most of the latter are beautifully symmetrical; they owe their origin to eruptions of volcanic mud without any lava: it is a remarkable circumstance that every one of the twenty-eight tuff-craters which were examined, had their southern sides either much lower than the other sides, or quite broken down and removed. As all these craters apparently have been formed when standing in the sea, and as the waves from the trade wind and the swell from the open Pacific here unite their forces on the southern coasts of all the islands, this singular uniformity in the broken state of the craters, composed of the soft and yielding tuff, is easily explained.

Considering that these islands are placed directly under the equator, the climate is far from being excessively hot; this seems chiefly caused by the singularly low temperature of the surrounding water, brought here by the great southern Polar current. Excepting during one short season, very little rain falls, and even then it is irregular; but the clouds generally hang low. Hence, whilst the lower parts of the islands are very sterile, the upper parts, at a height of a thousand feet and upwards, possess a damp climate and a tolerably luxuriant vegetation. This is especially the case on the windward sides of the islands, which first receive and condense the moisture from the atmosphere.

In the morning (17th) we landed on Chatham Island, which, like the others, rises with a tame and rounded outline, broken here and there by scattered hillocks, the remains of former craters. Nothing could be less inviting than the first appearance. A broken field of black basaltic lava, thrown into the most rugged waves, and crossed by great fissures, is everywhere covered by stunted, sun-burnt brushwood, which shows little signs of life. The dry and parched surface, being heated by the noon-day sun, gave to the air a close and sultry feeling, like that from a stove: we fancied even that the bushes smelt unpleasantly. Although I diligently tried to collect as many plants as possible, I succeeded in getting very few; and such wretched-looking little weeds would have better become an arctic than an equatorial Flora. The brushwood appears, from a short distance, as leafless as our trees during winter; and it was some time before I discovered that not only almost every plant was now in full leaf, but that the greater number were in flower. The commonest bush is one of the Euphorbiaceae: an acacia and a great odd-looking cactus are the only trees which afford any shade. After the season of heavy rains, the islands are said to appear for a short time partially green. The volcanic island of Fernando Noronha, placed in many respects under nearly similar conditions, is the only other country where I have seen a vegetation at all like this of the Galapagos Islands.

The *Beagle* sailed round Chatham Island, and anchored in several bays. One night I slept on shore on a part of the island, where black truncated cones were extraordinarily numerous: from one small eminence I counted sixty of them, all surmounted by craters more or less perfect. The greater number consisted merely of a ring of red scoriae or slags, cemented together: and their height above the plain of lava was not more than from fifty to a hundred feet; none had been very lately active. The entire surface of this part of the island seems to have been permeated, like a sieve, by the subterranean vapors: here and there the lava, whilst soft, has been blown into great bubbles; and in other parts, the tops of caverns similarly formed have fallen in, leaving circular pits with steep sides. From the regular form of the many craters, they gave to the country an artificial appearance, which vividly reminded me of those parts of Staffordshire, where the great iron-foundries are most numerous. The day was glowing hot, and the scrambling over the rough surface and through the intricate thickets, was very fatiguing; but I was well repaid by the strange Cyclopean scene. As I was walking along I met two large tortoises, each of which must have weighed at least two hundred pounds: one was eating a piece of cactus, and as I approached, it stared at me and slowly walked away; the other gave a deep hiss, and drew in its head. These huge reptiles, surrounded by the black lava, the leafless shrubs, and large cacti, seemed to my fancy like some antediluvian animals. The few dull-coloured birds cared no more for me than they did for the great tortoises.

23rd—The *Beagle* proceeded to Charles Island. This archipelago has long been frequented, first by the bucaniers, and latterly by whalers, but it is only within the last six years, that a small colony has been established here. The inhabitants are between two and three hundred in number; they are nearly all people of colour, who have been banished for political crimes from the Republic of the Equator, of which Quito is the capital. The settlement is placed about four and a half miles inland, and at a height probably of a thousand feet. In the first part of the road we passed through leafless thickets, as in Chatham Island. Higher up, the woods gradually became greener; and as soon as we crossed the ridge of the island, we were cooled by a fine southerly breeze, and our sight refreshed by a green and thriving vegetation. In this upper region coarse grasses and ferns abound; but there are no tree-ferns: I saw nowhere any member of the palm family, which is the more singular, as 360 miles northward, Cocos Island takes its name from the number of cocoa-nuts. The houses are irregularly scattered over a flat space of ground, which is cultivated with sweet potatoes and bananas. It will not easily be imagined how pleasant the sight of black mud was to us, after having been so long, accustomed to the parched soil of Peru and northern Chile. The inhabitants, although complaining of poverty, obtain, without much trouble, the means of subsistence. In the woods there are many wild pigs and goats; but the staple article of animal food is supplied by the tortoises. Their numbers have of course been greatly reduced in this island, but the people yet count on two days' hunting giving them food for the rest of the week. It is said that formerly single vessels have taken away as many as seven hundred, and that the ship's company of a frigate some years since brought down in one day two hundred tortoises to the beach.

September 29th—We doubled the south-west extremity of Albemarle Island, and the next day were nearly becalmed between it and Narborough Island. Both are covered with immense deluges of black naked lava, which have flowed either over the rims of the great caldrons, like pitch over the rim of a pot in which it has been boiled, or have burst forth from smaller orifices on the flanks; in their descent they have spread over miles of the sea-coast. On both of these islands, eruptions are known to have taken place; and in Albemarle, we saw a small jet of smoke curling from the summit of one of the great craters. In the evening we anchored in Bank's Cove, in Albemarle Island. The next morning I went out walking. To the south of the broken tuff-crater, in which the *Beagle* was anchored, there was another beautifully symmetrical one of an elliptic form; its longer axis was a little less than a mile, and its depth about 500 feet. At its bottom there was a shallow lake, in the middle of which a tiny crater formed an islet. The day was overpoweringly hot, and the lake looked clear and blue: I hurried down the cindery slope, and, choked with dust, eagerly tasted the water—but, to my sorrow, I found it salt as brine.

The rocks on the coast abounded with great black lizards, between three and four feet long; and on the hills, an ugly yellowish-brown species was equally common. We saw many of this latter kind, some clumsily running out of the way, and others shuffling into their burrows. I shall presently describe in more detail the habits of both these reptiles. The whole of this northern part of Albemarle Island is miserably sterile.

October 8th—We arrived at James Island: this island, as well as Charles Island, were long since thus named after our kings of the Stuart line. Mr. Bynoe, myself, and our servants were left here for a week, with provisions and a tent, whilst the *Beagle* went for water. We found here a party of Spaniards, who had been sent from Charles Island to dry fish, and to salt tortoise-meat. About six miles inland, and at the height of nearly 2000 feet, a hovel had been built in which two men lived, who were employed in catching tortoises, whilst the others were fishing on the coast. I paid this party two visits, and slept there one night. As in the other islands, the lower region was covered by nearly leafless bushes, but the trees were here of a larger growth than elsewhere, several being two feet and some even two feet nine inches in diameter. The upper region being kept damp by the clouds, supports a green and flourishing vegetation. So damp was the ground, that there were large beds of a coarse cyperus, in which great numbers of a very small water-rail lived and bred. While staying in this upper region, we lived entirely upon tortoise-meat: the breast-plate roasted (as the Gauchos do *carne con cuero*), with the flesh on it, is very good; and the young tortoises make excellent soup; but otherwise the meat to my taste is indifferent.

One day we accompanied a party of the Spaniards in their whale-boat to a salina, or lake from which salt is procured. After landing, we had a very rough walk over a rugged field of recent lava, which has almost surrounded a tuff-crater, at the bottom of which the salt-lake lies. The water is only three or four inches deep, and rests on a layer of beautifully crystallized, white salt. The lake is quite circular, and is fringed with a border of bright green succulent plants; the almost precipitous walls of the crater are clothed with wood, so that the scene was altogether both picturesque and curious. A few years since, the sailors belonging to a sealing-vessel murdered their captain in this quiet spot; and we saw his skull lying among the bushes.

During the greater part of our stay of a week, the sky was cloudless, and if the trade-wind failed for an hour, the heat became very oppressive. On two days, the thermometer within the tent stood for some hours at 93 degrees; but in the open air, in the wind and sun, at only 85 degrees. The sand was extremely hot; the thermometer placed in some of a brown colour immediately rose to 137 degrees, and how much above that it would have risen, I do not know, for it was not graduated any higher. The black sand felt much hotter, so that even in thick boots it was quite disagreeable to walk over it.

The natural history of these islands is eminently curious, and well deserves attention. Most of the organic productions are aboriginal creations, found nowhere else; there is even a difference between the inhabitants of the different islands; yet all show a marked relationship with those of America, though separated from that continent by an open space of ocean, between 500 and 600 miles in width. The archipelago is a little world within itself, or rather a satellite attached to America, whence it has derived a few stray colonists, and has received the general character of its indigenous productions. Considering the small size of the islands, we feel the more astonished at the number of their aboriginal beings, and at their confined range. Seeing every height crowned with its crater, and the boundaries of most of the lava-streams still distinct, we are led to believe that within a period geologically recent the unbroken ocean was here spread out. Hence, both in space and time, we seem to be brought somewhat near to that great fact—that mystery of mysteries—the first appearance of new beings on this earth.

Of terrestrial mammals, there is only one which must be considered as indigenous, namely, a mouse (*Mus galapagoensis*), and this is confined, as far as I could ascertain, to Chatham Island, the most easterly island of the group. It belongs, as I am informed by Mr. Waterhouse, to a division of the family of mice characteristic of America. At James Island, there is a rat sufficiently distinct from the common kind to have been named and described by Mr. Waterhouse; but as it belongs to the old-world division of the family, and as this island has been frequented by ships for the last hundred and fifty years, I can hardly doubt that this rat is merely a variety produced by the new and peculiar climate, food, and soil, to which it has been subjected. Although no one has a right to speculate without distinct facts, yet even with respect to the Chatham Island mouse, it should be borne in mind, that it may possibly be an American species imported here; for I have seen, in a most unfrequented part of the Pampas, a native mouse living in the roof of a newly built hovel, and therefore its transportation in a vessel is not improbable: analogous facts have been observed by Dr. Richardson in North America.

Of land-birds I obtained twenty-six kinds, all peculiar to the group and found nowhere else, with the exception of one lark-like finch from North America (*Dolichonyx oryzivorus*), which ranges on that continent as far north as 54 degrees, and generally frequents marshes. The other twenty-five birds consist, firstly, of a hawk, curiously intermediate in structure between a buzzard and the American group of carrion-feeding Polybori; and with these latter birds it agrees most closely in every habit and even tone of voice. Secondly, there are two owls, representing the short-eared and white barn-owls of Europe. Thirdly, a wren, three tyrant-flycatchers (two of them species of Pyrocephalus, one or both of which would be ranked by some ornithologists as only varieties), and a dove—all analogous to, but distinct from, American species. Fourthly, a swallow, which though differing from the Progne purpurea of both Americas, only in being rather duller colored, smaller, and slenderer, is considered by Mr. Gould as specifically distinct. Fifthly, there are three species of mocking thrush—a form highly characteristic of America. The remaining land-birds form a most singular group of finches, related to each other in the structure of their beaks, short tails, form of body and plumage: there are thirteen species, which Mr. Gould has divided into four subgroups. All these species are peculiar to this archipelago; and so is the whole group, with the exception of one species of the sub-group Cactornis, lately brought from Bow Island, in the Low Archipelago. Of Cactornis, the two species may be often seen climbing about the flowers of the great cactus-trees; but all the other species of this group of finches, mingled together in flocks, feed on the dry and sterile ground of the lower districts. The males of all, or certainly of the greater number, are jet black; and the females (with perhaps one or two exceptions) are brown. The most curious fact is the perfect gradation in the size of the beaks in the different species of Geospiza, from one as large as that of a hawfinch to that of a chaffinch, and (if Mr. Gould is right in including his sub-group, Certhidea, in the main group) even to that of a warbler. The largest beak in the genus Geospiza is shown in Fig. 1, and the smallest in Fig. 3; but instead of there being only one intermediate species, with a beak of the size shown in Fig. 2, there are no less than six species with insensibly graduated beaks. The beak of the sub-group Certhidea, is shown in Fig. 4. The beak of Cactornis is somewhat like that of a starling, and that of the fourth subgroup, Camarhynchus, is slightly parrot-shaped. Seeing this gradation and diversity of structure in one small, intimately related group of birds, one might really fancy that from an original paucity of birds in this archipelago, one species had been taken and modified for different ends. In a like manner it might be fancied that a bird originally a buzzard, had been induced here to undertake the office of the carrion-feeding Polybori of the American continent.

SOURCE: Darwin, Charles. *The Voyage of the Beagle.* London: J.M. Dent, 1839.

CHARLES DARWIN, ON *THE ORIGIN OF SPECIES* (1859)

Darwin expressed at the beginning of this excerpt a belief that the process of natural selection could work relentlessly across landscapes and time to perfect constantly varying living things. He had not set out on his quest to understand the mechanism of evolution to eliminate the need for a creator, but earlier philosophers and naturalists had inspired him to search for natural and material causes. He was determined not to rely on supernatural explanations in formulating his scientific theory. As a consequence, he reached a conclusion that essentially denied a place for a creator in natural history, and that conclusion caused him great consternation. Near the end of the book, Darwin acknowledged the attractiveness of an explanation that maintained the role of a designer, given that the human imagination much more readily recognizes complexity as the product of design. Yet, by laying down the simple steps and logical links of natural selection, he proposed an indisputable mechanism for change that could lead to complex forms, and not just complex but beautiful and wonderful. Throughout his exploration of change, Darwin referred only once to evolution in the book, in the very last word.

Chapter 4

Charles Darwin [Library of Congress Prints and Photographs Department, LC-USZ63-52389].

It may be said that natural selection is daily and hourly scrutinising, throughout the world, every variation, even the slightest; rejecting that which is bad, preserving and adding up all that is good; silently and insensibly working, whenever and wherever opportunity offers, at the improvement of each organic being in relation to its organic and inorganic conditions of life. We see nothing of these slow changes in progress, until the hand of time has marked the long lapse of ages, and then so imperfect is our view into long past geological ages, that we only see that the forms of life are now different from what they formerly were.

Although natural selection can act only through and for the good of each being, yet characters and structures, which we are apt to consider as of very trifling importance, may thus be acted on. When we see leaf-eating insects green, and bark-feeders mottled-grey; the alpine ptarmigan white in winter, the red-grouse the colour of heather, and the black-grouse that of peaty earth, we must believe that these tints are of service to these birds and insects in preserving them from danger. Grouse, if not destroyed at some period of their lives, would increase in countless numbers; they are known to suffer largely from birds of prey; and hawks are guided by eyesight to their prey,—so much so, that on parts of the Continent persons are warned not to keep white pigeons, as being the most liable to destruction. Hence I can see no reason to doubt that natural selection might be most effective in giving the proper colour to each kind of grouse, and in keeping that colour, when once acquired, true and constant. Nor ought we to think that the occasional destruction of an animal of any particular colour would produce little effect: we should remember how essential it is in a flock of white sheep to destroy every lamb with the faintest trace of black. In plants the down on the fruit and the colour of the flesh are considered by botanists as characters of the most trifling importance: yet we hear from an excellent horticulturist, Downing, that in the United States smooth-skinned fruits suffer far more from a beetle, a curculio, than those with down; that purple plums suffer far more from a certain disease than yellow plums; whereas another disease attacks yellow-fleshed peaches far more than those with other coloured flesh. If, with all the aids of art, these slight differences make a great difference in cultivating the several varieties, assuredly, in a state of nature, where the trees would have to struggle with other trees and with a host of enemies, such differences would effectually settle which variety, whether a smooth or downy, a yellow or purple fleshed fruit, should succeed.

In looking at many small points of difference between species, which, as far as our ignorance permits us to judge, seem to be quite unimportant, we must not forget that climate, food, &c., probably produce some slight and direct effect. It is, however, far more necessary to bear in mind that there are many unknown laws of correlation of growth, which, when one part of the organisation is modified through variation, and the modifications are accumulated by natural selection for the good of the being, will cause other modifications, often of the most unexpected nature.

As we see that those variations which under domestication appear at any particular period of life, tend to reappear in the offspring at the same period;—for instance, in the seeds of the many varieties of our culinary and agricultural plants; in the caterpillar and cocoon stages of the varieties of the silkworm; in the eggs of poultry, and in the colour of the down of their chickens; in the horns of our sheep and cattle when nearly adult;—so in a state of nature, natural selection will be enabled to act on and modify organic beings at any age, by the accumulation of profitable variations at that age, and by their inheritance at a corresponding age. If it profit a plant to have its seeds more and more widely disseminated by the wind, I can see no greater difficulty in this being effected through natural selection, than in the cotton-planter increasing and improving by selection the down in the pods on his cotton-trees. Natural selection may modify and adapt the larva of an insect to a score of contingencies, wholly different from those which concern the mature insect. These modifications will no doubt affect, through the laws of correlation, the structure of the adult; and probably in the case of those insects which live only for a few hours, and which never feed, a large part of their structure is merely the correlated result of successive changes in the structure of their larvæ. So, conversely, modifications in the adult will probably often affect the structure of the larva; but in all cases natural selection will ensure that modifications consequent on other modifications at a different period of life, shall not be in the least degree injurious: for if they became so, they would cause the extinction of the species.

Natural selection will modify the structure of the young in relation to the parent, and of the parent in relation to the young. In social animals it will adapt the structure of each individual for the benefit of the community; if each in consequence profits by the selected change. What natural selection cannot do, is to modify the structure of one species, without giving it any advantage, for the good of another species; and though statements to this effect may be found in works of natural history, I cannot find one case which will bear investigation. A structure used only once in an animal's whole life, if of high importance to it, might be modified to any extent by natural selection; for instance, the great jaws possessed by certain insects, and used exclusively for opening the cocoon or the hard tip to the beak of nestling birds, used for breaking the egg. It has been asserted, that of the best short-beaked tumbler-pigeons more perish in the egg than are able to get out of it; so that fanciers assist in the act of hatching. Now, if nature had to make the beak of a full-grown pigeon very short for the bird's own advantage, the process of modification would be very slow, and there would be simultaneously the most rigorous selection of the young birds within the egg, which had the most powerful and hardest beaks, for all with weak beaks would inevitably perish: or, more delicate and more easily broken shells might be selected, the thickness of the shell being known to vary like every other structure. . ..

Illustrations of the action of Natural Selection—In order to make it clear how, as I believe, natural selection acts, I must beg permission to give one or two imaginary illustrations. Let us take the case of a wolf, which preys on various animals, securing some by craft, some by strength, and some by fleetness; and let us suppose that the fleetest prey, a deer for instance, had from any change in the country increased in numbers, or that other prey had decreased in numbers, during that season of the year when the wolf is hardest pressed for food. I can under such circumstances see no reason to doubt that the swiftest and slimmest wolves would have the best chance of surviving, and so be preserved or selected,—provided always that they retained strength to master their prey at this or at some other period of the year, when they might be compelled to prey on other animals. I can see no more reason to doubt this, than that man can improve the fleetness of his greyhounds by careful and methodical selection, or by that unconscious selection which results from each man trying to keep the best dogs without any thought of modifying the breed.

Even without any change in the proportional numbers of the animals on which our wolf preyed, a cub might be born with an innate tendency to pursue certain kinds of prey. Nor can this be thought very improbable; for we often observe great differences in the natural tendencies of our domestic animals; one cat, for instance, taking to catch rats, another mice; one cat, according to Mr. St. John, bringing home winged game, another hares or rabbits, and another hunting on marshy ground and almost nightly catching woodcocks or snipes. The tendency to catch rats rather than mice is known to be inherited. Now, if any slight innate change of habit or of structure benefited an individual wolf, it would have the best chance of surviving and of leaving offspring. Some of its young would probably inherit the same habits or structure, and by the repetition of this process, a new variety might be formed which would either supplant or coexist with the parent-form of wolf. Or, again, the wolves inhabiting a mountainous district, and those frequenting the lowlands, would naturally be forced to hunt different prey; and from the continued preservation of the individuals best fitted for the two sites, two varieties might slowly be formed. These varieties would cross and blend where they met; but to this subject of intercrossing we shall soon have to return. I may add, that, according to Mr. Pierce, there are two varieties of the wolf inhabiting the Catskill Mountains in the United States, one with a light greyhound-like form, which pursues deer, and the other more bulky, with shorter legs, which more frequently attacks the shepherd's flocks.

Let us now take a more complex case. Certain plants excrete a sweet juice, apparently for the sake of eliminating something injurious from their sap: this is effected by glands at the base of the stipules in some Leguminosæ, and at the back of the leaf of the common laurel. This juice, though small in quantity, is greedily sought by insects. Let us now suppose a little sweet juice or nectar to be excreted by the inner bases of the petals of a flower. In this case insects in seeking the nectar would get dusted with pollen, and would certainly often transport the pollen from one flower to the stigma of another flower. The flowers of two distinct individuals of the same species would thus get crossed; and the act of crossing, we have good reason to believe (as will hereafter be more fully alluded to), would produce very vigorous seedlings, which consequently would have the best chance of flourishing and surviving. Some of these seedlings would probably inherit the nectar-excreting power. Those individual flowers which had the largest glands or nectaries, and which excreted most nectar, would be oftenest visited by insects, and would be oftenest crossed; and so in the long-run would gain the upper hand. Those flowers, also, which had their stamens and pistils placed, in relation to the size and habits of the particular insects which visited them, so as to favour in any degree the transportal of their pollen from flower to flower, would likewise be favoured or selected. We might have taken the case of insects visiting flowers for the sake of collecting pollen instead of nectar; and as pollen is formed for the sole object of fertilisation, its destruction appears a simple loss to the plant; yet if a little pollen were carried, at first occasionally and then habitually, by the pollen-devouring insects from flower to flower, and a cross thus effected, although nine-tenths of the pollen were destroyed, it might still be a great gain to the plant; and those individuals which produced more and more pollen, and had larger and larger anthers, would be selected.

When our plant, by this process of the continued preservation or natural selection of more and more attractive flowers, had been rendered highly attractive to insects, they would, unintentionally on their part, regularly carry pollen from flower to flower; and that they can most effectually do this, I could easily show by many striking instances. I will give only one—not as a very striking case, but as likewise illustrating one step in the separation of the sexes of plants, presently to be alluded to. Some holly-trees bear only male flowers, which have four stamens producing rather a small quantity of pollen, and a rudimentary pistil; other holly-trees bear only female flowers; these have a full-sized pistil, and four stamens with shrivelled anthers, in which not a grain of pollen can be detected. Having found a female tree exactly sixty yards from a male tree, I put the stigmas of twenty flowers, taken from different branches, under the microscope, and on all, without exception, there were pollen-grains, and on some a profusion of pollen. As the wind had set for several days from the female to the male tree, the pollen could not thus have been carried. The weather had been cold and boisterous, and therefore not favourable to bees, nevertheless every female flower which I examined had been effectually fertilised by the bees, accidentally dusted with pollen, having flown from tree to tree in search of nectar. But to return to our imaginary case: as soon as the plant had been rendered so highly attractive to insects that pollen was regularly carried from flower to flower, another process might commence. No naturalist doubts the advantage of what has been called the "physiological division of labour;" hence we may believe that it would be advantageous to a plant to produce stamens alone in one flower or on one whole plant, and pistils alone in another flower or on another plant. In plants under culture and placed under new conditions of life, sometimes the male organs and sometimes the female organs become more or less impotent; now if we suppose this to occur in ever so slight a degree under nature, then as pollen is already carried regularly from flower to flower, and as a more complete separation of the sexes of our plant would be advantageous on the principle of the division of labour, individuals with this tendency more and more increased, would be continually favoured or selected, until at last a complete separation of the sexes would be effected.

Let us now turn to the nectar-feeding insects in our imaginary case: we may suppose the plant of which we have been slowly increasing the nectar by continued selection, to be a common plant; and that certain insects depended in main part on its nectar for food. I could give many facts, showing how anxious bees are to save time; for instance, their habit of cutting holes and sucking the nectar at the bases of certain flowers, which they can, with a very little more trouble, enter by the mouth. Bearing such facts in mind, I can see no reason to doubt that an accidental deviation in the size and form of the body, or in the curvature and length of the proboscis, &c., far too slight to be appreciated by us, might profit a bee or other insect, so that an individual so characterised would be able to obtain its food more quickly, and so have a better chance of living and leaving descendants. Its descendants would probably inherit a tendency to a similar slight deviation of structure. The tubes of the corollas of the common red and incarnate clovers (Trifolium pratense and incarnatum) do not on a hasty glance appear to differ in length; yet the hive-bee can easily suck the nectar out of the incarnate clover, but not out of the common red clover, which is visited by humble-bees alone; so that whole fields of the red clover offer in vain an abundant supply of precious nectar to the hive-bee. Thus it might be a great advantage to the hive-bee to have a slightly longer or differently constructed proboscis. On the other hand, I have found by experiment that the fertility of clover greatly depends on bees visiting and moving parts of the corolla, so as to push the pollen on to the stigmatic surface. Hence, again, if humble-bees were to become rare in any country, it might be a great advantage to the red clover to have a shorter or more deeply divided tube to its corolla, so that the hive-bee could visit its flowers. Thus I can understand how a flower and a bee might slowly become, either simultaneously or one after the other, modified and adapted in the most perfect manner to each other, by the continued preservation of individuals presenting mutual and slightly favourable deviations of structure.

I am well aware that this doctrine of natural selection, exemplified in the above imaginary instances, is open to the same objections which were at first urged against Sir Charles Lyell's noble views on "the modern changes of the earth, as illustrative of geology"; but we now very seldom hear the action, for instance, of the coast-waves, called a trifling and insignificant cause, when applied to the excavation of gigantic valleys or to the formation of the longest lines of inland cliffs. Natural selection can act only by the preservation and accumulation of infinitesimally small inherited modifications, each profitable to the preserved being; and as modern geology has almost banished such views as the excavation of a great valley by a single diluvial wave, so will natural selection, if it be a true principle, banish the belief of the continued creation of new organic beings, or of any great and sudden modification in their structure. . . .

Chapter 14

As this whole volume is one long argument, it may be convenient to the reader to have the leading facts and inferences briefly recapitulated.

That many and grave objections may be advanced against the theory of descent with modification through natural selection, I do not deny. I have endeavoured to give to them their full force. Nothing at first can appear more difficult to believe than that the more complex organs and instincts should have been perfected, not by means superior to, though analogous with, human reason, but by the accumulation of innumerable slight variations, each good for the individual possessor. Nevertheless, this difficulty, though appearing to our imagination insuperably great, cannot be considered real if we admit the following propositions, namely,—that gradations in the perfection of any organ or instinct, which we may consider, either do now exist or could have existed, each good of its kind,—that all organs and instincts are, in ever so slight a degree, variable,—and, lastly, that there is a struggle for existence leading to the preservation of each profitable deviation of structure or instinct. The truth of these propositions cannot, I think, be disputed.

It is, no doubt, extremely difficult even to conjecture by what gradations many structures have been perfected, more especially amongst broken and failing groups of organic beings; but we see so many strange gradations in nature, as is proclaimed by the canon, "Natura non facit saltum," that we ought to be extremely cautious in saying that any organ or instinct, or any whole being, could not have arrived at its present state by many graduated steps. There are, it must be admitted, cases of special difficulty on the theory of natural selection; and one of the most curious of these is the existence of two or three defined castes of workers or sterile females in the same community of ants; but I have attempted to show how this difficulty can be mastered.

With respect to the almost universal sterility of species when first crossed, which forms so remarkable a contrast with the almost universal fertility of varieties when crossed, I must refer the reader to the recapitulation of the facts given at the end of the eighth chapter, which seem to me conclusively to show that this sterility is no more a special endowment than is the incapacity of two trees to be grafted together, but that it is incidental on constitutional differences in the reproductive systems of the intercrossed species. We see the truth of this conclusion in the vast difference in the result, when the same two species are crossed reciprocally; that is, when one species is first used as the father and then as the mother.

The fertility of varieties when intercrossed and of their mongrel offspring cannot be considered as universal; nor is their very general fertility surprising when we remember that it is not likely that either their constitutions or their reproductive systems should have been profoundly modified. Moreover, most of the varieties which have been experimentised on have been produced under domestication; and as domestication apparently tends to eliminate sterility, we ought not to expect it also to produce sterility.

The sterility of hybrids is a very different case from that of first crosses, for their reproductive organs are more or less functionally impotent; whereas in first crosses the organs on both sides are in a perfect condition. As we continually see that organisms of all kinds are rendered in some degree sterile from their constitutions having been disturbed by slightly different and new conditions of life, we need not feel surprise at hybrids being in some degree sterile, for their constitutions can hardly fail to have been disturbed from being compounded of two distinct organisations. This parallelism is supported by another parallel, but directly opposite, class of facts; namely, that the vigour and fertility of all organic beings are increased by slight changes in their conditions of life, and that the offspring of slightly modified forms or varieties acquire from being crossed increased vigour and fertility. So that, on the one hand, considerable changes in the conditions of life and crosses between greatly modified forms, lessen fertility; and on the other hand, lesser changes in the conditions of life and crosses between less modified forms, increase fertility.

Turning to geographical distribution, the difficulties encountered on the theory of descent with modification are grave enough. All the individuals of the same species, and all the species of the same genus, or even higher group, must have descended from common parents; and therefore, in however distant and isolated parts of the world they are now found, they must in the course of successive generations have passed from some one part to the others. We are often wholly unable even to conjecture how this could have been effected. Yet, as we have reason to believe that some species have retained the same specific form for very long periods, enormously long as measured by years, too much stress ought not to be laid on the occasional wide diffusion of the same species; for during very long periods of time there will always be a good chance for wide migration by many means. A broken or interrupted range may often be accounted for by the extinction of the species in the intermediate regions. It cannot be denied that we are as yet very ignorant of the full extent of the various climatal and geographical changes which have affected the earth during modern periods; and such changes will obviously have greatly facilitated migration. As an example, I have attempted to show how potent has been the influence of the Glacial period on the distribution both of the same and of representative species throughout the world. We are as yet profoundly ignorant of the many occasional means of transport. With respect to distinct species of the same genus inhabiting very distant and isolated regions, as the process of modification has necessarily been slow, all the means of migration will have been possible during a very long period; and consequently the difficulty of the wide diffusion of species of the same genus is in some degree lessened.

As on the theory of natural selection an interminable number of intermediate forms must have existed, linking together all the species in each group by gradations as fine as our present varieties, it may be asked, Why do we not see these linking forms all around us? Why are not all organic beings blended together in an inextricable chaos? With respect to existing forms, we should remember that we have no right to expect (excepting in rare cases) to discover *directly* connecting links between them, but only between each and some extinct and supplanted form. Even on a wide area, which has during a long period remained continuous, and of which the climate and other conditions of life change insensibly in going from a district occupied by one species into another district occupied by a closely allied species, we have no just right to expect often to find intermediate varieties in the intermediate zone. For we have reason to believe that only a few species are undergoing change at any one period; and all changes are slowly effected. I have also shown that the intermediate varieties which will at first probably exist in the intermediate zones, will be liable to be supplanted by the allied forms on either hand; and the latter, from existing in greater numbers, will generally be modified and improved at a quicker rate than the intermediate varieties, which exist in lesser numbers; so that the intermediate varieties will, in the long run, be supplanted and exterminated.

On this doctrine of the extermination of an infinitude of connecting links, between the living and extinct inhabitants of the world, and at each successive period between the extinct and still older species, why is not every geological formation charged with such links? Why does not every collection of fossil remains afford plain evidence of the gradation and mutation of the forms of life? We meet with no such evidence, and this is the most obvious and forcible of the many objections which may be urged against my theory. Why, again, do whole groups of allied species appear, though certainly they often falsely appear, to have come in suddenly on the several geological stages? Why do we not find great piles of strata beneath the Silurian system, stored with the remains of the progenitors of the Silurian groups of fossils? For certainly on my theory such strata must somewhere have been deposited at these ancient and utterly unknown epochs in the world's history.

I can answer these questions and grave objections only on the supposition that the geological record is far more imperfect than most geologists believe. It cannot be objected that there has not been time sufficient for any amount of organic change; for the lapse of time has been so great as to be utterly inappreciable by the human intellect. The number of specimens in all our museums is absolutely as nothing compared with the countless generations of countless species which certainly have existed. We should not be able to recognise a species as the parent of any one or more species if we were to examine them ever so closely, unless we likewise possessed many of the intermediate links between their past or parent and present states; and these many links we could hardly ever expect to discover, owing to the imperfection of the geological record. Numerous existing doubtful forms could be named which are probably varieties; but who will pretend that in future ages so many fossil links will be discovered, that naturalists will be able to decide, on the common view, whether or not these doubtful forms are varieties? As long as most of the links between any two species are unknown, if any one link or intermediate variety be discovered, it will simply be classed as another and distinct species. Only a small portion of the world has been geologically explored. Only organic beings of certain classes can be preserved in a fossil condition, at least in any great number. Widely ranging species vary most, and varieties are often at first local,—both causes rendering the discovery of intermediate links less likely. Local varieties will not spread into other and distant regions until they are considerably modified and improved; and when they do spread, if discovered in a geological formation, they will appear as if suddenly created there, and will be simply classed as new species. Most formations have been intermittent in their accumulation; and their duration, I am inclined to believe, has been shorter than the average duration of specific forms. Successive formations are separated from each other by enormous blank intervals of time; for fossiliferous formations, thick enough to resist future degradation, can be accumulated only where much sediment is deposited on the subsiding bed of the sea. During the alternate periods of elevation and of stationary level the record will be blank. During these latter periods there will probably be more variability in the forms of life; during periods of subsidence, more extinction.

With respect to the absence of fossiliferous formations beneath the lowest Silurian strata, I can only recur to the hypothesis given in the ninth chapter. That the geological record is imperfect all will admit; but that it is imperfect to the degree which I require, few will be inclined to admit. If we look to long enough intervals of time, geology plainly declares that all species have changed; and they have changed in the manner which my theory requires, for they have changed slowly and in a graduated manner. We clearly see this in the fossil remains from consecutive formations invariably being much more closely related to each other, than are the fossils from formations distant from each other in time.

Such is the sum of the several chief objections and difficulties which may justly be urged against my theory; and I have now briefly recapitulated the answers and explanations which can be given to them. I have felt these difficulties far too heavily during many years to doubt their weight. But it deserves especial notice that the more important objections relate to questions on which we are confessedly ignorant; nor do we know how ignorant we are. We do not know all the possible transitional gradations between the simplest and the most perfect organs; it cannot be pretended that we know all the varied means of Distribution during the long lapse of years, or that we know how imperfect the Geological Record is. Grave as these several difficulties are, in my judgment they do not overthrow the theory of descent with modification. . ..

It is interesting to contemplate an entangled bank, clothed with many plants of many kinds, with birds singing on the bushes, with various insects flitting about, and with worms crawling through the damp earth, and to reflect that these elaborately constructed forms, so different from each other, and dependent on each other in so complex a manner, have all been produced by laws acting around us. These laws, taken in the largest sense, being Growth with Reproduction; Inheritance which is almost implied by reproduction; Variability from the indirect and direct action of the external conditions of life, and from use and disuse; a Ratio of Increase so high as to lead to a Struggle for Life, and as a consequence to Natural Selection, entailing Divergence of Character and the Extinction of less-improved forms. Thus, from the war of nature, from famine and death, the most exalted object which we are capable of conceiving, namely, the production of the higher animals, directly follows. There is grandeur in this view of life, with its several powers, having been originally breathed into a few forms or into one; and that, whilst this planet has gone cycling on according to the fixed law of gravity, from so simple a beginning endless forms most beautiful and most wonderful have been, and are being, evolved.

SOURCE: Darwin, Charles. *On the Origin of Species by Means of Natural Selection.* London: Murray, 1859.

ALFRED RUSSEL WALLACE, "LETTERS AND REMINISCENCES" (1916)

Alfred Russel Wallace (1823–1913) provided insights regarding biodiversity built upon his own travels and accounts of other explorers, including Darwin. The parallels of time and space in the way species were distributed provided Wallace evidence to propose a law that suggested species appeared in close association with others, at a particular time and place. This law directed him to speculate on an analogy between the appearance of species as branches on an ancestral tree. Darwin had reached a similar conclusion. Wallace openly suggested that his conclusions, as hypotheses, were open to testing. He welcomed challenges from other naturalists who could weigh the validity of his ideas against the facts found in nature. Significantly, he distinguished those who would test his ideas with facts explicitly from those who might bring arguments formulated without examination of nature itself. This piece, published after his death, recounts the emergence of his contribution to the theory of evolution by natural selection.

Every naturalist who has directed his attention to the subject of the geographical distribution of animals and plants, must have been interested in the singular facts which it presents. Many of these facts are quite different from what would have been anticipated, and have hitherto been considered as highly curious, but quite inexplicable. None of the explanations attempted from the time of Linnaeus are now considered at all satisfactory; none of them have given a cause sufficient to account for the facts known at the time, or comprehensive enough to include all the new facts which have since been, and are daily being added. Of late years, however, a great light has been thrown upon the subject by geological investigations, which have shown that the present state of the earth, and the organisms now inhabiting it, are but the last stage of a long and uninterrupted series of changes which it has undergone, and consequently, that to endeavour to explain and account for its present condition without any reference to those changes (as has frequently been done) must lead to very imperfect and erroneous conclusions. . . . The following propositions in Organic Geography and Geology give the main facts on which the hypothesis is founded.

Geography

(1) Large groups, such as classes and orders, are generally spread over the whole earth, while smaller ones, such as families and genera, are frequently confined to one portion, often to a very limited district.

(2) In widely distributed families the genera are often limited in range; in widely distributed genera, well-marked groups of species are peculiar to each geographical district.

(3) When a group is confined to one district, and is rich in species, it is almost invariably the case that the most closely allied species are found in the same locality or in closely adjoining localities, and that therefore the natural sequence of the species by affinity is also geographical.

(4) In countries of a similar climate, but separated by a wide sea or lofty mountains, the families, genera and species of the one are often represented by closely allied families, genera and species peculiar to the other.

Geology

(5) The distribution of the organic world in time is very similar to its present distribution in space.

(6) Most of the larger and some of the smaller groups extend through several geological periods.

(7) In each period, however, there are peculiar groups, found nowhere else, and extending through one or several formations.

(8) Species of one genus, or genera of one family, occurring in the same geological time are more closely allied than those separated in time.

(9) As generally in geography no species or genus occurs in two very distant localities without being also found in intermediate places, so in geology the life of a species or genus has not been interrupted. In other words, no group or species has come into existence twice.

(10) The following law may be deduced from these facts: *Every species has come into existence coincident both in time and space with a pre-existing closely allied species.*

This law agrees with, explains and illustrates all the facts connected with the following branches of the subject: 1st, the system of natural affinities; 2nd, the distribution of animals and plants in space; 3rd, the same in time, including all the phenomena of representative groups, and those which Prof. Forbes supposed to manifest polarity; 4th, the phenomena of rudimentary organs. We will briefly endeavour to show its bearing upon each of these.

If [this] law be true, it follows that the natural series of affinities will also represent the order in which the several species came into existence, each one having had for its immediate antetype a clearly allied species existing at the time of its origin . . . if two or more species have been independently formed on the plan of a common antetype, then the series of affinities will be compound, and can only be represented by a forked or many-branched line . . . . Sometimes the series of affinities can be well represented for a space by a direct progression from species to species or from group to group, but it is generally found impossible so to continue. There constantly occur two or more modifications of an organ or modifications of two distinct organs, leading us on to two distinct series of species, which at length differ, so much from each other as to form distinct genera or families. These are the parallel series or representative groups and they often occur in different countries, or are found fossil in different formations . . . . We thus see how difficult it is to determine in every case whether a given relation is an analogy or an affinity, for it is evident that as we go back along the parallel or divergent series, towards the common antetype, the analogy which existed between the two groups becomes an affinity . . . . Again, if we consider that we have only the fragments of this vast system, the stems and main branches being represented by extinct species of which we have no knowledge, while a vast mass of limbs and boughs and minute twigs and scattered leaves is what we have to place in order, and determine the true position each originally occupied with regard to the others, the whole difficulty of the true Natural System of classification becomes apparent to us.

We shall thus find ourselves obliged to reject all those systems of classification which arrange species or groups in circles, as well as those which fix a definite number for the division of each group . . . . We have . . . never been able to find a case in which the circle has been closed by a direct affinity. In most cases a palpable analogy has been substituted, in others the affinity is very obscure or altogether doubtful . . . .

If we now consider the geographical distribution of animals and plants upon the earth, we shall find all the facts beautifully in accordance with, and readily explained by, the present hypothesis. A country having species, genera, and whole families peculiar to it, will be the necessary result of its having been isolated for a long period sufficient for many series of species to have been created on the type of pre-existing ones, which, as well as many of the earlier-formed species, have become extinct, and made the groups appear isolated . . . .

Such phenomena as are exhibited by the Galapagos Islands, which contain little groups of plants and animals peculiar to themselves, but most nearly allied to those of South America, have not hitherto received any, even a conjectural explanation. The Galapagos are a volcanic group of high antiquity and have probably never been more closely connected with the continent than they are at present. . ..

The question forces itself upon every thinking mind-why are these things so? They could not be as they are, had no law regulated their creation and dispersion. The law here enunciated not merely explains, but necessitates the facts we see to exist, while the vast and long-continued geological changes of the earth readily account for the exceptions and apparent discrepancies that here and there occur. The writer's object in putting forward his views in the present imperfect manner is to submit them to the tests of other minds, and to be made aware of all the facts supposed to be inconsistent with them. As his hypothesis is one which claims acceptance solely as explaining and connecting facts which exist in nature, he expects facts alone to be brought forward to disprove it, not a priori arguments against its probability. . . . .

To discover how the extinct species have from time to time been replaced by new ones down to the very latest geological period, is the most difficult, and at the same time the most interesting problem in the natural history of the earth. The present inquiry, which seeks to eliminate from known facts a law which has determined, to a certain degree, what species could and did appear at a given epoch, may, it is hoped, be considered as one step in the right direction towards a complete solution of it . . . . Admitted facts seem to show . . . a general, but not a detailed progression . . . . It is, however, by no means difficult to show that a real progression in the scale of organisation is perfectly consistent with all the appearances, and even with apparent retrogression should such occur.

SOURCE: Wallace, Alfred Russel. *Letters and Reminiscences*. Ed. by J. Marchant. London: Harper and Brothers, 1916.

THOMAS H. HUXLEY, "ON A PIECE OF CHALK" (1868)

Thomas Huxley (1825–1895) wrote a broad argument in support of Darwinian evolution, examining natural evidence in support of a less controversial scientific hypothesis. He then used that argument to demonstrate the equally solid evidence leading to natural selection as a mechanism of evolution that could not reasonably be dismissed. He began with a quite detailed description of the geological and geographical significance of chalk formations that extend across broad stretches of oceanic and continental landmasses. Readers were especially familiar with the chalk cliff at the edges of continents, and would be expected to infer the extension of the chalk into the ocean and beneath their own countryside homes. Huxley then engaged his audience with the questions of that chalk's composition and from whence it came. He suggested that the answers to those questions were based on the most substantial physical evidence of any subject known to science. Providing more detail, some of it technical, he demonstrated the depth of understanding naturalists enjoyed with respect to this topic. Moreover, he insisted that this understanding was available to any reader who willingly studied it. As his argument proceeded, Huxley illustrated themes reminiscent of his good friend Darwin, since they logically based their observations on the same facts of nature. In describing the composition and origin of the chalk, Huxley noted which conclusions surprised him, which were based on incontrovertible facts, which were open to debate and were based on a level of uncertainty. More to the point, however, he also could indicate how clearly the chalk provided evidence of the depths of time, how it suggested a history that predated all human existence, and how it aligned with the remote past wherein dwelled extinct and different living forms. Much more than Darwin, Huxley articulated the ways in which his argument contradicted the account of origins provided in the biblical story of Genesis. Chalk, as a material substance, underlay the area described as Eden in a way that would have taken eons to form before any garden could exist there. He admitted that the facts as any naturalist might present them could not adequately answer certain questions of causation that many readers would hasten to raise. Yet he answered such skepticism with a challenge to provide a better account of how so many different species could be found to exist, over such a long reach of time, which might also account for the six days of creation described in Genesis. Huxley would tolerate no prevarication on the facts of nature, especially in service to biblical history. As such, Huxley, more than Darwin, became the focus of deliberate debate over evolution.

If a well were sunk at our feet in the midst of the city of Norwich, the diggers would very soon find themselves at work in that white substance almost too soft to be called rock, with which we are all familiar as "chalk."

Not only here, but over the whole country of Norfolk, the well-sinker might carry his shaft down many hundred feet without coming to the end of the chalk; and, on the sea-coast, where the waves have pared away the face of the land which breasts them, the scarped faces of the high cliffs are often wholly formed of the same material. Northward, the chalk may be followed as far as Yorkshire; on the south coast it appears abruptly in the picturesque western bays of Dorset, and breaks into the Needles of the Isle of Wight; while on the shores of Kent it supplies that long line of white cliffs to which England owes her name of Albion.

Were the thin soil which covers it all washed away, a curved band of white chalk, here broader, and there narrower, might be followed diagonally across England from Lulworth in Dorset, to Flamborough Head in Yorkshire—a distance of over 280 miles as the crow flies. From this band to the North Sea, on the east, and the Channel, on the south, the chalk is largely hidden by other deposits; but, except in the Weald of Kent and Sussex, it enters into the very foundation of all the southeastern counties.

Attaining, as it does in some places, a thickness of more than a thousand feet, the English chalk must be admitted to be a mass of considerable magnitude. Nevertheless, it covers but an insignificant portion of the whole area occupied by the chalk formation of the globe, much of which has the same general characters as ours, and is found in detached patches, some less, and others more extensive, than the English. Chalk occurs in north-west Ireland; it stretches over a large part of France, the chalk which underlies Paris being, in fact, a continuation of that of the London basin; it runs through Denmark and Central Europe, and extends southward to North Africa; while eastward, it appears in the Crimea and in Syria, and may be traced as far as the shores of the Sea of Aral, in Central Asia. If all the points at which true chalk occurs were circumscribed, they would lie within an irregular oval about 3,000 miles in long diameter—the area of which would be as great as that of Europe, and would many times exceed that of the largest existing inland sea—the Mediterranean.

Thus the chalk is no unimportant element in the masonry of the earth's crust, and it impresses a peculiar stamp, varying with the conditions to which it is exposed, on the scenery of the districts in which it occurs. The undulating downs and rounded coombs, covered with sweet-grassed turf, of our inland chalk country, have a peacefully domestic and mutton-suggesting prettiness, but can hardly be called either grand or beautiful. But on our southern coasts, the wall-sided cliffs, many hundred feet high, with vast needles and pinnacles standing out in the sea, sharp and solitary enough to serve as perches for the wary cormorant, confer a wonderful beauty and grandeur upon the chalk headlands. And, in the East, chalk has its share in the formation of some of the most venerable of mountain ranges, such as the Lebanon.

What is this wide-spread component of the surface of the earth? and whence did it come?

You may think this no very hopeful inquiry. You may not unnaturally suppose that the attempt to solve such problems as these can lead to no result, save that of entangling the inquirer in vague speculations, incapable of refutation and of verification. If such were really the case, I should have selected some other subject than a "piece of chalk" for my discourse. But, in truth, after much deliberation, I have been unable to think of any topic which would so well enable me to lead you to see how solid is the foundation upon which some of the most startling conclusions of physical science rest.

A great chapter of the history of the world is written in the chalk. Few passages in the history of man can be supported by such an overwhelming mass of direct and indirect evidence as that which testifies to the truth of the fragment of the history of the globe, which I hope to enable you to read, with your own eyes, tonight. Let me add, that few chapters of human history have a more profound significance for ourselves. I weigh my words well when I assert, that the man who should know the true history of the bit of chalk which every carpenter carries about in his breeches-pocket, though ignorant of all other history, is likely, if he will think his knowledge out to its ultimate results, to have a truer, and therefore a better, conception of this wonderful universe, and of man's relation to it, than the most learned student who is deep-read in the records of humanity and ignorant of those of Nature.

The language of the chalk is not hard to learn, not nearly so hard as Latin, if you only want to get at the broad features of the story it has to tell; and I propose that we now set to work to spell that story out together....

Almost the whole of the bottom of this central plain [of the North Atlantic] (which extends for many hundred miles in a north and south direction) is covered by a fine mud, which, when brought to the surface, dries into a greyish white friable substance. You can write with this on a blackboard, if you are so inclined; and, to the eye, it is quite like very soft, greyish chalk. Examined chemically, it proves to be composed almost wholly of carbonate of lime; and if you make a section of it, in the same way as that of the piece of chalk was made, and view it with the microscope, it presents innumerable *Globigerinæ* imbedded in a granular matrix. Thus this deep-sea mud is substantially chalk. I say substantially, because there are a good many minor differences; but as these have no bearing on the question immediately before us,—which is the nature of the *Globigerinæ* of the chalk,—it is unnecessary to speak of them.

Globigerinæ of every size, from the smallest to the largest, are associated together in the Atlantic mud, and the chambers of many are filled by a soft animal matter. This soft substance is, in fact, the remains of the creature to which the *Globigerina* shell, or rather skeleton, owes its existence and which is an animal of the simplest imaginable description. It is, in fact, a mere particle of living jelly, without defined parts of any kind—without a mouth, nerves, muscles, or distinct organs, and only manifesting its vitality to ordinary observation by thrusting out and retracting from all parts of its surface, long filamentous processes, which serve for arms and legs. Yet this amorphous particle, devoid of everything which, in the higher animals, we call organs, is capable of feeding, growing, and multiplying; of separating from the ocean the small proportion of carbonate of lime which is dissolved in sea-water; and of building up that substance into a skeleton for itself, according to a pattern which can be imitated by no other known agency.

The notion that animals can live and flourish in the sea, at the vast depths from which apparently living *Globigerinæ* have been brought up, does not agree very well with our usual conceptions respecting the conditions of animal life; and it is not so absolutely impossible as it might at first sight appear to be, that the *Globigerinæ* of the Atlantic sea-bottom do not live and die where they are found.

As I have mentioned, the soundings from the great Atlantic plain are almost entirely made up of *Globigerinæ* with the granules which have been mentioned, and some few other calcareous shells; but a small percentage of the chalky mud—perhaps at most some five per cent of it is of a different nature, and consists of shells and skeletons composed of silex, or pure flint. These silicious bodies belong partly to the lowly vegetable organisms which are called *Diatomaceæ*, and partly to the minute, and extremely simple, animals, termed *Radiolaria*. It is quite certain that these creatures do not live at the bottom of the ocean, but at its surface—where they may be obtained in prodigious numbers by the use of a properly constructed net. Hence it follows that these silicious organisms, though they are not heavier than the lightest dust, must have fallen, in some cases, through fifteen thousand feet of water, before they reached their final restingplace on the ocean floor. And considering how large a surface these bodies expose in proportion to their weight, it is probable that they occupy a great length of time in making their burial journey from the surface of the Atlantic to the bottom.

But if the *Radiolaria* and Diatoms are thus rained upon the bottom of the sea, from the superficial layer of its waters in which they pass their lives, it is obviously possible that the *Globigerinæ* may be similarly derived; and if they were so, it would be much more easy to understand how they obtain their supply of food than it is at present. Nevertheless, the positive and negative evidence all points the other way. The skeletons of the full-grown, deep-sea *Globigerinæ* are so remarkably solid and heavy in proportion to their surface as to seem little fitted for floating; and, as a matter of fact, they are not to be found along with the Diatoms and *Radiolaria* in the uppermost stratum of the open ocean. It has been observed, again, that the abundance of *Globigerinæ*, in proportion to other organisms, of like kind, increases with the depth of the sea; and that deep-water *Globigerinæ* are larger than those which live in shallower parts of the sea; and such facts negative the supposition that these organisms have been swept by currents from the shallows into the deeps of the Atlantic. It therefore seems to be hardly doubtful that these wonderful creatures live and die at the depths in which they are found.

However, the important points for us are, that the living *Globigerinæ* are exclusively marine animals, the skeletons of which abound at the bottom of deep seas; and that there is not a shadow of reason for believing that the habits of the *Globigerinæ* of the chalk differed from those of the existing species. But if this be true, there is no escaping the conclusion that the chalk itself is the dried mud of an ancient deep sea. . . .

Thus there is a writing upon the wall of cliffs at Cromer, and whoso runs may read it. It tells us, with an authority which cannot be impeached, that the ancient sea bed of the chalk sea was raised up, and remained dry land, until it was covered with forest, stocked with the great game the spoils of which have rejoiced your geologists. How long it remained in that condition cannot be said; but, "the whirligig of time brought its revenges" in those days as in these. That dry land, with the bones and teeth of generations of long-lived elephants, hidden away among the gnarled roots and dry leaves of its ancient trees, sank gradually to the bottom of the icy sea, which covered it with huge masses of drift and boulder clay. Sea-beasts, such as the walrus now restricted to the extreme north, paddled about where birds had twittered among the topmost twigs of the fir-trees. How long this state of things endured we know not, but at length it came to an end.

The upheaved glacial mud hardened into the soil of modern Norfolk. Forests grew once more, the wolf and the beaver replaced the reindeer and the elephant; and at length what we call the history of England dawned.

Thus you have, within the limits of your own county, proof that the chalk can justly claim a very much greater antiquity than even the oldest physical traces of mankind. But we may go further and demonstrate, by evidence of the same authority as that which testifies to the existence of the father of men, that the chalk is vastly older than Adam himself. The Book of Genesis informs us that Adam, immediately upon his creation, and before the appearance of Eve, was placed in the Garden of Eden. The problem of the geographical position of Eden has greatly vexed the spirits of the learned in such matters, but there is one point respecting which, so far as I know, no commentator has ever raised a doubt. This is, that of the four rivers which are said to run out of it, Euphrates and Hiddekel are identical with the rivers now known by the names of Euphrates and Tigris. But the whole country in which these mighty rivers take their origin, and through which they run, is composed of rocks which are either of the same age as the chalk, or of later date. So that the chalk must not only have been formed, but, after its formation, the time required for the deposit of these later rocks, and for their upheaval into dry land, must have elapsed, before the smallest brook which feeds the swift stream of "the great river, the river of Babylon" began to flow.

Thus, evidence which cannot be rebutted, and which need not be strengthened, though if time permitted I might indefinitely increase its quantity, compels you to believe that the earth, from the time of the chalk to the present day, has been the theatre of a series of changes as vast in their amount, as they were slow in their progress. The area on which we stand has been first sea and then land, for at least four alternations; and has remained in each of these conditions for a period of great length.

Nor have these wonderful metamorphoses of sea into land, and of land into sea, been confined to one corner of England. During the chalk period, or "cretaceous epoch," not one of the present great physical features of the globe was in existence. Our great mountain ranges, Pyrenees, Alps, Himalayas, Andes, have all been upheaved since the chalk was deposited, and the cretaceous sea flowed over the sites of Sinai and Ararat. All this is certain, because rocks of cretaceous, or still later, date have shared in the elevatory movements which gave rise to these mountain chains; and may be found perched up, in some cases, many thousand feet high upon their flanks. And evidence of equal cogency demonstrates that, though, in Norfolk, the forest-bed rests directly upon the chalk, yet it does so, not because the period at which the forest grew immediately followed that at which the chalk was formed, but because an immense lapse of time, represented elsewhere by thousands of feet of rock, is not indicated at Cromer.

I must ask you to believe that there is no less conclusive proof that a still more prolonged succession of similar changes occurred, before the chalk was deposited. Nor have we any reason to think that the first term in the series of these changes is known. The oldest sea-beds preserved to us are sands, and mud, and pebbles, the wear and tear of rocks which were formed in still older oceans.

But, great as is the magnitude of these physical changes of the world, they have been accompanied by a no less striking series of modifications in its living inhabitants. All the great classes of animals, beasts of the field, fowls of the air, creeping things, and things which dwell in the waters, flourished upon the globe long ages before the chalk was deposited. Very few, however, if any, of these ancient forms of animal life were identical with those which now live. Certainly not one of the higher animals was of the same species as any of those now in existence. The beasts of the field, in the days before the chalk, were not our beasts of the field, nor the fowls of the air such as those which the eye of man has seen flying, unless his antiquity dates infinitely further back than we at present surmise. If we could be carried back into those times, we should be as one suddenly set down in Australia before it was colonized. We should see mammals, birds, reptiles, fishes, insects, snails, and the like, clearly recognizable as such, and yet not one of them would be just the same as those with which we are familiar, and many would be extremely different. . . .

Up to this moment I have stated, so far as I know, nothing but well-authenticated facts, and the immediate conclusions which they force upon the mind. But the mind is so constituted that it does not willingly rest in facts and immediate causes, but seeks always after a knowledge of the remoter links in the chain of causation.

Taking the many changes of any given spot of the earth's surface, from sea to land and from land to sea, as an established fact, we cannot refrain from asking ourselves how these changes have occurred. And when we have explained them—as they must be explained—by the alternate slow movements of elevation and depression which have affected the crust of the earth, we go still further back, and ask, Why these movements?

I am not certain that any one can give you a satisfactory answer to that question. Assuredly I cannot. All that can be said, for certain, is, that such movements are part of the ordinary course of nature, inasmuch as they are going on at the present time. Direct proof may be given, that some parts of the land of the northern hemisphere are at this moment insensibly rising and others insensibly sinking; and there is indirect, but perfectly satisfactory, proof, that an enormous area now covered by the Pacific has been deepened thousands of feet, since the present inhabitants of that sea came into existence. Thus there is not a shadow of a reason for believing that the physical changes of the globe, in past times, have been affected by other than natural causes. Is there any more reason for believing that the concomitant modifications in the forms of the living inhabitants of the globe have been brought about in other ways?

Before attempting to answer this question, let us try to form a distinct mental picture of what has happened in some special case. The crocodiles are animals which, as a group, have a very vast antiquity. They abounded ages before the chalk was deposited; they throng the rivers in warm climates, at the present day. There is a difference in the form of the joints of the back-bone, and in some minor particulars, between the crocodiles of the present epoch and those which lived before the chalk; but, in the cretaceous epoch, as I have already mentioned, the crocodiles had assumed the modern type of structure. Notwithstanding this, the crocodiles of the chalk are not identically the same as those which lived in the times called "older tertiary," which succeeded the cretaceous epoch; and the crocodiles of the older tertiaries are not identical with those of the newer tertiaries, nor are these identical with existing forms. I leave open the question whether particular species may have lived on from epoch to epoch. But each epoch has had its peculiar crocodiles; though all, since the chalk, have belonged to the modern type, and differ simply in their proportions, and in such structural particulars as are discernible only to trained eyes.

How is the existence of this long succession of different species of crocodiles to be accounted for? Only two suppositions seem to be open to us—Either each species of crocodile has been specially created, or it has arisen out of some pre-existing form by the operation of natural causes. Choose your hypothesis; I have chosen mine. I can find no warranty for believing in the distinct creation of a score of successive species of crocodiles in the course of countless ages of time. Science gives no countenance to such a wild fancy; nor can even the perverse ingenuity of a commentator pretend to discover this sense, in the simple words in which the writer of Genesis records the proceedings of the fifth and sixth days of the Creation.

On the other hand, I see no good reason for doubting the necessary alternative, that all these varied species have been evolved from pre-existing crocodilian forms, by the operation of causes as completely a part of the common order of nature as those which have effected the changes of the inorganic world. Few will venture to affirm that the reasoning which applies to crocodiles loses its force among other animals, or among plants. If one series of species has come into existence by the operation of natural causes, it seems folly to deny that all may have arisen in the same way.

A small beginning has led us to a great ending. If I were to put the bit of chalk with which we started into the hot but obscure flame of burning hydrogen, it would presently shine like the sun. It seems to me that this physical metamorphosis is no false image of what has been the result of our subjecting it to a jet of fervent, though nowise brilliant, thought to-night. It has become luminous, and its clear rays, penetrating the abyss of the remote past, have brought within our ken some stages of the evolution of the earth. And in the shifting "without haste, but without rest" of the land and sea, as in the endless variation of the forms assumed by living beings, we have observed nothing but the natural product of the forces originally possessed by the substance of the universe.

SOURCE: Huxley, Thomas H. "On a Piece of Chalk." *Macmillan's Magazine*, 1868, Collected Essays VIII.

CHARLES DARWIN, *THE DESCENT OF MAN* (1871)

Darwin struggled with the notion of instinct as an explanation for behavior, and not just in humans. He recognized the role of memory in influencing the actions of individuals in all social animals. An animal might respond to events with choices informed by prior experience in similar or analogous situations. As such, intentions could be influence by factors beyond the biological. A key question about the motivation to give aid to other individuals formed the basis of studies of altruism for over a century. With memory and a compulsion to give aid came the question of moral conscience. A parallel question, apparently unique to humankind, focused on the existence of God as well as the many spiritual beings of less advanced societies. Darwin viewed human uniqueness with respect to this ability to believe as a key distinction separating this one species from all of the lower animals. Since he therefore concluded that there was such a distinction, he defended his view of humans as descended from other species as no more nor less irreligious than the conclusion that the birth of an individual resulted from biological events. His point here was hardly the last word on that subject. Sexual selection represented another major theme in this book, wherein Darwin noted how significant the choice of a mate can be, not just for an individual but for an entire species. From such recognitions arose the eugenic hopes that many of his followers developed into social policies. Here, Darwin restricted his comments to the suggestion that potentially unfit parents should refrain from having offspring. He further qualified this suggestion by noting that naturalists still understood only vaguely the laws of inheritance. And yet, once such laws could be well understood, Darwin expected advanced governments would have to step in to ensure that the reckless not supplant the prudent in choosing to reproduce.

Chapter 21

A brief summary will be sufficient to recall to the reader's mind the more salient points in this work. Many of the views which have been advanced are highly speculative, and some no doubt will prove erroneous; but I have in every case given the reasons which have led me to one view rather than to another. It seemed worth while to try how far the principle of evolution would throw light on some of the more complex problems in the natural history of man. False facts are highly injurious to the progress of science, for they often endure long; but false views, if supported by some evidence, do little harm, for every one takes a salutary pleasure in proving their falseness: and when this is done, one path towards error is closed and the road to truth is often at the same time opened.

The main conclusion here arrived at, and now held by many naturalists who are well competent to form a sound judgment is that man is descended from some less highly organised form. The grounds upon which this conclusion rests will never be shaken, for the close similarity between man and the lower animals in embryonic development, as well as in innumerable points of structure and constitution, both of high and of the most trifling importance,—the rudiments which he retains, and the abnormal reversions to which he is occasionally liable,—are facts which cannot be disputed. They have long been known, but until recently they told us nothing with respect to the origin of man. Now when viewed by the light of our knowledge of the whole organic world, their meaning is unmistakable. The great principle of evolution stands up clear and firm, when these groups or facts are considered in connection with others, such as the mutual affinities of the members of the same group, their geographical distribution in past and present times, and their geological succession. It is incredible that all these facts should speak falsely. He who is not content to look, like a savage, at the phenomena of nature as disconnected, cannot any longer believe that man is the work of a separate act of creation. He will be forced to admit that the close resemblance of the embryo of man to that, for instance, of a dog—the construction of his skull, limbs and whole frame on the same plan with that of other mammals, independently of the uses to which the parts may be put—the occasional re-appearance of various structures, for instance of several muscles, which man does not normally possess, but which are common to the Quadrumana—and a crowd of analogous facts—all point in the plainest manner to the conclusion that man is the co-descendant with other mammals of a common progenitor.

We have seen that man incessantly presents individual differences in all parts of his body and in his mental faculties. These differences or variations seem to be induced by the same general causes, and to obey the same laws as with the lower animals. In both cases similar laws of inheritance prevail. Man tends to increase at a greater rate than his means of subsistence; consequently he is occasionally subjected to a severe struggle for existence, and natural selection will have effected whatever lies within its scope. A succession of strongly-marked variations of a similar nature is by no means requisite; slight fluctuating differences in the individual suffice for the work of natural selection; not that we have any reason to suppose that in the same species, all parts of the organization tend to vary to the same degree. We may feel assured that the inherited effects of the long-continued use or disuse of parts will have done much in the same direction with natural selection. Modifications formerly of importance, though no longer of any special use, are long-inherited. When one part is modified, other parts change through the principle of correlation, of which we have instances in many curious cases of correlated monstrosities. Something may be attributed to the direct and definite action of the surrounding conditions of life, such as abundant food, heat or moisture; and lastly, many characters of slight physiological importance, some indeed of considerable importance, have been gained through sexual selection.

No doubt man, as well as every other animal, presents structures, which seem to our limited knowledge, not to be now of any service to him, nor to have been so formerly, either for the general conditions of life, or in the relations of one sex to the other. Such structures cannot be accounted for by any form of selection, or by the inherited effects of the use and disuse of parts. We know, however, that many strange and strongly-marked peculiarities of structure occasionally appear in our domesticated productions, and if their unknown causes were to act more uniformly, they would probably become common to all the individuals of the species. We may hope hereafter to understand something about the causes of such occasional modifications, especially through the study of monstrosities: hence the labours of experimentalists such as those of M. Camille Dareste, are full of promise for the future. In general we can only say that the cause of each slight variation and of each monstrosity lies much more in the constitution of the organism, than in the nature of the surrounding conditions; though new and changed conditions certainly play an important part in exciting organic changes of many kinds.

Through the means just specified, aided perhaps by others as yet undiscovered, man has been raised to his present state. But since he attained to the rank of manhood, he has diverged into distinct races, or as they may be more fitly called sub-species. Some of these, such as the Negro and European, are so distinct that, if specimens had been brought to a naturalist without any further information, they would undoubtedly have been considered by him as good and true species. Nevertheless all the races agree in so many unimportant details of structure and in so many mental peculiarities that these can be accounted for only by inheritance from a common progenitor; and a progenitor thus characterised would probably deserve to rank as man.

It must not be supposed that the divergence of each race from the other races, and of all from a common stock, can be traced back to any one pair of progenitors. On the contrary, at every stage in the process of modification, all the individuals which were in any way better fitted for their conditions of life, though in different degrees, would have survived in greater numbers than the less well-fitted. The process would have been like that followed by man, when he does not intentionally select particular individuals, but breeds from all the superior individuals, and neglects the inferior. He thus slowly but surely modifies his stock, and unconsciously forms a new strain. So with respect to modifications acquired independently of selection, and due to variations arising from the nature of the organism and the action of the surrounding conditions, or from changed habits of life, no single pair will have been modified much more than the other pairs inhabiting the same country, for all will have been continually blended through free intercrossing.

By considering the embryological structure of man,—the homologies which he presents with the lower animals,—the rudiments which he retains,—and the reversions to which he is liable, we can partly recall in imagination the former condition of our early progenitors; and can approximately place them in their proper place in the zoological series. We thus learn that man is descended from a hairy, tailed quadruped, probably arboreal in its habits, and an inhabitant of the Old World. This creature, if its whole structure had been examined by a naturalist, would have been classed amongst the Quadrumana, as surely as the still more ancient progenitor of the Old and New World monkeys. The Quadrumana and all the higher mammals are probably derived from an ancient marsupial animal, and this through a long series of diversified forms, from some amphibian-like creature, and this again from some fish-like animal. In the dim obscurity of the past we can see that the early progenitor of all the Vertebrata must have been an aquatic animal provided with branchiae, with the two sexes united in the same individual, and with the most important organs of the body (such as the brain and heart) imperfectly or not at all developed. This animal seems to have been more like the larvæ of the existing marine. Ascidians than any other known form.

The high standard of our intellectual powers and moral disposition is the greatest difficulty which presents itself, after we have been driven to this conclusion on the origin of man. But every one who admits the principle of evolution, must see that the mental powers of the higher animals, which are the same in kind with those of man, though so different in degree, are capable of advancement. Thus the interval between the mental powers of one of the higher apes and of a fish, or between those of an ant and scale-insect, is immense; yet their development does not offer any special difficulty; for with our domesticated animals, the mental faculties are certainly variable, and the variations are inherited. No one doubts that they are of the utmost importance to animals in a state of nature. Therefore the conditions are favourable for their development through natural selection. The same conclusion may be extended to man; the intellect must have been all-important to him, even at a very remote period, as enabling him to invent and use language, to make weapons, tools, traps, &c., whereby with the aid of his social habits, he long ago became the most dominant of all living creatures.

A great stride in the development of the intellect will have followed, as soon as the half-art and half-instinct of language came into use; for the continued use of language will have reacted on the brain and produced an inherited effect; and this again will have reacted on the improvement of language. As Mr. Chauncey Wright has well remarked, the largeness of the brain in man relatively to his body, compared with the lower animals, may be attributed in chief part to the early use of some simple form of language,—that wonderful engine which affixes signs to all sorts of objects and qualities, and excites trains of thought which would never arise from the mere impression of the sense, or if they did arise could not be followed out. The higher intellectual powers of man, such as those of ratiocination, abstraction, self-consciousness, &c., probably follow from the continued improvement and exercise of the other mental faculties.

The development of the moral qualities is a more interesting problem. The foundation lies in the social instincts, including under this term the family ties. These instincts are highly complex, and in the case of the lower animals give special tendencies towards certain definite actions; but the more important elements are love, and the distinct emotion of sympathy. Animals endowed with the social instincts take pleasure in one another's company, warn one another of danger, defend and aid one another in many ways. These instincts do not extend to all the individuals of the species, but only to those of the same community. As they are highly beneficial to the species, they have in all probability been acquired through natural selection.

A moral being is one who is capable of reflecting on his past actions and their motives—of approving of some and disapproving of others; and the fact that man is the one being who certainly deserves this designation, is the greatest of all distinctions between him and the lower animals. But in the fourth chapter I have endeavoured to shew that the moral sense follows, firstly, from the enduring and ever-present nature of the social instincts; secondly, from man's appreciation of the approbation and disapprobation of his fellows; and thirdly, from the high activity of his mental faculties, with past impressions extremely vivid; and in these latter respects he differs from the lower animals. Owing to this condition of mind, man cannot avoid looking both backwards and forwards, and comparing past impressions. Hence after some temporary desire or passion has mastered his social instincts, he reflects and compares the now weakened impression of such past impulses with the ever-present social instincts; and he then feels that sense of dissatisfaction which all unsatisfied instincts leave behind them, he therefore resolves to act differently for the future,—and this is conscience. Any instinct, permanently stronger or more enduring than another, gives rise to a feeling which we express by saying that it ought to be obeyed. A pointer dog, if able to reflect on his past conduct, would say to himself, I ought (as indeed we say of him) to have pointed at that hare and not have yielded to the passing temptation of hunting it.

Social animals are impelled partly by a wish to aid the members of their community in a general manner, but more commonly to perform certain definite actions. Man is impelled by the same general wish to aid his fellows; but has few or no special instincts. He differs also from the lower animals in the power of expressing his desires by words, which thus become a guide to the aid required and bestowed. The motive to give aid is likewise much modified in man: it no longer consists solely of a blind instinctive impulse, but is much influenced by the praise or blame of his fellows. The appreciation and the bestowal of praise and blame both rest on sympathy; and this emotion, as we have seen, is one of the most important elements of the social instincts. Sympathy, though gained as an instinct, is also much strengthened by exercise or habit. As all men desire their own happiness, praise or blame is bestowed on actions and motives, according as they lead to this end; and as happiness is an essential part of the general good, the greatest-happiness principle indirectly serves as a nearly safe standard of right and wrong. As the reasoning powers advance and experience is gained, the remoter effects of certain lines of conduct on the character of the individual, and on the general good, are perceived; and then the self-regarding virtues come within the scope of public opinion, and receive praise, and their opposites blame. But with the less civilised nations reason often errs, and many bad customs and base superstitions come within the same scope, and are then esteemed as high virtues, and their breach as heavy crimes.

The moral faculties are generally and justly esteemed as of higher value than the intellectual powers. But we should bear in mind that the activity of the mind in vividly recalling past impressions is one of the fundamental though secondary bases of conscience. This affords the strongest argument for educating and stimulating in all possible ways the intellectual faculties of every human being. No doubt a man with a torpid mind, if his social affections and sympathies are well developed, will be led to good actions, and may have a fairly sensitive conscience. But whatever renders the imagination more vivid and strengthens the habit of recalling and comparing past impressions, will make the conscience more sensitive, and may even somewhat compensate for weak social affections and sympathies.

The moral nature of man has reached its present standard, partly through the advancement of his reasoning powers and consequently of a just public opinion, but especially from his sympathies having been rendered more tender and widely diffused through the effects of habit, example, instruction, and reflection. It is not improbable that after long practice virtuous tendencies may be inherited. With the more civilised races, the conviction of the existence of an all-seeing Deity has had a potent influence on the advance of morality. Ultimately man does not accept the praise or blame of his fellows as his sole guide, though few escape this influence, but his habitual convictions, controlled by reason, afford him the safest rule. His conscience then becomes the supreme judge and monitor. Nevertheless the first foundation or origin of the moral sense lies in the social instincts, including sympathy; and these instincts no doubt were primarily gained, as in the case of the lower animals, through natural selection.

The belief in God has often been advanced as not only the greatest, but the most complete of all the distinctions between man and the lower animals. It is however impossible, as we have seen, to maintain that this belief is innate or instinctive in man. On the other hand a belief in all-pervading spiritual agencies seems to be universal; and apparently follows from a considerable advance in man's reason, and from a still greater advance in his faculties of imagination, curiosity and wonder. I am aware that the assumed instinctive belief in God has been used by many persons as an argument for His existence. But this is a rash argument, as we should thus be compelled to believe in the existence of many cruel and malignant spirits, only a little more powerful than man; for the belief in them is far more general than in a beneficent Deity. The idea of a universal and beneficent Creator does not seem to arise in the mind of man, until he has been elevated by long-continued culture.

He who believes in the advancement of man from some low organised form, will naturally ask how does this bear on the belief in the immortality of the soul. The barbarous races of man, as Sir J. Lubbock has shewn, possess no clear belief of this kind; but arguments derived from the primeval beliefs of savages are, as we have just seen, of little or no avail. Few persons feel any anxiety from the impossibility of determining at what precise period in the development of the individual, from the first trace of a minute germinal vesicle, man becomes an immortal being; and there is no greater cause for anxiety because the period cannot possibly be determined in the gradually ascending organic scale.

I am aware that the conclusions arrived at in this work will be denounced by some as highly irreligious; but he who denounces them is bound to shew why it is more irreligious to explain the origin of man as a distinct species by descent from some lower form, through the laws of variation and natural selection, than to explain the birth of the individual through the laws of ordinary reproduction. The birth both of the species and of the individual are equally parts of that grand sequence of events, which our minds refuse to accept as the result of blind chance. The understanding revolts at such a conclusion, whether or not we are able to believe that every slight variation of structure,—the union of each pair in marriage,—the dissemination of each seed,—and other such events, have all been ordained for some special purpose.

Sexual selection has been treated at great length in this work; for, as I have attempted to shew, it has played an important part in the history of the organic world. I am aware that much remains doubtful, but I have endeavoured to give a fair view of the whole case. In the lower divisions of the animal kingdom, sexual selection seems to have done nothing: such animals are often affixed for life to the same spot, or have the sexes combined in the same individual, or what is still more important, their perceptive and intellectual faculties are not sufficiently advanced to allow of the feelings of love and jealousy, or of the exertion of choice. When, however, we come to the Arthropoda and Vertebrata, even to the lowest classes in these two great Sub-Kingdoms, sexual selection has effected much.

In the several great classes of the animal kingdom,—in mammals, birds, reptiles, fishes, insects, and even crustaceans,—the differences between the sexes follow nearly the same rules. The males are almost always the wooers; and they alone are armed with special weapons for fighting with their rivals. They are generally stronger and larger than the females, and are endowed with the requisite qualities of courage and pugnacity. They are provided, either exclusively or in a much higher degree than the females, with organs for vocal or instrumental music, and with odoriferous glands. They are ornamental with infinitely diversified appendages, and with the most brilliant or conspicuous colours, often arranged in elegant patterns, whilst the females are unadorned. When the sexes differ in more important structures, it is the male which is provided with special sense-organs for discovering the female, with locomotive organs for reaching her, and often with prehensile organs for holding her. These various structures for charming or securing the female are often developed in the male during only part of the year, namely the breeding-season. They have in many cases been more or less transferred to the females; and in the latter case they often appear in her as mere rudiments. They are lost or never gained by the males after emasculation. Generally they are not developed in the male during early youth, but appear a short time before the age for reproduction. Hence in most cases the young of both sexes resemble each other; and the female somewhat resembles her young offspring throughout life. In almost every great class a few anomalous cases occur, where there has been an almost complete transposition of the characters proper to the two sexes; the females assuming characters which properly belong to the males. This surprising uniformity in the laws regulating the differences between the sexes in so many and such widely separated classes, is intelligible if we admit the action of one common cause, namely sexual selection.

Sexual selection depends on the success of certain individuals over others of the same sex, in relation to the propagation of the species; whilst natural selection depends on the success of both sexes, at all ages, in relation to the general conditions of life. The sexual struggle is of two kinds; in the one it is between individuals of the same sex, generally the males, in order to drive away or kill their rivals, the females remaining passive; whilst in the other, the struggle is likewise between the individuals of the same sex, in order to excite or charm those of the opposite sex, generally the females, which no longer remain passive, but select the more agreeable partners. This latter kind of selection is closely analogous to that which man unintentionally, yet effectually, brings to bear on his domesticated productions, when he preserves during a long period the most pleasing or useful individuals, without any wish to modify the breed.

The laws of inheritance determine whether characters gained through sexual selection by either sex shall be transmitted to the same sex, or to both; as well as the age at which they shall be developed. It appears that variations arising late in life are commonly transmitted to one and the same sex. Variability is the necessary basis for the action of selection, and is wholly independent of it. It follows from this, that variations of the same general nature have often been taken advantage of and accumulated through sexual selection in relation to the propagation of the species, as well as through natural selection in relation to the general purposes of life. Hence secondary sexual characters, when equally transmitted to both sexes can be distinguished from ordinary specific characters only by the light of analogy. The modifications acquired through sexual selection are often so strongly pronounced that the two sexes have frequently been ranked as distinct species, or even as distinct genera. Such strongly-marked differences must be in some manner highly important; and we know that they have been acquired in some instances at the cost not only of inconvenience, but of exposure to actual danger.

The belief in the power of sexual selection rests chiefly on the following considerations. Certain characters are confined to one sex; and this alone renders it probable that in most cases they are connected with the act of reproduction. In innumerable instances these characters are fully developed only at maturity, and often during only a part of the year, which is always the breeding-season. The males (passing over a few exceptional cases) are the more active in courtship; they are the better armed, and are rendered the more attractive in various ways. It is to be especially observed that the males display their attractions with elaborate care in the presence of the females; and that they rarely or never display them excepting during the season of love. It is incredible that all this should be purposeless. Lastly we have distinct evidence with some quadrupeds and birds, that the individuals of one sex are capable of feeling a strong antipathy or preference for certain individuals of the other sex.

Bearing in mind these facts, and the marked results of man's unconscious selection, when applied to domesticated animals and cultivated plants, it seems to me almost certain that if the individuals of one sex were during a long series of generations to prefer pairing with certain individuals of the other sex, characterised in some peculiar manner, the offspring would slowly but surely become modified in this same manner. I have not attempted to conceal that, excepting when the males are more numerous than the females, or when polygamy prevails, it is doubtful how the more attractive males succeed in leaving a large number of offspring to inherit their superiority in ornaments or other charms than the less attractive males; but I have shewn that this would probably follow from the females,—especially the more vigorous ones, which would be the first to breed,—preferring not only the more attractive but at the same time the more vigorous and victorious males.

Although we have some positive evidence that birds appreciate bright and beautiful objects, as with the bower-birds of Australia, and although they certainly appreciate the power of song, yet I fully admit that it is astonishing that the females of many birds and some mammals should be endowed with sufficient taste to appreciate ornaments, which we have reason to attribute to sexual selection; and this is even more astonishing in the case of reptiles, fish, and insects. But we really know little about the minds of the lower animals. It cannot be supposed, for instance, that male birds of paradise or peacocks should take such pains in erecting, spreading, and vibrating their beautiful plumes before the females for no purpose. We should remember the fact given on excellent authority in a former chapter, that several peahens, when debarred from an admired male, remained widows during a whole season rather than pair with another bird.

Nevertheless I know of no fact in natural history more wonderful than that of the female Argus pheasant should appreciate the exquisite shading of the ball-and-socket ornaments and the elegant patterns on the wing-feathers of the male. He who thinks that the male was created as he now exists must admit that the great plumes, which prevent the wings from being used for flight, and which are displayed during courtship and at no other time in a manner quite peculiar to this one species, were given to him as an ornament. If so, he must likewise admit that the female was created and endowed with the capacity of appreciating such ornaments. I differ only in the conviction that the male Argus pheasant acquired his beauty gradually, through the preference of the females during many generations for the more highly ornamented males; the aesthetic capacity of the females having been advanced through exercise or habit, just as our own taste is gradually improved. In the male through the fortunate chance of a few feathers, being left unchanged, we can distinctly trace how simple spots with a little fulvous shading on one side may have been developed by small steps into the wonderful ball-and-socket ornaments; and it is probable that they were actually thus developed.

Everyone who admits the principle of evolution, and yet feels great difficulty in admitting that female mammals, birds, reptiles, and fish, could have acquired the high taste implied by the beauty of the males, and which generally coincides with our own standard, should reflect that the nerve-cells of the brain in the highest as well as in the lowest members of the Vertebrate series, are derived from those of the common progenitor of this great Kingdom. For we can thus see how it has come to pass that certain mental faculties, in various and widely distinct groups of animals, have been developed in nearly the same manner and to nearly the same degree.

The reader who has taken the trouble to go through the several chapters devoted to sexual selection, will be able to judge how far the conclusions at which I have arrived are supported by sufficient evidence. If he accepts these conclusions he may, I think, safely extend them to mankind; but it would be superfluous here to repeat what I have so lately said on the manner in which sexual selection apparently has acted on man, both on the male and female side, causing the two sexes to differ in body and mind, and the several races to differ from each other in various characters, as well as from their ancient and lowly-organised progenitors.

He who admits the principle of sexual selection will be led to the remarkable conclusion that the nervous system not only regulates most of the existing functions of the body, but has indirectly influenced the progressive development of various bodily structures and of certain mental qualities. Courage, pugnacity, perseverance, strength and size of body, weapons of all kinds, musical organs, both vocal and instrumental, bright colours and ornamental appendages, have all been indirectly gained by the one sex or the other, through the exertion of choice, the influence of love and jealousy, and the appreciation of the beautiful in sound, colour or form; and these powers of the mind manifestly depend on the development of the brain.

Man scans with scrupulous care the character and pedigree of his horses, cattle, and dogs before he matches them; but when he comes to his own marriage he rarely, or never, takes any such care. He is impelled by nearly the same motives as the lower animals, when they are left to their own free choice, though he is in so far superior to them that he highly values mental charms and virtues. On the other hand he is strongly attracted by mere wealth or rank. Yet he might by selection do something not only for the bodily constitution and frame of his offspring, but for their intellectual and moral qualities. Both sexes ought to refrain from marriage if they are in any marked degree inferior in body or mind; but such hopes are Utopian and will never be even partially realised until the laws of inheritance are thoroughly known. Everyone does good service, who aids towards this end. When the principles of breeding and inheritance are better understood, we shall not hear ignorant members of our legislature rejecting with scorn a plan for ascertaining whether or not consanguineous marriages are injurious to man.

The advancement of the welfare of mankind is a most intricate problem: all ought to refrain from marriage who cannot avoid abject poverty for their children; for poverty is not only a great evil, but tends to its own increase by leading to recklessness in marriage. On the other hand, as Mr. Galton has remarked, if the prudent avoid marriage, whilst the reckless marry, the inferior members tend to supplant the better members of society. Man, like every other animal, has no doubt advanced to his present high condition through a struggle for existence consequent on his rapid multiplication; and if he is to advance still higher, it is to be feared that he must remain subject to a severe struggle. Otherwise he would sink into indolence, and the more gifted men would not be more successful in the battle of life than the less gifted. Hence our natural rate of increase, though leading to many and obvious evils, must not be greatly diminished by any means. There should be open competition for all men; and the most able should not be prevented by laws or customs from succeeding best and rearing the largest number of offspring. Important as the struggle for existence has been and even still is, yet as far as the highest part of man's nature is concerned there are other agencies more important. For the moral qualities are advanced, either directly or indirectly, much more through the effects of habit, the reasoning powers, instruction, religion, &c., than through natural selection; though to this latter agency may be safely attributed the social instincts, which afforded the basis for the development of the moral sense.

The main conclusion arrived at in this work, namely, that man is descended from some lowly organised form, will, I regret to think, be highly distasteful to many. But there can hardly be a doubt that we are descended from barbarians. The astonishment which I felt on first seeing a party of Fuegians on a wild and broken shore will never be forgotten by me, for the reflection at once rushed into my mind—such were our ancestors. These men were absolutely naked and bedaubed with paint, their long hair was tangled, their mouths frothed with excitement, and their expression was wild, startled, and distrustful.

They possessed hardly any arts, and like wild animals lived on what they could catch; they had no government, and were merciless to every one not of their own small tribe. He who has seen a savage in his native land will not feel much shame, if forced to acknowledge that the blood of some more humble creature flows in his veins. For my own part I would as soon be descended from that heroic little monkey, who braved his dreaded enemy in order to save the life of his keeper, or from that old baboon, who descending from the mountains, carried away in triumph his young comrade from a crowd of astonished dogs—as from a savage who delights to torture his enemies, offers up bloody sacrifices, practises infanticide without remorse, treats his wives like slaves, knows no decency, and is haunted by the grossest superstitions.

Man may be excused for feeling some pride at having risen, though not through his own exertions, to the very summit of the organic scale; and the fact of his having thus risen, instead of having been aboriginally placed there, may give him hope for a still higher destiny in the distant future. But we are not here concerned with hopes or fears, only with the truth as far as our reason permits us to discover it; and I have given the evidence to the best of my ability. We must, however, acknowledge, as it seems to me, that man with all his noble qualities, with sympathy which feels for the most debased, with benevolence which extends not only to other men but to the humblest living creature, with his god-like intellect which has penetrated into the movements and constitution of the solar system—with all these exalted powers—Man still bears in his bodily frame the indelible stamp of his lowly origin.

SOURCE: Darwin, Charles. *The Descent of Man*. London: Murray, 1871.

TERMS

Fossil—remains indicating that species that had existed in the past were not identical to species represented by living organisms. While the differences provided clues to the kinds of changes that had taken place, the similarities to living forms also demonstrated the continuity of life. As naturalists explored the unity and diversity of organisms living in far-flung parts of the world, they could add to those examples in every category an ever-increasing catalogue of fossil forms.

Geologic Timescale—a concept that provided a key to naturalists who began to think about biological change. While geologists debated the merits of explanations that involved uniform, steady state change over eons versus catastrophic and global change that might have occurred recently, Charles Lyell provided overwhelming evidence for uniformitarianism, as the concept came to be known. Biologists could adopt this concept, and the corresponding timeframe, in order to accommodate the long reach of time needed to see changes of the sort indicated by explorers and naturalists collecting around the world in the nineteenth century.

Natural Selection—the unique contribution of Charles Darwin and Alfred Russel Wallace, representing a specific process or mechanism by which populations could evolve. Before Darwin and Wallace, exploration of the idea of evolution remained highly speculative, lacking a natural mechanism. With natural selection, populations of varying individuals could be examined to find those individuals with advantageous survival characteristics. Darwin referred to the "struggle for existence" as a source of pressure on populations that removed the weak and unfit, and allowed the fit to emerge as parents of the next generation. Over time, with each generation producing an increased proportion of individuals with advantageous variations, populations shifted to include a preponderance of those individuals with adaptations uniquely suited to the current environment.

Sexual Selection—a concept introduced by Charles Darwin as a corollary to natural selection, where reproductive advantages, rather than survival, provided the key to increased fitness. As such, individuals who were likely to find greater mating success would be more likely to pass their characteristics on to the next generation. Even if the traits that assisted in attracting mates carried a cost to survival, additional matings would compensate the attractive individuals in terms of offspring. Some examples of sexual selection involve competition among members of the same sex of a species, although many involve mate choice alone.

Variation—refers to the differences among individuals as they struggled to survive and reproduce. Natural selection could amplify any difference between individuals of a species, no matter how minute or seemingly insignificant. In the lifetime of a single individual, a given variation might produce no effect in the next generation, as random factors may often govern the survival of one or many individuals. But taken as a population, and considering the effects over many generations, each variation became the source of potential selective action, producing adaptations beneficial to survival among many individuals and even distinguishing subpopulations from others in the development of new species.

3

REVIEWS OF DARWIN'S WORK

Publication of Darwin's *On the Origin of Species* brought immediate response from scientists and a wide range of educated British and English readers. In a sense, the responses closely resembled the author's worst fears. Some indicated flaws in logic, others asked for more evidence, and still others identified the unwelcome theological implications. For the most part, Darwin retreated from the public debate, continuing to work at his theory and letting his supporters defend the publication he had considered premature.

Darwin regarded Asa Gray not only as his strongest American supporter, but also as the person who best understood his formulation of natural selection. Gray held different philosophical and theological views, even if he concurred with Darwin in describing the mechanism of selection. Gray explicitly and doggedly retained a role for a divine creator in providing the ultimate design of the universe. He reflected in depth upon the ways that Darwin's work challenged and engaged religious and spiritual notions, and how critics and supporters alike took seriously the scientific evidence that supported ideas new and old.

Gray and others regularly made reference to William Paley's arguments in *Natural Theology*, specifically reconsidering the human eye as an object of fascination to explain. Although supporting the notion of a designer, Gray advanced Darwin's theory of natural selection over the static view of natural theology. The evidence for change undermined Paley's argument for stasis. On this level, Gray embraced evolution as a means for understanding nature even if it meant rethinking the role of the Creator, and in this regard Darwin cherished his American advocate.

Even before Darwin's book appeared, Louis Agassiz suggested that the evidence of nature pointed to the unity of a plan originally conceived by a source of intelligence greater than any known in the physical universe. The unity of plan could be invoked as a means to understanding the relationship of one species to another, and of one group to another. The basis for such relationships in many ways remained obscure to science, and Agassiz could see no greater mystery to be solved than to reveal the principles of the unity of plan that could also account for the diversity of life. Near the end of his life, Agassiz became increasingly critical and more desperate to defend a position that continued to oppose Darwin. By then, most scientists had embraced evolution. Agassiz continued to consider the evidence and reflected on new discoveries in ways that provided evolutionists distinctive insights. He became more entrenched in the view that, whatever the evidence or argument, species could not change and thus evolution as a precept could not be supported.

Throughout the exchanges between Agassiz and his evolutionary opponents, Agassiz upheld the highest standards of scientific discourse. He had earned the respect of his colleagues and treated their ideas with due consideration. He often complimented Darwin on the method and quality of his work, and he took seriously the challenge to understand and answer the arguments put forth. From his position at Harvard University, he helped found the Museum of Comparative Zoology, an institution that eventually served as a center of evolutionary study.

As one who inspired Darwin in the formative years of his career as a naturalist, Charles Lyell provided major philosophical and scientific foundations for natural selection. Lyell's notion of uniformitarian or steady state change in geology became the basis for Darwin's gradualism in organic evolution. Lyell himself wrote extensively on change in organic forms, although he did not provide any explanation for how such change might take place. Upon seeing Darwin's mechanism, he generally accepted the process as both reasonable from a philosophical standpoint and adequate from a scientific standpoint.

While Lyell criticized some of Darwin's logic, he judged more harshly those "opponents of transmutation" who could not see the utility of natural selection as a guide to exploring populations and species. The facts, he pointed out, were well established, even if some questions remained unanswered. Those unanswered questions, even more importantly, provided clues to the mysteries of human history. Through a series of analogies, Lyell showed how humans might be considered the product of natural selection, even explaining intellect and the capacity for belief. Those products suggested the role of that "Author of Nature" rather than inevitable consequences of natural history and selection. This point, that evolution was not progressive, became a significant point of debate, and Lyell sided with Darwin. Together they argued against the rise of Man to a pinnacle of nature. From there, each questioned the possible source and explanation of human self-awareness. Lyell held to belief in a designer, while Darwin explored the topic more deeply in *The Descent of Man*, published shortly before Lyell's death in 1875.

Some of Darwin's critics provided important perspective on questions that remained to be answered, even when they were not naturalists themselves. A Scottish engineer, Fleeming Jenkin claimed neither background as a naturalist nor religious motivations, but provided one of the most thorough critiques of Darwin's theory of natural selection. He carefully outlined one problem recognized by Darwin himself, blending inheritance. The appearance of unique variations that might benefit an individual and serve as the basis of adaptation within a population of offspring would in almost every case, according to the hypothesis of blending inheritance, be obliterated by the much more common characters of every available mate. In generation after generation, natural selection seemed unlikely to identify effectively those few unique characters or "sports" and preserve them in abundance. Darwin wondered at this as well, and only after the basic principles of genetics became widely recognized after 1900 could biologists effectively answer this criticism.

Similarly, Jenkin knew enough of the physical laws that governed the heating and cooling of matter to demonstrate that evidence from geology was inadequate to explain the apparent age of the Earth and solar system. Essentially, the heat from creation, divine or otherwise, must be diminishing throughout the ages. While a biblical figure of 6,000 years would be inadequate to account for current conditions, the assumptions of geologists about the age of the Earth proved incompatible with the still-warm physical material of the planet. Again, new information from physics eventually obviated this criticism as well.

Like other critics of Darwin, Jenkin also suggested that the theory of natural selection lacked the creativity that allowed many theorists to imagine a Creator that could form plants and animals in countless ways. Jenkin did not confine himself to a biblical account, but dismissed Darwin's single-minded approach to assigning cause in nature. He thus provided a logical critique that troubled Darwin, and many of his points remained unanswered at the heart of criticisms from others that followed for decades.

Samuel Wilberforce faced Thomas Huxley in a widely recounted debate that pitted religion against science. Wilberforce, the Bishop of Oxford, defended religion and opposed Darwinian evolution, but caricature's that suggest his defense of religion equate with literal reading of the Bible or other fundamentalists views misrepresent him. In this excerpt, Wilberforce clearly expressed his admiration of Darwin's scientific work, even assuring readers that Darwin must be counted as a Christian.

Whatever Darwin's scientific faults, the greatest problem encountered by Wilberforce revealed an inability of the naturalist to accept and marvel at the mysteries of nature. Speculation that attempted to account for the currently available evidence left little room for admiration of the Creator who had assembled nature. Taking the Creator as a starting point, Wilberforce saw no need to explain the creation on its own terms. The bishop could thus demonstrate that the faults of Darwin's theory corresponded with other naturalists and philosophers who had come before. Readers could see these faults and dismiss Darwin, just as they had dismissed the others, and return to their focus on the creation as the product of divine goodness.

ASA GRAY, "DESIGN VERSUS NECESSITY: DISCUSSION BETWEEN TWO READERS OF DARWIN'S TREATISE ON *THE ORIGIN OF SPECIES*, UPON ITS NATURAL THEOLOGY" (1860)

Asa Gray (1810–1888) published an account by his antagonist, identified here as "D.T.," that examined the issue of design as contrasted with "necessity," where necessity described the outcome of a set of events as prescribed by natural laws. Design occurred where a source of intelligence intended an outcome and set in motion events that would tend to produce that outcome. When other actors intervened, it could be said that design was thwarted, but that necessity would produce other outcomes according to the combination of actions. Taken at this level, natural selection acted independently of design and by necessity to produce species with characteristics as scientists find them. This could not be used as an argument to support or deny the existence of a Designer in nature, but where necessity alone was adequate to supply an explanation, science could reasonably forego any further search for the supernatural. In response, Gray challenged any narrow meaning of design, proposing instead that a Designer would encompass the broader scenario, consider various contingencies, and work to achieve a goal. It was the intention, rather than the mere existence, of the designer that made the difference.

D. T.—Is Darwin's theory atheistic or pantheistic? Or, does it tend to atheism or pantheism? Before attempting any solution of this question, permit me to say a few words tending to obtain

American naturalists worked to identify the taxonomic relationships among species along Darwinian lines throughout the late nineteenth and early twentieth centuries [Comstock, Anna Botsford. (1893) Moths. From John Henry Comstock, *Evolution and Taxonomy* (Ithaca, New York). Plate 1: 23 cm × 15 cm. Image courtesy of Comstock Memorial Library of Entomology, Cornell University].

a definite conception of *necessity* and *design*, as the sources from which events may originate, each independent of the other; and we shall, perhaps, best attain a clear understanding of each, by the illustration of an example in which simple human designers act upon the physical powers of common matter.

Suppose, then, a square billiard-table to be placed with its corners directed to the four cardinal points. Suppose a player, standing at the north corner, to strike a red ball directly to the south, his design being to lodge the ball in the south pocket; which design, if not interfered with, must, of course be accomplished. Then suppose another player, standing at the east corner, to direct a white ball to the west corner. This design also, if not interfered with, must be accomplished. Next suppose both players to strike their balls at the same instant, with like forces, in the directions before given. In this case the balls would not pass as before, namely, the red ball to the south, and the white ball to the west, but they must both meet and strike each other in the centre of the table, and, being perfectly elastic, the red ball must pass to the west pocket, and the white ball to the south pocket. We may suppose that the players acted wholly without concert with each other, indeed, they may be ignorant of each other's design, or even of each other's existence; still we know that the events must happen as herein described. Now, the first half of the course of these two balls is from an impulse, or proceeds from a power, acting from design. Each player has the design of driving his ball across the table in a diagonal line to accomplish its lodgment at the opposite corner of the table. Neither designed that his ball should be deflected from that course and pass to another corner of the table. The *direction* of this second part of the motion must be referred entirely to *necessity*, which directly interferes with the purpose of him who designed the rectilinear direction. We are not, in this case, to go back to find design in the creation of the powers or laws of inertia and elasticity, after the order of which the deflection, at the instant of collision, necessarily takes place. We know that these powers were inherent in the balls, and were not created to answer this special deflection. We are required, by the hypothesis, to confine attention in point of time, from the instant preceding the impact of the balls, to the time of their arrival at the opposite corners of the table. The cues are moved by design. The impacts are acts from design. The first half of the motion of each ball is under the direction of design. We mean by this the particular design of each player. But, at the instant of the collision of the balls upon each other, direction from design ceases, and the balls no longer obey the particular designs of the players, the ends or purposes intended by them are not accomplished, but frustrated, by *necessity*, or by the necessary action of the powers of inertia and elasticity, which are inherent in matter, and are not made by any design of a Creator for this special action, or to serve this special purpose, but would have existed in the materials of which the balls were made, although the players had never been born.

I have thus stated, by a simple example in physical action, what is meant by design and what by necessity; and that the latter may exist without any dependence upon the former. If I have given the statement with what may be thought, by some, unnecessary prolixity, I have only to say that I have found many minds to have a great difficulty in conceiving of necessity as acting altogether independent of design.

Let me now trace these principles as sources of action in Darwin's work or theory. Let us see how much there is of design acting to produce a foreseen end, and thus proving a reasoning and self-conscious Creator; and how much of mere blind power acting without rational design, or without a specific purpose or conscious foresight. Mr. Darwin has specified in a most clear and unmistakable manner the operation of his three great powers, or rather, the three great laws by which the organic power of life acts in the formation of an eye. Following the method he has pointed out, we will take a number of animals of the same species, in which the eye is not developed. They may have all the other senses, with the organs of nutrition, circulation, respiration, and locomotion. They all have a brain and nerves, and some of these nerves may be sensitive to light; but have no combination of retina, membranes, humors, etc., by which the distinct image of an object may be formed and conveyed by the optic nerve to the cognizance of the internal perception, or the mind. The animal in this case would be merely sensible of the difference between light and darkness. He would have no power of discriminating form, size, shape, or color, the difference of objects, and to gain from these a knowledge of their being useful or hurtful, friends or enemies. Up to this point there is no appearance of *necessity* upon the scene. The billiard-balls have not yet struck together, and we will suppose that none of the arguments that may be used to prove, from this organism, thus existing, that it could not have come into form and being without a creator acting to this end with intelligence and design, are opposed by anything that can be found in Darwin's theory; for, so far, Darwin's laws are supposed not to have come into operation. Give the animals, thus organized, food and room, and they may go on, from generation to generation, upon the same organic level. Those individuals that, from natural variation, are born with *light-nerves* a little more sensitive to light than their parents, will cross or interbreed with those who have the same organs a little less sensitive, and thus the mean standard will be kept up without any advancement. If our billiard-table were sufficiently extensive, i.e., infinite, the balls rolled from the corners would never meet, and the *necessity* which we have supposed to deflect them would never act.

The moment, however, that the want of space or food commences *natural selection* begins. Here the balls meet, and all future action is governed by *necessity*. The best forms, or those nerves most sensitive to light, connected with incipient membranes and humors for corneas and lenses, are picked out and preserved by natural selection, of necessity. All cannot live and propagate, and it is a necessity, obvious to all, that the weaker must perish, if the theory be true. Working on, in this way, through countless generations, the eye is at last formed in all its beauty and excellence. It must (always assuming that this theory is true) result from this combined action of natural variation, the struggle for life, and natural selection, with as much certainty as the balls, after collision, must pass to corners of the table different from those to which they were directed, and so far forth as the eye is formed by these laws, acting upward from the nerve merely sensitive to light, we can no more infer design, and from design a designer, than we can infer design in the direction of the billiard-balls after the collision. Both are sufficiently accounted for by blind powers acting under a blind necessity. Take away the struggle for life from the one, and the collision of the balls from the other—and neither of these was designed—and the animal would have gone on without eyes. The balls would have found the corners of the table to which they were first directed.

While, therefore, it seems to me clear that one who can find no proof of the existence of an intelligent Creator except through the evidence of design in the organic world, can find no evidence of such design in the construction of the eye, if it were constructed under the operation of Darwin's laws, I shall not for one moment contend that these laws are *incompatible* with design and a self-conscious, intelligent Creator. Such design might, indeed, have coexisted with the necessity or natural selection; and so the billiard-players might have designed the collision of their balls; but neither the formation of the eye, nor the path of the balls after collision, furnishes any sufficient proof of such design in either case.

One, indeed, who believes, from revelation or any other cause, in the existence of such a Creator, the fountain and source of all things in heaven above and in the earth beneath, will see in natural variation, the struggle for life, and natural selection, only the order or mode in which this Creator, in his own perfect wisdom, sees fit to act. Happy is he who can thus see and adore. But how many are there who have no such belief from intuition, or faith in revelation; but who have by careful and elaborate search in the physical, and more especially in the organic world, inferred, by induction, the existence of God from what has seemed to them the wonderful adaptation of the different organs and parts of the animal body to its, apparently, designed ends! Imagine a mind of this skeptical character, in all honesty and under its best reason, after finding itself obliged to reject the evidence of revelation, to commence a search after the Creator, in the light of natural theology. He goes through the proof for final cause and design, as given in a summary though clear, plain, and convincing form, in the pages of Paley and the "Bridgewater Treatises." The eye and the hand, those perfect instruments of optical and mechanical contrivance and adaptation, without the least waste or surplusage—these, say Paley and Bell, certainly prove a designing maker as much as the palace or the watch proves an architect or a watchmaker. Let this mind, in this state, cross Darwin's work, and find that, after a sensitive nerve or a rudimentary hoof or claw, no design is to be found. From this point upward the development is the mere necessary result of natural selection; and let him receive this law of natural selection as true, and where does he find himself? Before, he could refer the existence of the eye, for example, only to design, or chance. There was no other alternative. He rejected chance, as impossible. It must then be a design. But Darwin brings up another power, namely, natural selection, in place of this impossible chance. This not only may, but, according to Darwin, must of necessity produce an eye. It may indeed coexist with design, but it must exist and act and produce its results, even without design. Will such a mind, under such circumstances, infer the existence of the designer—God—when he can, at the same time, satisfactorily account for the thing produced, by the operation of this natural selection? It seems to me, therefore, perfectly evident that the substitution of natural selection, by necessity, for design in the formation of the organic world, is a step decidedly atheistical. It is in vain to say that Darwin takes the creation of organic life, in its simplest forms, to have been the work of the Deity. In giving up design in these highest and most complex forms of organization, which have always been relied upon as the crowning proof of the existence of an intelligent Creator, without whose intellectual power they could not have been brought into being, he takes a most decided step to banish a belief in the intelligent action of God from the organic world. The lower organisms will go next.

The atheist will say, Wait a little. Some future Darwin will show how the simple forms came necessarily from inorganic matter. This is but another step by which, according to Laplace, "The discoveries of science throw final causes further back."

A. G.—It is conceded that, if the two players in the supposed case were ignorant of each other's presence, the designs of both were frustrated, and from necessity. Thus far it is not needful to inquire whether this necessary consequence is an unconditional or a conditioned necessity, nor to require a more definite statement of the meaning attached to the word *necessity* as a supposed third alternative.

But, if the players knew of each other's presence, we could not infer from the result that the design of both or of either was frustrated. One of them may have intended to frustrate the other's design, and to effect his own. Or both may have been equally conversant with the properties of the matter and the relation of the forces concerned (whatever the cause, origin, or nature, of these forces and properties), and the result may have been according to the designs of both.

As you admit that they might or might not have designed the collision of their balls and its consequences, the question arises whether there is any way of ascertaining which of the two conceptions we may form about it is the true one. Now, let it be remarked that *design* can never be *demonstrated*. Witnessing the act does not make known the *design*, as we have seen in the case assumed for the basis of the argument. The word of the actor is not proof; and that source of evidence is excluded from the cases in question. The only way left, and the only possible way in cases where testimony is out of the question, is to infer the design from the result, or from arrangements which strike us as *adapted* or *intended* to produce a certain result, which affords a presumption of design. The strength of this presumption may be zero, or an even chance, as perhaps it is in the assumed case; but the probability of design will increase with the particularity of the act, the specialty of the arrangement or machinery, and with the number of identical or yet more of similar and analogous instances, until it rises to a moral certainty—i.e., to a conviction which practically we are as unable to resist as we are to deny the cogency of a mathematical demonstration. A single instance, or set of instances, of a comparatively simple arrangement might suffice. For instance, we should not doubt that a pump was designed to raise water by the moving of the handle. Of course, the conviction is the stronger, or at least the sooner arrived at, where we can imitate the arrangement, and ourselves produce the result at will, as we could with a pump, and also with the billiard-balls.

And here I would suggest that your billiard-table, with the case of collision, answers well to a machine. In both a result is produced by indirection—by applying a force out of line of the ultimate direction. And, as I should feel as confident that a man intended to raise water who was working a pump-handle, as if he were bringing it up in pailfuls from below by means of a ladder, so, after due examination of the billiard-table and its appurtenances, I should probably think it likely that the effect of the rebound was expected and intended no less than that of the immediate impulse. And a similar inspection of arrangements and results in Nature would raise at least an equal presumption of design.

You allow that the rebound might have been intended, but you require proof that it was. We agree that a single such instance affords no evidence either way. But how would it be if you saw the men doing the same thing over and over? And if they varied it by other arrangements of the balls or of the blow, and these were followed by analogous results? How if you at length discovered a profitable end of the operation, say the winning of a wager? So in the counterpart case of natural selection: must we not infer intention from the arrangements and the results? But I will take another case of the very same sort, though simpler, and better adapted to illustrate natural selection; because the change of direction—your necessity—acts gradually or successively, instead of abruptly.

Suppose I hit a man standing obliquely in my rear, by throwing forward a crooked stick, called a boomerang. How could he know whether the blow was intentional or not? But suppose I had been known to throw boomerangs before; suppose that, on different occasions, I had before wounded persons by the same, or other indirect and apparently aimless actions; and suppose that an object appeared to be gained in the result—that definite ends were attained—would it not at length be inferred that my assault, though indirect, or apparently indirect, was designed?

To make the case more nearly parallel with those it is brought to illustrate, you have only to suppose that, although the boomerang thrown by me went forward to a definite place, and at least appeared to subserve a purpose, and the bystanders, after a while, could get traces of the mode or the empirical law of its flight, yet they could not themselves do anything with it. It was quite beyond their power to use it. Would they doubt, or deny *my* intention, on that account? No: They would insist that design on my part must be presumed from the nature of the results; that, though design may have been wanting in any one case, yet the repetition of the result, and from different positions and under varied circumstances, showed that there *must* have been design.

Moreover, in the way your case is stated, it seems to concede the most important half of the question, and so affords a presumption for the rest, on the side of design. For you seem to assume an actor, a designer, accomplishing his design in the first instance. You—a bystander—infer that the player effected his design in sending the first ball to the pocket before him. You infer this from observation alone. Must you not from a continuance of the same observation equally infer a common design of the two players in the complex result, or a design of one of them to frustrate the design of the other? If you grant a designing actor, the presumption of design is as strong, or upon continued observation of instances soon becomes as strong, in regard to the deflection of the balls, or variation of the species, as it was for the result of the first impulse or for the production of the original animal, etc.

But, in the case to be illustrated, we do not see the player. We see only the movement of the balls. Now, if the contrivances and adaptations referred to really do "prove a designer as much as the palace or the watch proves an architect or a watchmaker"—as Paley and Bell argue, and as your skeptic admits, while the alternative is between design and chance—then they prove it with all the proof the case is susceptible of, and with complete conviction. For we cannot doubt that the watch had a watchmaker. And if they prove it on the supposition that the unseen operator acted *immediately*—i.e., that the player directly impelled the balls in the directions we see them moving, I insist that this proof is not impaired by our ascertaining that he acted *mediately*—i.e., that the present state or form of the plants or animals, like the present position of the billiard-balls, resulted from the collision of the individuals with one another, or with the surroundings. The original impulse, which we once supposed was in the line of the observed movement, only proves to have been in a different direction; but the series of movements took place with a series of results, each and all of them none the less determined, none the less designed.

Wherefore, when, at the close, you quote Laplace, that "the discoveries of science throw final causes farther back," the most you can mean is, that they constrain us to look farther back for the impulse. They do not at all throw *the argument for design* farther back, in the sense of furnishing evidence or presumption that only the primary impulse was designed, and that all the rest followed from chance or necessity.

Evidence of design, I think you will allow, everywhere is drawn from the observation of adaptations and of results, and has really nothing to do with anything else, except where you can take the *word* for the *will*. And in that case you have not *argument for design*, but *testimony*. In Nature we have no testimony; but the argument is overwhelming.

Now, note that the argument of the olden time—that of Paley, etc., which your skeptic found so convincing—was always the argument for design in the movement of the balls *after deflection*. For it was drawn from animals produced by generation, not by creation, and through a long succession of generations or deflections. Wherefore, if the argument for design is perfect in the case of an animal derived from a long succession of individuals as nearly alike as offspring is generally like parents and grandparents, and if this argument is not weakened when a variation, or series of variations, has occurred in the course, as great as any variations we know of among domestic cattle, how then is it weakened by the supposition, or by the likelihood, that the variations have been twice or thrice as great as we formerly supposed, or because the variations have been "picked out," and a few of them preserved as breeders of still other variations, by natural selection?

Finally let it be noted that your element of *necessity* has to do, so far as we know, only with the picking out and preserving of certain changing forms, i.e., with the natural selection. This selection, you may say, must happen under the circumstances. This is a necessary result of the collision of the balls; and these results can be predicted. If the balls strike so and so, they will be deflected so and so. But the *variation* itself is of the nature of an origination. It answers well to the original impulse of the balls, or to a series of such impulses. We cannot predict what particular new variation will occur from any observation of the past. Just as the first impulse was given to the balls at a point out of sight, so the impulse which resulted in the variety or new form was given at a point beyond observation, and is equally mysterious or unaccountable, except on the supposition of an ordaining will. The parent had not the peculiarity of the variety, the progeny has. Between the two is the din or obscure region of the formation of a new individual, in some unknown part of which, and in some wholly unknown way, the difference is intercalated. To introduce necessity here is gratuitous and unscientific; but here you must have it to make your argument valid.

I agree that, judging from the past, it is not improbable that variation itself may be hereafter shown to result from physical causes. When it is so shown, you may extend your necessity into this region, but not till then. But the whole course of scientific discovery goes to assure us that the discovery of the cause of variation will be only a resolution of variation into two factors: one, the immediate secondary cause of the changes, which so far explains them; the other an unresolved or unexplained phenomenon, which will then stand just where the product, variation, stands now, only that it will be one step nearer to the efficient cause.

This line of argument appears to me so convincing that I am bound to suppose that it does not meet your case. Although you introduced players to illustrate what design is, it is probable that you did not intend, and would not accept, the parallel which your supposed case suggested. When you declare that the proof of design in the eye and the hand, as given by Paley and Bell, was convincing, you mean, of course, that it was convincing, so long as the question was between *design* and *chance*, but that now another alternative is offered, one which obviates the force of those arguments, and may account for the actual results without design. I do not clearly apprehend this third alternative.

Will you be so good, then, as to state the grounds upon which you conclude that the supposed proof of design from the eye, or the hand, as it stood before Darwin's theory was promulgated, would be invalidated by the admission of this new theory?

D. T.—As I have ever found you, in controversy, meeting the array of your opponent fairly and directly, without any attempt to strike the body of his argument through an unguarded joint in the phraseology, I was somewhat surprised at the course taken in your answer to my statement on Darwin's theory. You there seem to suppose that I instanced the action of the billiard balls and players as a parallel, throughout, to the formation of the organic world. Had it occurred to me that such an application might be supposed to follow legitimately from my introduction of this action, I should certainly have stated that I did not intend, and should by no means accede to, that construction. My purpose in bringing the billiard-table upon the scene was to illustrate, by example, *design* and *necessity*, as different and independent sources from which results, it might indeed be identical results, may be derived. All the conclusions, therefore, that you have arrived at through this misconception or misapplication of my illustration, I cannot take as an answer to the matter stated or intended to be stated by me. Again, following this misconception, you suppose the skeptic (instanced by me as revealing through the evidence of design, exhibited in the structure of the eye, for its designer, God) as bringing to the examination a belief in the existence of design in the construction of the animals as they existed up to the moment when the eye was, according to my supposition, added to the heart, stomach, brain, etc. By skeptic I, of course, intended one who doubted the existence of design in every organic structure, or at least required proof of such design. Now, as the watch may be instanced as a more complete exhibition of design than a flint knife or an hour-glass, I selected, after the example of Paley, the eye, as exhibiting by its complex but harmonious arrangements a higher evidence of design and a designer than is to be found in a nerve sensitive to light, or any mere rudimentary part or organ. I could not mean by skeptic one who believed in design so far as a claw, or a nerve sensitive to light, was concerned, but doubted all above. For one who believes in design at all will not fail to recognize it in a hand or an eye. But I need not extend these remarks, as you acknowledge in the sequel to your argument that you may not have suited it to the case as I have stated it.

You now request me to "state the grounds upon which I conclude that the supposed proof of design from the eye and the hand, as it stood before Darwin's theory was promulgated, is invalidated by the admission of that theory." It seems to me that a sufficient answer to this question has already been made in the last part of my former paper; but, as you request it, I will go over the leading points as there given, with more minuteness of detail.

Let us, then, suppose a skeptic, one who is yet considering and doubting of the existence of God, having already concluded that the testimony from any and all revelation is insufficient, and having rejected what is called the *a priori* arguments brought forward in natural theology, and pertinaciously insisted upon by Dr. Clark and others, turning as a last resource to the argument from design in the organic world. Voltaire tells him that a palace could not exist without an architect to design it. Dr. Paley tells him that a watch proves the design of a watchmaker. He thinks this very reasonable, and, although he sees a difference between the works of Nature and those of mere human art, yet if he can find in any organic body, or part of a body, the same adaptation to its use that he finds in a watch, this truth will go very far toward proving, if it is not entirely conclusive, that, in making it, the powers of life by which it grew were directed by an intelligent, reasoning master. Under the guidance of Paley he takes an eye, which, although an optical, and not a mechanical instrument like the watch, is as well adapted to testify to design. He sees, first, that the eye is transparent when every other part of the body is opaque. Was this the result of a mere Epicurean or Lucretian "fortuitous concourse" of living "atoms"? He is not yet certain it might not be so. Next he sees that it is spherical, and that this convex form alone is capable of changing the direction of the light which proceeds from a distant body, and of collecting it so as to form a distinct image within its globe. Next he sees at the exact place where this image must be formed a curtain of nerve-work, ready to receive and convey it, or excite from it, in its own mysterious way, an idea of it in the mind. Last of all, he comes to the crystalline lens. Now, he has before learned that without this lens an eye would by the aqueous and vitreous humors alone form an image upon the retina, but this image would be indistinct from the light not being sufficiently refracted, and likewise from having a colored fringe round its edges. This last effect is attributable to the refrangibility of light, that is, to some of the colors being more refracted than others. He likewise knows that more than a hundred years ago Mr. Dollond having found out, after many experiments, that some kinds of glass have the power of dispersing light, for each degree of its refraction, much more than other kinds, and that on the discovery of this fact he contrived to make telescopes in which he passed the light through two object-glasses successively, one of which he made of crown and one of flint glass, so ground and adapted to each other that the greater dispersion produced by the substance of one should be corrected by the smaller dispersion of the other. This contrivance corrected entirely the colored images which had rendered all previous telescopes very imperfect. He finds in this invention all the elements of design, as it appeared in the thought and action of a human designer. First, conjecture of certain laws or facts in optics. Then, experiment proving these laws or facts. Then, the contrivance and formation of an instrument by which those laws or facts must produce a certain sought result.

Thus enlightened, our skeptic turns to his crystalline lens to see if he can discover the work of a Dollond in this. Here he finds that an eye, having a crystalline lens placed between the humors, not only refracts the light more than it would be refracted by the humors alone, but that, in this combination of humors and lens, the colors are as completely corrected as in the combination of Dollond's telescope. Can it be that there was no design, no designer, directing the powers of life in the formation of this wonderful organ? Our skeptic is aware that, in the arts of man, great aid has been, sometimes, given by chance, that is, by the artist or workman observing some fortuitous combination, form, or action, around him. He has heard it said that the chance arrangement of two pairs of spectacles, in the shop of a Dutch optician, gave the direction for constructing the first telescope. Possibly, in time, say a few geological ages, it might in some optician's shop have brought about a combination of flint and crown glass which, together, should have been achromatic. But the space between the humors of the eye is not an optician's shop where object-glasses of all kinds, shapes, and sizes, are placed by chance, in all manner of relations and positions. On the hypothesis under which our skeptic is making his examination—the eye having been completed in all but the formation of the lens—the place which the lens occupies when completed was filled with parts of the humors and plane membrane, homogeneous in texture and surface, resenting, therefore, neither the variety of the materials nor forms which are contained in the optician's shop for chance to make its combinations with. How, then, could it be cast of a combination not before used, and fashioned to a shape different from that before known, and placed in exact combination with all the parts before enumerated, with many others not even mentioned? He sees no parallelism of condition, then, by which chance could act in forming a crystalline lens, which answers to the condition of an optician's shop, where it might be possible in many ages for chance to combine existing forms into an achromatic object-glass.

Considering, therefore, the eye thus completed and placed in its bony case and provided with its muscles, its lids, its tear-ducts, and all its other elaborate and curious appendages, and, a thousand times more wonderful still, without being encumbered with a single superfluous or useless part, can he say that this could be the work of chance? The improbability of this is so great, and consequently the evidence of design is so strong, that he is about to seal his verdict in favor of design, when he opens Mr. Darwin's book.

There he finds that an eye is no more than a vital aggregation or growth, directed, not by design nor chance, but moulded by natural variation and natural selection, through which it must, necessarily, have been developed and formed. Particles or atoms being aggregated by the blind powers of life, must become under the given conditions, by natural variation and natural selection, eyes, without design, as certainly as the red billiard-ball went to the west pocket, by the powers of inertia and elasticity, without the design of the hand that put it in motion.

Let us lay before our skeptic the way in which we may suppose that Darwin would trace the operation of life, or the vital force conforming to these laws. In doing this we need not go through with the formation of the several membranes, humors, etc., but take the crystalline lens as the most curious and nicely arranged and adapted of all the parts, and as giving, moreover, a close parallel, in the end produced, to that produced by design, by a human designer, Dollond, in forming his achromatic object-glass. If it can be shown that natural variation and natural selection were capable of forming the crystalline lens, it will not be denied that they were capable of forming the iris, the sclerotica, the aqueous humors, or any and all the other parts. Suppose, then, that we have a number of animals, with eyes yet wanting the crystalline. In this state the animals can see, but dimly and imperfectly, as a man sees after having been couched. Some of the offspring of these animals have, by *natural variation*, merely a portion of the membrane which separates the aqueous from the vitreous humor a little thickened in its middle part, a little swelled out. This refracts the light a little more than it would be refracted by a membrane in which no such swelling existed, and not only so, but, in combination with the humors, it corrects the errors of dispersion and makes the image somewhat more colorless. All the young animals that have this swelled membrane see more distinctly than their parents or brethren. They, therefore, have an advantage over them in the *struggle for life*. They can obtain food more easily; can find their prey, and escape from their enemies with greater facility than their kindred. This thickening and rounding of the membrane goes on from generation to generation by natural variation; natural selection all the while "picking out with unerring skill all the improvements, through countless generations," until at length it is found that the membrane has become a perfect crystalline lens. Now, where is the design in all this? The membrane was not thickened and rounded to the end that the image should be more distinct and colorless; but, being thickened and rounded by the operation of natural variation, *inherent* in generation, natural selection of *necessity* produced the result that we have seen. The same result was thus produced of necessity, in the eye, that Dollond came at, in the telescope, with design, through painful guessing, reasoning, experimenting, and forming.

Suppose our skeptic to believe in all this power of natural selection; will he now seal up his verdict for design, with the same confidence that he would before he heard of Darwin? If not, then "the supposed proof from design is invalidated by Darwin's theory."

A. G.—Waiving incidental points and looking only to the gist of the question, I remark that the argument for design as against chance, in the formation of the eye, is most convincingly stated in your argument. Upon this and upon numerous similar arguments the whole question we are discussing turns. So, if the skeptic was about to seal his verdict in favor of design, and a designer, when Darwin's book appeared, why should his verdict now be changed or withheld? All the facts about the eye, which convinced him that the organ was designed, remain just as they were. His conviction was not produced through testimony or eyewitness, but design was irresistibly inferred from the evidence of contrivance in the eye itself.

Now, if the eye as it is, or has become, so convincingly argued design, why not each particular step or part of this result? If the production of a perfect crystalline lens in the eye— you know not how—as much indicated design as did the production of a Dollond achromatic lens—you understand how—then why does not "the swelling out" of a particular portion of the membrane behind the iris—caused you know not how—which, by "correcting the errors of dispersion and making the image somewhat more colorless," enabled the "young animals to see more distinctly than their parents or brethren," equally indicate design—if not as much as a perfect crystalline, or a Dollond compound lens, yet as much as a common spectacle-glass?

Darwin only assures you that what you may have thought was done directly and at once was done indirectly and successively. But you freely admit that indirection and succession do not invalidate design, and also that Paley and all the natural theologians drew the arguments which convinced your skeptic wholly from eyes indirectly or naturally produced.

Recall a woman of a past generation and show her a web of cloth; ask her how it was made, and she will say that the wool or cotton was carded, spun, and woven by hand. When you tell her it was not made by manual labor, that probably no hand has touched the materials throughout the process, it is possible that she might at first regard your statement as tantamount to the assertion that the cloth was made without design. If she did, she would not credit your statement. If you patiently explained to her the theory of carding-machines, spinning-jennies, and power-looms, would her reception of your explanation weaken her conviction that the cloth was the result of design? It is certain that she would believe in design as firmly as before, and that this belief would be attended by a higher conception and reverent admiration of a wisdom, skill, and power greatly beyond anything she had previously conceived possible.

Wherefore, we may insist that, for all that yet appears, the argument for design, as presented by the natural theologians, is just as good now, if we accept Darwin's theory, as it was before that theory was promulgated; and that the skeptical juryman, who was about to join the other eleven in a unanimous verdict in favor of design, finds no good excuse for keeping the court longer waiting.

SOURCE: Gray, Asa. "Design versus Necessity: Discussion between Two Readers of Darwin's Treatise on the Origin of Species, upon Its Natural Theology," in *Darwiniana*. Ed. A. Hunter Dupree. Cambridge: Belknap Press, 1963 (1860), pp. 51–71.

LOUIS AGASSIZ, ESSAY ON CLASSIFICATION (1851)

Among the best-known critics of evolution in the nineteenth century, Swiss-born naturalist Louis Agassiz (1807–1873) worked diligently to demonstrate the role of the Creator in nature. This essay, first published while Darwin still pondered his evolutionary theory, contained the essence of an argument Agassiz advanced for decades afterward. He described evidence from the natural world in terms that clearly illustrated the importance of intelligence in designing the diverse and wonderful forms that naturalists worked to categorize. The implications for classification that evolution suggested vexed Agassiz more than the plain suggestion that evolution could explain natural forms without reference to a creator. His position with respect to the ideas he opposed remained unique. Louis Agassiz embraced much of Darwin's descriptive work, and frequently applauded the British naturalist's botanical, zoological, paleontological, and geological publications. He could point to the concept of homology, just as Darwin did, to support his broader philosophical conclusions. The similarities in bone structure among vertebrate animals—even radically different vertebrates birds, horses, and fish—demonstrated for Agassiz the unity of plan he professed. Other similarities in other groups counted equally as homologies. The limits of homology, however, illustrated the limits of Darwin's theory. Where the gaps between structures of different groups remained large, Agassiz insisted that evolution could not provide reasonable explanations. To insist upon such explanations would deny the creativity of a divine intelligence.

Section IV: Unity of Plan in Otherwise Highly Diversified Types

Nothing is more striking throughout the animal and vegetable kingdoms than the unity of plan in the structure of the most diversified types. From pole to pole, in every longitude, mammalia, birds, reptiles, and fishes exhibit one and the same plan of structure, involving abstract conceptions of the highest order, far transcending the broadest generalizations of man, for it is only after the most laborious investigations man has arrived at an imperfect understanding of this plan. Other plans, equally wonderful, may be traced in Articulata, in Mollusks, in Radiata, and in the various types of plants. And yet the logical connection, these beautiful harmonies, this infinite diversity in unity are represented by some as the result of forces exhibiting no trace of intelligence, no power of thinking, no faculty of combination, no knowledge of time and space. If there is anything which places man above all other beings in nature, it is precisely the circumstance that he possesses those noble attributes without which, in their most exalted excellence and perfection, not one of these general traits of relationship so characteristic of the great types of the animal and vegetable kingdoms can be understood or even perceived. How, then, could these relations have been devised without similar powers? If all these relations are almost beyond the reach of the mental powers of man, and if man himself is part and parcel of the whole system, how could this system have been called into existence if there does not exist One Supreme Intelligence as the Author of all things?

Louis Agassiz [Library of Congress Prints and Photographs Department, LC-USZ62-103949].

Section V: Correspondence in the Details of Structure in Animals Otherwise Entirely Disconnected

During the first decade of this century naturalists began to study relations among animals which had escaped almost entirely the attention of earlier observers. Though Aristotle knew already that the scales of fishes correspond to the feathers of birds, it is but recently that anatomists have discovered the close correspondence which exists between all the parts of all animals belonging to the same type, however different they may appear at first sight. Not only is the wing of the bird identical in its structure with the arm of man or the fore leg of a quadruped, it agrees quite as closely, with the fin of the whale or the pectoral fin of the fish, and all these together correspond in the same manner with their hind extremities. Quite as striking a coincidence is observed between the solid skull-box, the immovable bones of the face and the lower jaw of man and the other mammalia, and the structure of the bony frame of the head of birds, turtles, lizards, snakes, frogs, and fishes. But this correspondence is not limited to the skeleton; every other system of organs exhibits in these animals the same relations, the same identity in plan and structure, whatever be the differences in the form of the parts, in their number, and even in their functions. Such an agreement in the structure of animals is called their homology and is more or less close in proportion as the animals in which it is traced are more or less nearly related.

The same agreement exists between the different systems and their parts in Articulata, in Mollusks, and in Radiata, only that their structure is built up upon respectively different plans, though in these three types the homologies have not yet been traced to the same extent as among Vertebrata. There is, therefore, still a wide field open for investigations in this most attractive branch of Zoology. So much, however, is already plain from what has been done in this department of our science, that the identity of structure among animals does not extend to all the four branches of the animal kingdom; that, on the contrary, every great type is constructed upon a distinct plan, so peculiar, indeed, that homologies cannot be extended from one type to the other but are strictly limited within each of them. The more remote resemblance which may be traced between representatives of different types is founded upon analogy and not upon affinity. While, for instance, the head of fishes exhibits the most striking homology with that of reptiles, birds, and mammalia, as a whole, as well as in all its parts, that of Articulata is only analogous to it and to its part. What is commonly called head in Insects is not a head like that of Vertebrata; it has not a distinct cavity for the brain, separated from that, which communicates below the neck with the chest and abdomen; its solid envelope does not consist of parts of an internal skeleton, surrounded by flesh, but is formed of external rings, like those of the body, soldered together; it contains but one cavity, which includes the cephalic ganglion, as well as the organs of the mouth and all the muscles of the head. The same may be said of the chest, the legs and wings, the abdomen, and all the parts they contain. The cephalic ganglion is not homologous to the brain, nor are the organs of senses homologous to those of Vertebrata, even though they perform the same functions. The alimentary canal is formed in a very different way in the embryos of the two types, as are also their respiratory organs, and it is as unnatural to identify them, as it would be still to consider gills and lungs as homologous among Vertebrata, now that Embryology has taught us that in different stages of growth these two kinds of respiratory organs exist in all Vertebrata in very different organic connections one from the other.

What is true of the branch of Articulata when compared to that of Vertebrata is equally true of the Mollusks and Radiata when, compared with one another or with the two other types, as might easily be shown by a fuller illustration of the correspondence of their structure within these limits. This inequality in the fundamental character of the structure of the four branches of the animal kingdom points to the necessity of a radical reform in the nomenclature of Comparative Anatomy. Some naturalists, however, have already extended such comparisons respecting the structure of animals beyond the limits pointed out by nature, when they have attempted to show that all structures may be reduced to one norm, and when they have maintained, for instance, that every bone existing in any Vertebrate must have its counterpart in every other species of that type. To assume such a uniformity among animals would amount to denying to the Creator even as much freedom in expressing his thoughts as man enjoys.

If it be true, as pointed out above, that all animals are constructed upon four different plans of structure, in such a manner that all the different kinds of animals are only different expressions of these fundamental formulae, we may well compare the whole animal kingdom to a work illustrating four great ideas, between which there is no other connecting link than the unity exhibited in the eggs in which their most diversified manifestations are first embodied in an embryonic form, to undergo a series of transformations, and appear in the end in that wonderful variety of independent living beings which inhabit our globe, or have inhabited it from the earliest period of the existence of life upon its surface.

The most surprising feature of the animal kingdom seems, however, to me to rest neither in its diversity, nor in the various degrees of complication of its structure, nor in the close affinity of some of its representatives while others are so different, nor in the manifold relations of all of them to one another and the surrounding world, but in the circumstances that beings, endowed with such different and such unequal gifts should nevertheless constitute an harmonious whole, intelligibly connected in all its parts.

SOURCE: Agassiz, Louis. *Essay on Classification.* Ed. Edward Lurie. Cambridge: Belknap Press, 1962 (1857), pp. 20–23.

CHARLES LYELL, *THE ANTIQUITY OF MAN* (1863)

For as much as Charles Lyell's (1797–1895) geological work was interwoven with Darwin's biology, Lyell held some different views about the ultimate implications of evolution for understanding the history of humankind. For that issue in particular, the geologist retained a distinctly creationist view. Although his beliefs were nothing like biblical creationism, he acknowledged the work of an "Author of Nature," a being that could create around a divine design. This acknowledgment relieved a scientist from insisting upon conclusions that required absurd logic. An example noted in this excerpt involved the success of species that arrive suddenly in a new region and exceed all expectations for expansion, despite their not having been adapted to the new area as Darwin's theory would suggest. For Lyell, that success related to the original design of a creator, while for Darwin it depended on chance.

When I formerly advocated the doctrine that species were primordial creations and not derivative, I endeavoured to explain the manner of their geographical distribution, and the affinity of living forms to the fossil types nearest akin to them in the Tertiary strata of the same part of the globe, by supposing that the creative power, which originally adapts certain types to aquatic and others to terrestrial conditions, has at successive geological epochs introduced new forms best suited to each area and climate, so as to fill the places of those which may have died out.

In that case, although the new species would differ from the old (for these would not be revived, having been already proved by the fact of their extinction to be incapable of holding their ground), still they would resemble their predecessors generically. For, as Mr. Darwin states in regard to new races, those of a dominant type inherit the advantages which made their parent species flourish in the same country, and they likewise partake in those general advantages which made the genus to which the parent species belonged a large genus in its own country.

We might therefore, by parity of reasoning, have anticipated that the creative power, adapting the new types to the new combination of organic and inorganic conditions of a given region, such as its soil, climate, and inhabitants, would introduce new modifications of the old types—marsupials, for

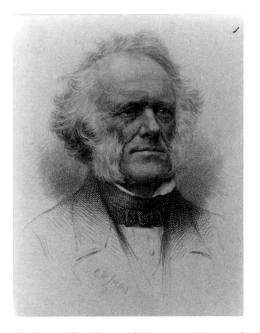

Charles Lyell [Library of Congress Prints and Photographs Department, LC-USZ62-123180].

example, in Australia, new sloths and armadilloes in South America, new heaths at the Cape, new roses in the northern and new calceolarias in the southern hemisphere. But to this line of argument Mr. Darwin and Dr. Hooker reply that when animals or plants migrate into new countries, whether assisted by man or without his aid, the most successful colonisers appertain by no means to those types which are most allied to the old indigenous species. On the contrary it more frequently happens that members of genera, orders, or even classes, distinct and foreign to the invaded country, make their way most rapidly and become dominant at the expense of the endemic species. Such is the case with the placental quadrupeds in Australia, and with horses and many foreign plants in the pampas of South America, and numberless instances in the United States and elsewhere which might easily be enumerated. Hence the transmutationists infer that the reason why these foreign types, so peculiarly fitted for these regions, have never before been developed there is simply that they were excluded by natural barriers. But these barriers of sea or desert or mountain could never have been of the least avail had the creative force acted independently of material laws or had it not pleased the Author of Nature that the origin of new species should be governed by some secondary causes analogous to those which we see preside over the appearance of new varieties, which never appear except as the offspring of a parent stock very closely resembling them. . . .

Chapter 24

Some of the opponents of transmutation, who are well versed in Natural History, admit that though that doctrine is untenable, it is not without its practical advantages as a "useful working hypothesis," often suggesting good experiments and observations and aiding us to retain in the memory a multitude of facts respecting the geographical distribution of genera and species, both of animals and plants, the succession in time of organic remains, and many other phenomena which, but for such a theory, would be wholly without a common bond of relationship.

It is in fact conceded by many eminent zoologists and botanists, as before explained, that whatever may be the nature of the species-making power or law, its effects are of such a character as to imitate the results which variation, guided by natural selection, would produce, if only we could assume with certainty that there are no limits to the variability of species. But as the anti-transmutationists are persuaded that such limits do exist, they regard the hypothesis as simply a provisional one, and expect that it will one day be superseded by another cognate theory, which will not require us to assume the former continuousness of the links which have connected the past and present states of the organic world, or the outgoing with the incoming species.

In like manner, many of those who hesitate to give in their full adhesion to the doctrine of progression, the other twin branch of the development theory, and who even object to it, as frequently tending to retard the reception of new facts supposed to militate against opinions solely founded on negative evidence, are nevertheless agreed that on the whole it is of great service in guiding our speculations. Indeed it cannot be denied that a theory which establishes a connection between the absence of all relics of vertebrata in the oldest fossiliferous rocks, and the presence of man's remains in the newest, which affords a more than plausible explanation of the successive appearance in strata of intermediate age of the fish, reptile, bird, and mammal, has no ordinary claims to our favour as comprehending the largest number of positive and negative facts gathered from all parts of the globe, and extending over countless ages, that science has perhaps ever attempted to embrace in one grand generalisation.

But will not transmutation, if adopted, require us to include the human race in the same continuous series of developments, so that we must hold that Man himself has been derived by an unbroken line of descent from some one of the inferior animals? We certainly cannot escape from such a conclusion without abandoning many of the weightiest arguments which have been urged in support of variation and natural selection considered as the subordinate causes by which new types have been gradually introduced into the earth. Many of the gaps which separate the most nearly allied genera and orders of mammalia are, in a physical point of view, as wide as those which divide Man from the mammalia most nearly akin to him, and the extent of his isolation, whether we regard his whole nature or simply his corporeal attributes, must be considered before we can discuss the bearing of transmutation upon his origin and place in the creation. . . .

The author of an elaborate review of Darwin's "Origin of Species," himself an accomplished geologist [William Hopkins], declares that if we embrace the doctrine of the continuous variation of all organic forms from the lowest to the highest, including Man as the last link in the chain of being, there must have been a transition from the instinct of the brute to the noble mind of Man; and in that case, "Where," he asks, "are the missing links, and at what point of his progressive improvement did Man acquire the spiritual part of his being, and become endowed with the awful attribute of immortality?"

Before we raise objections of this kind to a scientific hypothesis, it would be well to pause and inquire whether there are no analogous enigmas in the constitution of the world around us, some of which present even greater difficulties than that here stated. When we contemplate, for example, the many hundred millions of human beings who now people the earth, we behold thousands who are doomed to helpless imbecility, and we may trace an insensible gradation between them and the half-witted, and from these again to individuals of perfect understanding, so that tens of thousands must have existed in the course of ages, who in their moral and intellectual condition, have exhibited a passage from the irrational to the rational, or from the irresponsible to the responsible. Moreover we may infer from the returns of the Registrar General of Births and Deaths in Great Britain, and from Quetelet's statistics of Belgium, that one-fourth of the human race die in early infancy, nearly one-tenth before they are a month old; so that we may safely affirm that millions perish on the earth in every century, in the first few hours of their existence. To assign to such individuals their appropriate psychological place in the creation is one of the unprofitable themes on which theologians and metaphysicians have expended much ingenious speculation.

The philosopher, without ignoring these difficulties, does not allow them to disturb his conviction that "whatever is, is right," nor do they check his hopes and aspirations in regard to the high destiny of his species; but he also feels that it is not for one who is so often confounded by the painful realities of the present, to test the probability of theories respecting the past, by their agreement or want of agreement with some ideal of a perfect universe which those who are opposed to opinions may have pictured to themselves.

We may also demur to the assumption that the hypothesis of variation and natural selection obliges us to assume that there was an absolutely insensible passage from the highest intelligence of the inferior animals to the improvable reason of Man. The birth of an individual of transcendent genius, of parents who have never displayed any intellectual capacity above the average standard of their age or race, is a phenomenon not to be lost sight of, when we are conjecturing whether the successive steps in advance by which a progressive scheme has been developed may not admit of occasional strides, constituting breaks in an otherwise continuous series of psychical changes.

The inventors of useful arts, the poets and prophets of the early stages of a nation's growth, the promulgators of new systems of religion, ethics, and philosophy, or of new codes of laws, have often been looked upon as messengers from Heaven, and after their death have had divine honours paid to them, while fabulous tales have been told of the prodigies which accompanied their birth. Nor can we wonder that such notions have prevailed when we consider what important revolutions in the moral and intellectual world such leading spirits have brought about; and when we reflect that mental as well as physical attributes are transmissible by inheritance, so that we may possibly discern in such leaps the origin of the superiority of certain races of mankind. In our own time the occasional appearance of such extraordinary mental powers may be attributed to atavism; but there must have been a beginning to the series of such rare and anomalous events. If, in conformity with the theory of progression, we believe mankind to have risen slowly from a rude and humble starting point, such leaps may have successively introduced not only higher and higher forms and grades of intellect, but at a much remoter period may have cleared at one bound the space which separated the highest stage of the unprogressive intelligence of the inferior animals from the first and lowest form of improvable reason manifested by Man.

To say that such leaps constitute no interruption to the ordinary course of nature is more than we are warranted in affirming. In the case of the occasional birth of an individual of superior genius there is certainly no break in the regular genealogical succession; and when all the mists of mythological fiction are dispelled by historical criticism, when it is acknowledged that the earth did not tremble at the nativity of the gifted infant and that the face of heaven was not full of fiery shapes, still a mighty mystery remains unexplained, and it is the ORDER of the phenomena, and not their CAUSE, which we are able to refer to the usual course of nature.

Dr. Asa Gray, in the excellent essay already cited, has pointed out that there is no tendency in the doctrine of Variation and Natural Selection to weaken the foundations of Natural Theology, for, consistently with the derivative hypothesis of species, we may hold any of the popular views respecting the manner in which the changes of the natural world are brought about. We may imagine "that events and operations in general go on in virtue simply of forces communicated at the first, and without any subsequent interference, or we may hold that now and then, and only now and then, there is a direct interposition of the Deity; or, lastly, we may suppose that all the changes are carried on by the immediate orderly and constant, however infinitely diversified, action of the intelligent, efficient Cause." They who maintain that the origin of an individual, as well as the origin of a species or a genus, can be explained only by the direct action of the creative cause, may retain their favourite theory compatibly with the doctrine of transmutation.

Professor Agassiz, having observed that, "while human thought is consecutive, divine thought is simultaneous," Dr. Asa Gray has replied that, "if divine thought is simultaneous, we have no right to affirm the same of divine action."

The whole course of nature may be the material embodiment of a preconcerted arrangement; and if the succession of events be explained by transmutation, the perpetual adaptation of the organic world to new conditions leaves the argument in favour of design, and therefore of a designer, as valid as ever; "for to do any work by an instrument must require, and therefore presuppose, the exertion rather of more than of less power, than to do it directly."* (*Asa Gray, "Natural Selection Not Inconsistent with Natural Theology" Trubner & Co. London 1861 page 55.)

As to the charge of materialism brought against all forms of the development theory, Dr. Gray has done well to remind us that "of the two great minds of the seventeenth century, Newton and Leibnitz, both profoundly religious as well as philosophical, one produced the theory of gravitation, the other objected to that theory, that it was subversive of natural religion."* (*Ibid. page 31.)

It may be said that, so far from having a materialistic tendency, the supposed introduction into the earth at successive geological periods of life—sensation—instinct—the intelligence of the higher mammalia bordering on reason—and lastly the improvable reason of Man himself, presents us with a picture of the ever-increasing dominion of mind over matter.

SOURCE: Lyell, Charles. *The Antiquity of Man*. London: Murray, 1863.

FLEEMING JENKIN, "THE ORIGIN OF SPECIES [REVIEW ARTICLE]" (1867)

Fleeming Jenkin (1822–1885) recounted accurately the main points of the argument laid out in *On the Origin of Species*, and referred extensively to evidence presented in that book. He esteemed Darwin as an observer and contributor to the enterprise of science, but he withheld any praise that might relate to Darwin's efforts as a logician and philosopher. Jenkin sought to dismiss natural selection on points of logic and reduce its general applicability by demonstrating the triviality of observing the process in action. Jenkin organized his argument around five key elements. He questioned the extent to which natural variability among individuals could account for kinds of new features that must have appeared in populations. He questioned the efficiency with which natural selection might operate. He questioned the timeframe over which evolution was purported to have taken place. He also questioned the importance of solving problems of classification by claiming only that similarly classified species must share common ancestry. Finally, he questioned the significance of observations that served as minor facts in support of Darwin's broader claims. In this excerpt, each of these questions received treatment by Jenkin. Throughout, he insisted that his purpose was to defend the logic of truth, rather than to align himself with any religious cause or argument. He consistently accepted those observations and facts that any reasonable person could identify, and even defended those that required specialized training to understand. His point was not to deny evidence, but to examine the connections being made between facts and theoretical suppositions. Jenkin wanted to limit, wherever possible, the assumptions that led Darwin to conclude that all of nature could be understood as the product of natural selection operating over vast reaches of time.

The theory proposed by Mr. Darwin as sufficient to account for the origin of species has been received as probably, and even as certainly true, by many who from their knowledge of physiology, natural history, and geology, are competent to form an intelligent opinion.

The facts, they think, are consistent with the theory. Small differences are observed between animals and their offspring. Greater differences are observed between varieties known to be sprung from a common stock. The differences between what have been termed species are sometimes hardly greater in appearance than those between varieties owning a common origin. Even when species differ more widely, the difference they say, is one of degree only, not of kind. They can see no clear, definite distinction by which to decide in all cases, whether two animals have sprung from a common ancestor or not. They feel warranted in concluding, that for aught the structure of animals shows to the contrary, they may be descended from a few ancestors only,—nay, even from a single pair.

The most marked differences between varieties known to have sprung from one source have been obtained by artificial breeding. Men have selected, during many generations, those individuals possessing the desired attributes in the highest degree. They have thus been able to add, as it were, small successive differences, till they have at last produced marked varieties. Darwin shows that by a process, which he calls natural selection, animals more favourably constituted than their fellows will survive in the struggle for life, will produce descendants resembling themselves, of which the strong will live, the weak will die; and so, generation after generation, nature, by a metaphor, may be said to choose certain animals, even as man does when he desires to raise a special breed. The device of nature is based on the attributes most useful to the animal; the device of man on the attributes useful to man, or admired by him. All must agree that the process termed natural selection is in universal operation. The followers of Darwin believe that by that process differences might be added even as they are added by man's selection, though more slowly, and that this addition might in time be carried to so great an extent as to produce every known species of animal from one or two pairs, perhaps from organisms of the lowest type.

A very long time would be required to produce in this way the great differences observed between existing beings. Geologists say their science shows no ground for doubting that the habitable world has existed for countless ages. Drift and inundation, proceeding at the rate we now observe, would require cycles of ages to distribute the materials of the surface of the globe in their present form and order; and they add, for aught we know, countless ages of rest may at many places have intervened between the ages of action.

But if all beings are thus descended from a common ancestry, a complete historical record would show an unbroken chain of creatures, reaching from each one now known back to the first type, with each link differing from its neighbour by no more than the several offspring of a single pair of animals now differ. We have no such record; but geology can produce vestiges which may be looked upon as a few out of the innumerable links of the whole conceivable chain, and what, say the followers of Darwin, is more certain than that the record of geology must necessarily be imperfect? The records we have show a certain family likeness between the beings living at each epoch, and this is at least consistent with our views.

There are minor arguments in favour of the Darwinian hypothesis, but the main course of the argument has, we hope, been fairly stated. It bases large conclusions as to what has happened upon the observation of comparatively small facts now to be seen. The cardinal facts are the production of varieties by man, and the similarity of all existing animals. About the truth and extent of those facts none but men possessing a special knowledge of physiology and natural history have any right to an opinion; but the superstructure based on those facts enters the region of pure reason, and may be discussed apart from all doubt as to the fundamental facts.

Can natural selection choose special qualities, and so breed special varieties, as man does? Does it appear that man has the power indefinitely to magnify the peculiarities which distinguish his breeds from the original stock? Is there no other evidence than that of geology as to the age of the habitable earth? And what is the value of the geological evidence? How far, in the absence of other knowledge, does the mere difficulty in classifying organized beings justify us in expecting that they have had a common ancestor? And finally, what value is to be attached to certain minor facts supposed to corroborate the new theory? These are the main questions to be debated in the present essay, written with the belief that some of them have been unduly overlooked. The opponents of Darwin have been chiefly men having special knowledge similar to his own, and they have therefore naturally directed their attention to the cardinal facts of his theory. They have asserted that animals are not so similar but that specific differences can be detected, and that man can produce no varieties differing from the parent stock, as one species differs from another. They naturally neglect the deductions drawn from facts which they deny. If your facts were true, they say, perhaps nature would select varieties, and in endless time, all you claim might happen; but we deny the facts. You produce no direct evidence that your selection took place, claiming only that your hypothesis is not inconsistent with the teaching of geology. Perhaps not, but you only claim a 'may be,' and we attack the direct evidence you think you possess.

To an impartial looker-on the Darwinians seem rather to have had the best of the argument on this ground, and it is at any rate worthwhile to consider the question from the other point of view; admit the facts, and examine the reasoning. This we now propose to do, and for clearness will divide the subject into heads corresponding to the questions asked above, as to the extent of variability, the efficiency of natural selection, the lapse of time, the difficulty of classification, and the value of minor facts adduced in support of Darwin.

Some persons seem to have thought his theory dangerous to religion, morality, and what not. Others have tried to laugh it out of court. We can share neither the fears of the former nor the merriment of the latter; and, on the contrary, own to feeling the greatest admiration both for the ingenuity of the doctrine and for the temper in which it was broached, although, from a consideration of the following arguments, our opinion is adverse to its truth. . . .

Variability: Darwin says that in the struggle for life a grain may turn the balance in favour of a given structure, which will then be preserved. But one of the weights in the scale of nature is due to the number of a given tribe. Let there be 7000 A's and 7000 B's, representing two varieties of a given animal, and let all the B's, in virtue of a slight difference of structure, have the better chance of life by 1/7000th part. We must allow that there is a slight probability that the descendants of B will supplant the descendants of A; but let there be only 7001 A's against 7000 B's at first, and the chances are once more equal, while if there be 7002 A's to start, the odds would be laid on the A's. True, they stand a greater chance of being killed; but then they can better afford to be killed. The grain will only turn the scales when these are very nicely balanced, and an advantage in numbers counts for weight, even as an advantage in structure. As the numbers of the favoured variety diminish, so must its relative advantage increase, if the chance of its existence is to surpass the chance of its extinction, until hardly any conceivable advantage would enable the descendants of a single pair to exterminate the descendants of many thousands if they and their descendants are supposed to breed freely with the inferior variety, and so gradually lose their ascendancy. If it is impossible that any sport or accidental variation in a single individual, however favourable to life, should be preserved and transmitted by natural selection, still less can slight an imperceptible variations, occurring in single individuals be garnered up and transmitted to continually increasing numbers; for if a very highly-favoured white cannot blanch a nation of Negroes, it will hardly be contended that a comparatively very dull mulatto has a good chance of producing a tawny tribe; the idea, which seems almost absurd when presented in connexion with a practical case, rests on a fallacy of exceedingly common occurrence in mechanics and physics generally. When a man shows that a tendency to produce a given effect exists he often thinks he has proved that the effect must follow. He does not take into account the opposing tendencies, much less does he measure the various forces, with a view to calculate the result. For instance, there is a tendency on the part of a submarine cable to assume a catenary curve, and very high authorities once said it would; but, in fact, forces neglected by them utterly alter the curve from the catenary. There is a tendency on the part of the same cables, as usually made, to untwist entirely; luckily there are opposing forces, and they untwist very little. These cases will hardly seem obvious; but what should we say to a man who asserted that the centrifugal tendency of the earth must send it off in a tangent? One tendency is balanced or outbalanced by others; the advantage of structure possessed by an isolated specimen is enormously outbalanced by the advantage of numbers possessed by the others. . . .

Efficiency of Natural Selection—Those individual of any species which are most adapted to the life they lead, live on an average longer than those which are less adapted to the circumstances in which the species is placed. The individuals which live the longest will have the most numerous offspring, and as the offspring on the whole resemble their parents, the descendants from any given generation will on the whole resemble the more favoured rather than the less favoured individuals of the species. So much of the theory of natural selection will hardly be denied; but it will be worth while to consider how far this process can tend to cause a variation in some one direction. It is clear that it will frequently, and indeed generally, tend to prevent any deviation from the common type. The mere existence of a species is a proof that it is tolerably well adapted to the life it must lead; many of the variations which may occur will be variations for the worse, and natural selection will assuredly stamp these out. A white grouse in the heather, or a white hare on a fallow would be sooner detected by its enemies than one of the usual plumage or colour. Even so, any favourable deviation must, according to the very terms of the statement, give its fortunate possessor a better chance of life; but this conclusion differs widely from the supposed consequence that a whole species may or will gradually acquire some one new quality, or wholly change in one direction and in the same manner. In arguing this point, two distinct kinds of possible variation must be separately considered: *first*, that kind of common variation which must be conceived as not only possible, but inevitable, in each individual of the species, such as longer and shorter legs, better or worse hearing, etc.; and, *secondly*, that kind of variation which only occurs rarely, and may be called a sport of nature, or more briefly a 'sport,' as when a child is born with six fingers on each hand. The common variation is not limited to one part of any animal, but occurs in all; and when we say that on the whole the stronger live longer than the weaker, we mean that in some cases long life will have been due to good lungs, in others to good ears, in others to good legs. There are few cases in which one faculty is pre-eminently useful to an animal beyond all other faculties, and where that is not so, the effect of natural selection will simply be to kill the weakly, and insure a sound, healthy, well-developed breed. If we could admit the principle of a gradual accumulation of improvements, natural selection would gradually improve the breed of everything, making the hare of the present generation run faster, hear better, digest better, than his ancestors; his enemies, the weasels, greyhounds, etc., would have improved likewise, so that perhaps the hare would not be really better off; but at any rate the direction of the change would be from a war of pigmies to a war of Titans. Opinions may differ as to the evidence of this gradual perfectibility of all things, but it is beside the question to argue this point, as the origin of species requires not the gradual improvement of animals retaining the same habits and structure, but such modification of those habits and structure as will actually lead to the appearance of new organs. We freely admit, that if an accumulation of slight improvements be possible, natural selection might improve hares as hares, and weasels as weasels, that is to say, it might produce animals having every useful faculty and every useful organ of their ancestors developed to a higher degree; more than this, it may obliterate some once useful organs when circumstances have so changed that they are no longer useful, for since that organ will weigh for nothing in the struggle of life, the average animal must be calculated as though it did not exist.

We will even go further: if, owing to a change of circumstances some organ becomes pre-eminently useful, natural selection will undoubtedly produce a gradual improvement in that organ, precisely as man's selection can improve a special organ. In all cases the animals above the average live longer, those below the average die sooner, but in estimating the chance of life of a particular animal, one special organ may count much higher or lower according to circumstances, and will accordingly be improved or degraded. Thus it must apparently be conceded that natural selection is a true cause or agency whereby in some cases variations of special organs may be perpetuated and accumulated, but the importance of this admission is much limited by a consideration of the cases where it applies: first of all we have required that it should apply to variations which must occur in every individual, so that enormous numbers of individuals will exist, all having a little improvement in the same direction; as, for instance, each generation of hares will include an enormous number which have longer legs than the average of their parents although there may be an equally enormous number who have shorter legs; secondly, we require that the variation shall occur in an organ already useful owing to the habits of the animal. Such a process of improvement as is described could certainly never give organs of sight, smell or hearing to organisms which had never possessed them. It could not add a few legs to a hare, or produce a new organ, or even cultivate any rudimentary organ which was not immediately useful to any enormous majority of hares. No doubt half the hares which are born have longer tails than the average of their ancestors; but as no large number of hares hang by their tails, it is inconceivable that any change of circumstances should breed hares with prehensile tails; or, to take an instance less shocking in its absurdity, half the hares which are born may be presumed to be more like their cousins the rabbits in their burrowing organs than the average hare ancestor was; but this peculiarity cannot be improved by natural selection as described above, until a considerable number of hares begin to burrow, which we have as yet seen no likelihood of their doing. Admitting, therefore, that natural selection may improve organs already useful to great numbers of a species, does not imply an admission that it can create or develop new organs, and so original species. . . .

But this theory of the origin of species is surely not the Darwinian theory; it simply amounts to the hypothesis that, from time to time, an animal is born differing appreciably from its progenitors, and possessing the power of transmitting the difference to its descendants. What is this but stating that, from time to time, a new species is created? It does not, indeed, imply that the new specimen suddenly appears in full vigour, made out of nothing; but it offers no explanation of the cause of the divergence from the progenitors, and still less of the mysterious faculty by which the divergence is transmitted unimpaired to countless descendants. It is clear that every divergence is not thus transmitted, for otherwise one and the same animal might have to be big to suit its father and little to suit is mother, might require a long nose in virtue of its grandfather and a short one in virtue of its grandmother, in a word, would have to resume in itself the countless contradictory peculiarities of its ancestors, all in full bloom, and unmodified one by the other, which seems as impossible as at one time to be and not to be. The appearance of a new specimen capable of perpetuating its peculiarity is precisely what might be termed a creation, the word being used to express our ignorance of how the thing happened. The substitution of the new specimens, descendants from the old species, would then be simply an example of strong race supplanting a weak one, by a process known long before the term 'natural selection' was invented. Perhaps this is the way in which new species are introduced, but it does not express the Darwinian theory of the gradual accumulation of infinitely minute differences of every-day occurrence, and apparently fortuitous in their character. . . .

Lapse of Time: To resume the arguments in this chapter—Darwin's theory requires countless ages, during which the earth shall have been habitable, and he claims geological evidence as showing an inconceivably great lapse of time, and as not being in contradiction with inconceivably greater periods than are even geologically indicated,—periods of rest between formations, and periods anterior to our so-called first formations, during which the rudimentary organs of the early fossils became degraded from their primeval uses. In answer, it is shown that a general physical law obtains, irreconcilable with the persistence of active change at a constant rate; in any portion of the universe, however large, only a certain capacity for change exists, so that every change which occurs renders the possibility of future change less, and, on the whole, the rapidity or violence of changes tends to diminish. Not only would this law gradually entail in the future the death of all beings and cessation of all change in the planetary system, and in the past point to a state of previous violence equally inconsistent with life, if no energy were lost by the system, but this gradual decay from a previous state of violence is rendered far more rapid by the continual loss of energy going on by means of radiation. From this general conception pointing either to a beginning, or to the equally inconceivable idea of infinite energy in finite materials, we pass to the practical application of the law to the sun and earth, showing that their present state proves that they cannot remain for ever adapted to living beings, and that living beings can have existed on the earth only for a definite time, since in distant periods the earth must have been in fusion, and the sun must have been mere hot gas, or a group of distant meteors, so as to have been incapable of fulfilling its present functions as the comparatively small centre of the system. From the earth we have no very safe calculation of past time, but the sun gives five hundred million years as the time separating us from a condition inconsistent with life. We next argue that the time occupied in the arrangement of the geological formations need not have been longer than is fully consistent with this view, since the gradual dissipation of energy must have resulted in a gradual diminution of violence of all kinds, so that calculations of the time occupied by denudations or deposits based on the simple division of the total mass of a deposit, or denudation by the annual action now observed, are fallacious, and that even as the early geologists erred in attempting to compress all action into six thousand years, so later geologists have outstepped all bounds in their figures, by assuming that the world has always gone on much as it now does, and that the planetary system contains an inexhaustible motive power, by which the vast labour of the system has been, and can be maintained for ever. We have endeavoured to meet the main objections to these views, and conclude, that countless ages cannot be granted to the expounder of any theory of living beings, but that the age of the inhabited world is proved to have been limited to a period wholly inconsistent with Darwin's views. . . .

Difficulty of Classification: It appears that it is difficult to classify animals or plants, arranging them in groups as genera, species, and varieties; that the line of demarcation is by no means clear between species and sub-species, between sub-species and well-marked varieties, or between lesser varieties and individual differences; that these lines of demarcation, as drawn by different naturalists, vary much, being sometimes made to depend on this, sometimes on that organ, rather arbitrarily. This difficulty chiefly seems to have led men to devise theories of transmutation of species, and is the very starting point of Darwin's theory, which depicts the differences between various individuals of any one species as identical in nature with the differences between individuals of various species, and supposes all these differences, varying in degree only, to have been produced by the same causes; so that the subdivision into groups is, in this view, to a great extent arbitrary, but may be considered rational if the words variations, varieties, sub-species, species, and genera, be used to signify or be considered to express that the individuals included in these smaller or greater groups, have had a common ancestor very lately, some time since, within the later geological ages, or before the primary rocks. The common terms, explained by Darwin's principles, signify, in fact, the more or less close blood-relationship of the individuals. This, if it could be established, would undoubtedly afford a less arbitrary principle of classification than pitching on some organ in any degree similar. The application of the new doctrine might offer some difficulty, as it does not clearly appear what would be regarded as the sign of more or less immediate descent from a common ancestor, and perhaps each classifier would have pet marks by which to decide the question, in which case the new principle would not be of much practical use; yet if the theory were really true, in time the marks of common ancestry would probably come to be known with some accuracy, and meanwhile the theory would give an aim and meaning to classification, which otherwise might be looked upon as simply a convenient form of catalogue. . . .

It may perhaps be thought irreverent to hold an opinion that the Creator could not create animals of any shape and fashion whatever; undoubtedly we may conceive all rules and all laws as entirely self-imposed by him, as possibly quite different or non-existent elsewhere; but what we mean is this, that just as with the existing chemical laws of the world, the number of possible chemical combinations of a particular kind is limited, and not even the Creator could make more without altering the laws he has himself imposed, even so, if we imagine animals created or existing under some definite law, the number of species, and of possible varieties of one species, will be limited; and these varieties and species being definite arrangements of organic compounds, will as certainly be capable of arrangement in series as inorganic chemical compounds are. These views no more imply a limit to the power of God than the statement that the three angles of a triangle are necessarily equal to two right angles. . . .

Observed Facts Supposed to Support Darwin's Views: The chief argument used to establish the theory rest on conjecture. Beasts may have varied; variation may have accumulated; they may have become permanent; continents may have arisen or sunk, and seas and winds been so arranged as to dispose of animals just as we find them, now spreading a race widely, now confining it to one Galapagos island. There may be records of infinitely more animals than we know of in geological formations yet unexplored. Myriads of species differing little from those we know to have been preserved, may actually not have been preserved at all. There may have been an inhabited world for ages before the earliest known geological strata. The world may indeed have been inhabited for an indefinite time; even the geological observations may perhaps give most insufficient idea of the enormous times which separated one formation from another; the peculiarities of hybrids may result from accidental differences between the parents, not from what have been called specific differences.

We are asked to believe all these maybe's happening on an enormous scale, in order that we may believe the final Darwinian 'maybe,' as to the origin of species. The general form of his argument is as follows: All these things may have been, therefore my theory is possible, and since my theory is a possible one, all those hypotheses which it requires are rendered probable. There is little direct evidence that any these maybe's actually *have been*.

In this essay an attempt has been made to show that many of these assumed possibilities are actually impossibilities, or at the best have not occurred in this world, although it is proverbially somewhat difficult to prove a negative.

Let us now consider what direct evidence Darwin brings forward to prove that animals really are descended from a common ancestor. As direct evidence we may admit the possession of webbed feet by unplumed birds; the stripes observed on some kinds of horses and hybrids of horses, resembling not their parents, but other species of the genus; the generative variability of abnormal organs; the greater tendency to vary of widely diffused and widely ranging species, certain peculiarities of distribution. All these facts are consistent with Darwin's theory, and if it could be shown that they could not possibly have occurred except in consequence of natural selection, they would prove the truth of this theory. It would, however, clearly be impossible to prove that in no other way could these phenomena have been produced, and Darwin makes no attempt to prove this. He only says he cannot imagine why unplumed birds should have webbed feet, unless in consequence of their direct descent from web-footed ancestors who lived in the water; that he thinks it would in some way be derogatory to the Creator to let hybrids have stripes on their legs, unless some ancestors of theirs had stripes on his leg. He cannot imagine why abnormal organs and widely diffused genera should vary more than others, unless his views be true; and he says he cannot account for the peculiarities of distribution in any way but one. It is perhaps hardly necessary to combat these arguments, and to show that our inability to account for certain phenomena, in any way but one, is no proof of the truth of the explanation given, but simply is a confession of our ignorance. When a man says a glowworm must be on fire, and in answer to our doubts challenges us to say how it can give out light unless it be on fire, we do not admit his challenge as any proof of his assertion, and indeed we allow it no weight whatever as against positive proof we have that the glowworm is not on fire. We conceive Darwin's theory to be in exactly the same case; its untruth can, as we think, be proved, and his or our own inability to explain a few isolated facts consistent with his views would simply prove his and our ignorance of the true explanation. But although unable to give any certainly true explanations of the above phenomena, it is possible to suggest explanations perhaps as plausible as the Darwinian theory, and though the fresh suggestions may very probably not be correct, they may serve to show that at least more than one conceivable explanation may be given. . . .

We by no means wish to assert that we know the above suggestions to be the true explanation of the facts. We merely wish to show that other explanation than those given by Darwin are conceivable, although this is indeed not required by our argument, since, if his main assumptions can be proved false, his theory will derive no benefit from the few facts which may be allowed to be consistent with its truth. . . .

These arguments are cumulative. If it be true that no species can vary beyond defined limits, it matters little whether natural selection would be efficient in producing definite variations. If natural selection, though it does select the stronger average animals, and under peculiar circumstances may develop special organs already useful, can never select new imperfect organs such as are produced in sports, then, even though eternity were granted, and no limit assigned to the possible changes of animals, Darwin's cannot be the true explanation of the manner in which change has been brought about. Lastly, even if no limit be drawn to the possible difference between offspring and their progenitors, and if natural selection were admitted to be an efficient cause capable of building up even new senses, even then, unless time, vast time, be granted, the changes which might have been produced by the gradual selection of peculiar offspring have not really been so produced. Any one of the main pleas of our argument, if established, is fatal to Darwin's theory. What then shall we say if we believe that experiment has shown a sharp limit to the variation of every species, that natural selection is powerless to perpetuate new organs even should they appear, that countless ages of a habitable globe are rigidly proven impossible by the physical laws which forbid the assumption of infinite power in a finite mass? What can we believe but that Darwin's theory is an ingenious and plausible speculation, to which future physiologists will look back with the kind of admiration we bestow on the atoms of Lucretius, or the crystal spheres of Eudoxus, containing like these some faint half-truths, marking at once the ignorance of the age and the ability of the philosopher. Surely the time is past when a theory unsupported by evidence is received as probable, because in our ignorance we know not why it should be false, though we cannot show it to be true. Yet we have heard grave men gravely urge, that because Darwin's theory was the most plausible known, it should be believed. Others seriously allege that it is more consonant with a lofty idea of the Creator's action to suppose that he produced beings by natural selection, rather than by the finikin process of making each separate little race by the exercise of Almighty power. The argument such as it is, means simply that the user of it thinks that this is how he personally would act if possessed of almighty power and knowledge, but his speculations as to his probable feelings and actions, after such a great change of circumstances, are not worth much. If we are told that our experience shows that God works by laws, then we answer, 'Why the special Darwinian law?' A plausible theory should not be accepted while unproven; and if the arguments of this essay be admitted, Darwin's theory of the origin of species is not only without sufficient support from evidence, but is proved false by a cumulative proof.

SOURCE: Jenkin, Fleeming. "*The Origin of Species* [Review Article]." *The North British Review* 46 (1867).

SAMUEL WILBERFORCE, "REVIEW OF DARWIN'S *ORIGIN OF SPECIES*" (1860)

In examining Darwin's theory of natural selection, Samuel Wilberforce (1805–1873) remained circumspect. He understood Darwin's argument and provided an accurate account of certain evidence for his readers. He chose to criticize points where analogies overreached the evidence as most readers might interpret it. Where Darwin seemed determined to account for change, Wilberforce noted that most of nature exhibited remarkable stability. In domestic breeding, few species retained the hard-won characteristics breeders sought when individuals were allowed to breed without human guidance. Over thousands of years of human history, examples of the kind of change Darwin described seemed nonexistent. While the fossil record might be imperfect, as paleontologists generally argued, the lack of progressive recent evidence, for Wilberforce, undermined the notion of evolution. While praising Darwin for his skill and ingenuity in developing an account of evolutionary change, Wilberforce was much less generous in crediting the naturalist with original ideas. Instead, Darwin's contributions were compared to those of his grandfather, Erasmus Darwin, who had dabbled with evolutionary thinking, but produced no evidence of its action in nature. To dabble with ideas and meddle with evidence violated Wilberforce's clear picture of how science ought to be conducted. He noted how seriously geologists offended their science when they constructed explanations that required the shifting of strata, rather than examining more carefully the solid rock on which their theories must be built.

The Lord Bishop of Winchester, Samuel Wilberforce [Library of Congress Prints and Photographs Department, LC-USZ-62-97016].

Any contribution to our Natural History literature from the pen of Mr. C. Darwin is certain to command attention. His scientific attainments, his insight and carefulness as an observer, blended with no scanty measure of imaginative sagacity, and his clear and lively style, make all his writings unusually attractive. His present volume on the 'Origin of Species' is the result of many years of observation, thought, and speculation; and is manifestly regarded by him as the 'opus' upon which his future fame is to rest. It is true that he announces it modestly enough as the mere precursor of a mightier volume. But that volume is only intended to supply the facts which are to support the completed argument of the present essay. In this we have a specimen-collection of the vast accumulation; and, working from these as the high analytical mathematician may work from the admitted results of his conic sections, he proceeds to deduce all the conclusions to which he wishes to conduct his readers.

The essay is full of Mr. Darwin's characteristic excellences. It is a most readable book; full of facts in natural history, old and new, of his collecting and of his observing; and all of these are told in his own perspicuous language, and all thrown into picturesque language, and all sparkle with the colours of fancy and the lights of imagination. It assumes, too, the grave proportions of a sustained argument upon a matter of the deepest interest, not to naturalists only, or even to men of science exclusively, but to every one who is interested in the history of man and of the relations of nature around him to the history and plan of creation. . . .

Now, the main propositions by which Mr. Darwin's conclusion is attained are these:

1. That observed and admitted variations spring up in the course of descents from a common progenitor.
2. That many of these variations tend to an improvement upon the parent stock.
3. That, by a continued selection of these improved specimens as the progenitors of future stock, its powers may be unlimitedly increased.
4. And, lastly, that there is in nature a power continually and universally working out this selection, and so fixing and augmenting these improvements.

Mr. Darwin's whole theory rests upon the truth of these propositions, and crumbles utterly away if only one of them fail him. These therefore we must closely scrutinize. We will begin with the last in our series, both because we think it the newest and the most ingenious part of Mr. Darwin's whole argument, and also because, whilst we absolutely deny the mode in which he seeks to apply the existence of the power to help him in his argument, yet we think that he throws great and very interesting light upon the fact that such a self-acting power does actively and continuously work in all creation around us. . . .

Mr. Darwin begins by endeavoring to prove that such variations are produced under the selecting power of man amongst domestic animals. Now here we demur *in limine*. Mr. Darwin himself allows that there is a plastic habit amongst domesticated animals which is not found amongst them when in a state of nature. 'Under domestication, it may be truly said that the whole organization becomes in some degree plastic.' If so, it is not fair to argue, from the variations of the plastic nature, as to what he himself admits is the far more rigid nature of the undomesticated animal. But we are ready to give Mr. Darwin this point, and to join issue with him on the variations which he is able to adduce, as having been produced under circumstances the most favourable to change. He takes for this purpose the domestic pigeon, the most favourable specimen no doubt, for many reasons, which he could select, as being a race eminently subject to variation, the variations of which have been most carefully observed by breeders, and which, having been for some 4000 years domesticated, affords the longest possible period for the accumulation of variations. But with all this in his favour, what is he able to show? He writes a delightful chapter upon pigeons. Runts and fantails, short-faced tumblers and long-faced tumblers, long-beaked carriers and pouters, black barbs, Jacobins, and turbots, coo and tumble, inflate their oesophagi, and pout and spread out their tails before us. We learn that 'pigeons have been watched and tended with the utmost care, and loved by many people.' They have been domesticated for thousands of years in several quarters of the world. The earliest known record of pigeons is in the fifth Egyptian dynasty, about 3000 B.C., though 'pigeons are given in a bill of fare' (what an autograph would be that of the chef-de-cuisine of the day!) 'in the previous dynasty' and so we follow pigeons on down to the days of 'that most skilful breeder Sir John Sebright,' who 'used to say, with respect to pigeons, that "he would produce any given feather in three years, but it would take him six years to produce beak and head."'

Now all this is very pleasant writing, especially for pigeon-fanciers; but what step do we really gain in it all towards establishing the alleged fact that variations are but species in the act of formation, or in establishing Mr. Darwin's position that a well-marked variety may be called an incipient species? We affirm positively that no single *fact* tending even in that direction is brought forward. On the contrary, every one points distinctly towards the opposite conclusions; for with all the change wrought in appearance, with all the apparent variation in manners, there is not the faintest beginning of any such change in what that great comparative anatomist, Professor Owen, calls 'the characteristics of the skeleton or other parts of the frame upon which specific differences are founded.' There is no tendency to that great law of sterility which, in spite of Mr. Darwin, we affirm ever to mark the hybrid; for every variety of pigeon, and the descendants of every such mixture, breed as freely, and with as great fertility, as the original pair; nor is there the very first appearance of that power of accumulating variations until they grow into specific differences, which is essential to the argument for the transmutation of species; for, as Mr. Darwin allows, sudden returns in colour, and other most altered appearances, to the parent stock continually attest the tendency of variations not to become fixed, but to vanish, and manifest the perpetual presence of a principle which leads not to the accumulation of minute variations into well-marked species, but to return from the abnormal to the original type. So clear is this, that it is well known that any relaxation in the breeder's care effaces all the established points of difference, and the fancy-pigeon reverts again to the character of its simplest ancestor.

The same relapse may moreover be traced in still wider instances. There are many testimonies to the fact that domesticated animals, removed from the care and tending of man, lose rapidly the peculiar variations which domestication had introduced amongst them, and relapse into their old untamed condition. 'Plus,' says M.P.S. Pallas, 'je réfléchis, plus je suis dispose à croire que la race des chevaux sauvages que l'on trouve dans les landes baignées par le Jaik et le Don, et dans celles de Baraba, ne providnt que de chevaux Kirguis et Kalmouks devenus sauvages,' &c.; and he proceeds to show how far they have relapsed from the type of tame into that of wild horses. Prichard, in his 'Natural History of Man,' remarks that the present state of the escaped domesticated animals, which, since the discovery of the Western Continent by the Spaniards, have been transported from Europe to American, gives us an opportunity of seeing how soon the relapse may become almost complete. 'Many of these races have multiplied (he says) exceedingly on a soil and under a climate congenial to their nature. Several of them have run wild in the vast forests of America, and have lost all the most obvious appearances of domestication.' This he proceeds to prove to be more or less the case as to the hog, the horse, the ass, the sheep, the goat, the cow, the dog, the cat, and gallinaceous fowls.

Now, in all these instances we have the result of the power of selection exercised on the most favourable species for a very long period of time, in a race of that peculiarly plastic habit which is the result of long domestication; and that result is, to prove that there has been no commencement of any such mutation as could, if it was infinitely prolonged, become really a specific change. . . .

We come then to these conclusions. All the facts presented to us in the natural world tend to show that none of the variations produced in the fixed forms of animal life, when seen in its most plastic condition under domestication, give any promise of a true transmutation of species; first, from the difficulty of accumulating and fixing variations within the same species; secondly, from the fact that these variations, though most serviceable for man, have no tendency to improve the individual beyond the standard of his own specific type, and so to afford matter, even if they were infinitely produced, for the supposed power of natural selection on which to work; whilst all variation from the mixture of species are barred by the inexorable law of hybrid sterility. Further, the embalmed records of 3000 years show that there has been no beginning of transmutation in the species of our most familiar domesticated animals; and beyond this, that in the countless tribes of animal life around us, down to its lowest and most variable species, no one has ever discovered a single instance of such transmutation being now in prospect; no new organ has ever been known to be developed—no new natural instance to be formed—whilst, finally in the fast museum of departed animal life which the strata of the earth imbed for our examination, whilst they contain far too complete a representation of the past to be set aside as a mere imperfect record, yet afford no one instance of any such change as having ever been in progress, or give us anywhere the missing links of the assumed chain, or the remains which would enable now existing variations, by gradual approximations, to shade off into unity. . . .

There are no parts of Mr. Darwin's ingenious book in which he gives the reins more completely to his fancy than where he deals with the improvement of instinct by his principle of natural selection. We need but instance his assumption, without a fact on which to build it, that the marvelous skill of the honey-bee in constructing its cells is thus obtained, and the slave-making habits of the Formica Polyerges thus formed. There seems to be no limit here to the exuberance of his fancy, and we cannot but think that we detect one those hints by which Mr. Darwin indicates the application of his system from the lower animals to man himself, when he dwells so pointedly upon the fact that it is always the *black* ant which is enslaved by his other coloured and more fortunate brethren. "The slaves are black!' We believe that, if we had Mr. Darwin in the witness-box, and could subject him to a moderate cross-examination, we should find that he believed that the tendency of the lighter-coloured races of mankind to prosecute the Negro slave-trade was really a remains, in their more favoured condition, of the 'extraordinary and odious instinct' which had possessed them before they had been 'improved by natural selection' from Formica Polyerges into Homo. This at least is very much the way in which slips in quite incidentally the true identity of man with the horse, the bat, and the porpoise:

> The framework of bones being the same in the hand of a man, wing of a bat, fin of a porpoise, and leg of the horse, the same number of vertebrae forming the neck of the giraffe and of the elephant, and innumerable other such facts, at once explain themselves on the theory of descent with slow and slight successive modifications.

Such assumptions as these, we once more repeat, are most dishonourable and injurious to science; and though, out of respect to Mr. Darwin's high character and to the tone of his work, we have felt it right to weigh the 'argument' again set by him before us in the simple scales of logical examination, yet we must remind him that the view is not a new one, and that it has already been treated with admirable humour when propounded by another of his name and of his lineage. We do not think that, with all his matchless ingenuity, Mr. Darwin has found any instance which so well illustrates his own theory of the improved descendant under the elevating influences of natural selection exterminating the progenitor whose specialties he has exaggerated as he himself affords us in this work. For if we go back two generations we find the ingenious grandsire of the author of the "Origin of Species' speculating on the same subject, and almost in the same manner with his more daring descendant. . . .

Our readers will not have failed to notice that we have objected to the views with which we have been dealing solely on scientific grounds. We have done so from our fixed conviction that it is thus that the truth or falsehood of such arguments should be tried. We have no sympathy with those who object to any facts or alleged facts in nature, or to any inference logically deduced from them, because they believe them to contradict what it appears to them is taught be Revelation. We think that all such objections savour of a timidity which is really inconsistent with a firm and well-instructed faith:

'Let us for a moment,' profoundly remarks Professor Sedwick, 'suppose that there are some religious difficulties in the conclusions of geology. How, then, are we to solve them? Not by making a world after a pattern of our own—not by shifting a shuffling the solid strata of the earth, and then dealing them out in such a way as to play the game of an ignorant or dishonest hypothesis—not by shutting our eyes to facts, or denying the evidence of our senses—but by patient investigation, carried on in the sincere love of truth, and by learning to reject every consequence not warranted by physical evidence.'

He who is as sure as he is of his own existence that the God of Truth is at once the God of Nature and the God of Revelation, cannot believe it to be possible that His voice in either, rightly understood, can differ, or deceive His creatures. To oppose facts in the natural world because they seem to oppose Revelation, or to humour them so as to compel them to speak its voice, is, he knows, but another form of the ever-ready feebleminded dishonesty of lying for God, and trying by fraud or falsehood to do the work of the God of truth. It is with another and a nobler spirit that the true believer walks amongst the words of nature. The words graven on the everlasting rocks are the words of God, and they are graven by His hand. No more can they contradict His Word written in His hand on the stony tables contradict the writing of His hand in the volume of the new dispensation. There may be to man difficulty in reconciling all the utterances of the two voices. But what of that? He has learned already that here he knows only in part, and that the day of reconciling all apparent contradictions between what must agree is nigh at hand. He rests his mind in perfect quietness on this assurance, and rejoices in the gift of light without a misgiving as to what it may discover:

'A man of deep thought and great practical wisdom,' says Sedwick, 'one whose piety and benevolence have for many years been shining before the world, and of whose sincerity no scoffer (of whatever school) will dare to start a doubt, recorded his opinion in the great assembly of the men of science who during the past year were gathered from every corner of the Empire within the walls of this University, "that Christianity had everything to hope and nothing to fear from the advancement of philosophy."'

This is as truly the spirit of Christianity as it is that of philosophy. Few things have more deeply injured the cause of religion than the busy fussy energy with which men, narrow and feeble alike in faith and in science, have bustled forth to reconcile all new discoveries in physics with the word of inspiration. For it continually happens that some larger collection of facts, or some wider view of the phenomena of nature, alter the whole philosophic scheme; whilst Revelation has been committed to declare an absolute agreement with what turns out after all to have been a misconception or an error. We cannot, therefore, consent to test the truth of natural science by the Word of Revelation. But this does not make it the less important to point out on scientific grounds scientific errors, when those errors tend to limit God's glory in creation, or to gainsay the revealed relations of that creation to Himself. To both these classes of error, though, we doubt not, quite unintentionally on his part, we think that Mr. Darwin's speculations directly tend.

Mr. Darwin writes as a Christian, and we doubt not that he is one. We do not for a moment believe him to be one of those who retain in some corner of their hearts a secret unbelief which they dare not vent; and we therefore pray him to consider well the grounds on which we brand his speculations with the charge of such a tendency. First, then, he not obscurely declares that he applies his scheme of the action of the principle of natural selection to MAN himself, as well as to the animals around him. Now, we must say at once, and openly, that such a notion is absolutely incompatible not only with single expressions in the word of God on that subject of natural science with which it is not immediately concerned, but, which in our judgment is of far more importance, with the whole representation of that moral and spiritual condition of man which is its proper subject-matter. Man's derived supremacy over the earth; man's power of articulate speech; man's gift of reason; man's free-will and responsibility; man's fall and man's redemption; the incarnation of the Eternal Son; the indwelling of the Eternal Spirit,— all are equally and utterly irreconcilable with the degrading notion of the brute origin of him who was created in the image of God, and redeemed by the Eternal Son assuming to himself his nature. Equally inconsistent, too, not with any passing expressions, but with the whole scheme of God's dealings with man as recorded in His word, is Mr. Darwin's daring notion of man's further development into some unknown extent of powers, and shape, and size, through natural selection acting through that long vista of ages which he casts mistily over the earth upon the most favoured individuals of his species. We care not in these pages to push the argument further. We have done enough for our purpose in thus succinctly intimating its course. If any of our readers doubt what must be the result of such speculations carried to their logical and legitimate conclusion, let them turn to the pages of Oken, and see for themselves the end of that path the opening of which is decked out in these pages with the bright hues and seemingly innocent deductions of the transmutation-theory.

Nor can we doubt, secondly, that this view, which thus contradicts the revealed relation of creation to its Creator, is equally inconsistent with the fullness of His glory. It is, in truth, an ingenious theory for diffusing throughout creation the working and so the personality of the Creator. And thus, however unconsciously to him who holds them, such views really tend inevitably to banish from the mind most of the peculiar attributes of the Almighty.

How, asks Mr. Darwin, can we possibly account for the manifest plan, order, and arrangement which pervade creation, except we allow to it this self-developing power through modified descent? . . .

How can we account for all this? By the simplest and yet the most comprehensive answer. By declaring the stupendous fact that all creation is the transcript in matter of ideas eternally existing in the mid of the Most High—that order in the utmost perfectness of its relation pervades His works, because it exists as in its centre and highest fountainhead in Him the Lord of all. Here is the true account of the fact which has so utterly misled shallow observers, that Man himself, the Prince and Head of this creation, passes in the earlier stages of his being through phases of existence closely analogous, so far as his earthly tabernacle is concerned, to those in which the lower animals ever remain. At that point of being the development of the protozoa is arrested. Through it the embryo of their chief passes to the perfections of his earthly frame. But the types of those lower forms of being must be found in the animals which never advance beyond them—not in man for whom they are but the foundation for an after-development; whilst he too, Creation's crown and perfection, thus bears witness in his own frame to the law of order which pervades the universe.

In like manner, could we answer every other question as to which Mr. Darwin thinks all oracles are dumb unless they speak his speculation. He is, for instance, more than once troubled by what he considers imperfections in Nature's work. 'If,' he says, 'our reason leads us to admire with enthusiasm a multitude of inimitable contrivances in Nature, this same reason tells us that some other contrivances are less perfect.'

'Nor ought we to marvel if all the contrivances in nature be not, as far as we can judge, absolutely perfect; and if some of them be abhorrent to our idea fitness. We need not marvel at the sting of the bee causing the bee's own death; at drones being produced in such vast numbers for one single act, with the great majority slaughtered by their sterile sisters; at the astonishing waste of pollen by our fir-trees; at the instinctive hatred of the queen-bee for her own fertile daughters; at ichneumonidae feeding within the live bodies of caterpillars; and at other such cases. The wonder indeed is, on the theory of natural selection, that more cases of the want of absolute perfection have not been observed.'

We think that the real temper of this whole speculation as to nature itself may be read in these few lines. It is a dishonouring view of nature.

That reverence for the work of God's hands with which a true belief in the All-wise Worker fills the believer's heart is at the root of all great physical discovery; it is the basis of philosophy. He who would see the venerable features of Nature must not seek with the rudeness of a licensed roisterer violently to unmask her countenance; but must wait as a learner for her willing unveiling. There was more of the true temper of philosophy in the poetic fiction of the Pan-ic shriek, than in the atheistic speculations of Lucretius. But this temper must beset those who do in effect banish God from nature. And so Mr. Darwin not only finds in it these bungling contrivances which his own greater skill could amend, but he stands aghast at them until reconciled to their presence by his own theory that 'a ratio of increase so high as to lead to a struggle for life, and as a consequence to natural selection entailing divergence of character and the extinction of less improved forms, as decidedly followed by the most exalted object which we are capable of conceiving, namely, the production of the higher animals.' But we can give him a simpler solution still for the presence of these strange forms of imperfection and suffering amongst the works of God. . . .

It is by our deep conviction of the truth and importance of this view for the scientific mind of England that we have been led to treat at so much length Mr. Darwin's speculation. The contrast between the sober, patient, philosophical courage of our home philosophy, and the writings of Lamarack and his followers and predecessors, of M. M. Demailet, Bory de Saint Vincent, Virey, and Oken, is indeed most wonderful; and it is greatly owing to the noble tone which has been given by those great men whose words we have quoted to the school of British science. That Mr. Darwin should have wandered from this broad highway of nature's works into the jungle of fanciful assumption is no small evil. We trust that he is mistaken in believing that he may count Sir C. Lyell as one of his converts. We know indeed the strength of the temptations which he can bring to bear upon his geological brother. The Lyellian hypothesis, itself not free from some of Mr. Darwin's faults, stands eminently in need for its own support of some such new scheme of physical life as that propounded here. Yet no man has been more distinct and more logical in the denial of the transmutation of species than Sir C. Lyell, and that not in the infancy of his scientific life, but in its full vigour and maturity.

Sir C. Lyell devotes the 33rd to the 36th chapter of his 'Principles of Geology' to an examination of this question. He gives a clear account of the mode in which Lamarack supported his belief of the transmutation of species; he 'interrupts the author's argument to observe that no positive fact is cited to exemplify the substitution of some *entirely new* sense, faculty, or organ—because no examples were to be found; and remarks that when Lamarack talks' of 'the effects of internal sentiment.' &c., as causes whereby animals and plants may acquire *new organs*, he substitutes names for things, and with a disregard to the strict rules of induction resorts to fictions.

He shows the fallacy of Lamarck's reasoning, and by anticipation confutes the whole theory of Mr. Darwin, when gathering clearly up into a heads the recapitulation of the whole argument in favour of the reality of species in nature. He urges:

1. That there is capacity in all species to accommodate themselves to a certain extent to a change of external circumstances.
2. The entire variation from the original type. . .may usually be effected in a brief period of time, after which no further deviation can be obtained.
3. The intermixing distinct species is guarded against by the sterility of the mule offspring.

4. It appears that species have a real existence in nature, and that each was endowed at the time of its creation with the attributes and organization by which it is now distinguished.

We trust that Sir. C. Lyell abides still by these truly philosophical principles; and that with his help and with that of his brethren this flimsy speculation may be as completely put down as was what twin though less-instructed brother, the 'Vestiges of Creation.' In so doing they will assuredly provide for the strength and continually growing progress of British science.

Indeed, not only do all laws for the study of nature vanish when the great principle of order pervading and regulating all her processes is give up, but all that imparts the deepest interest in the investigation of her wonders will have departed too. Under such influences man soon goes back to the marveling stare of childhood at the centaurs and turn, he comes like Oken to write a scheme of creation under 'a sort inspiration;' but it is the frenzied inspiration of the inhaler of mephitic gas. The whole world of nature is laid for such a man under a fantastic law of glamour, and he becomes capable of believing anything: to him it is just as probably that Dr. Livingstone will find the next tribe of Negroes with their heads growing under their arms as fixed on the summit of the cervical vertebrae; and he is able, with a continually growing neglect of all the facts around him, with equal confidence and equal delusion, to look back to any past and to look on to any future.

SOURCE: Wilberforce, Samuel. "Review of Darwin's *Origin of Species.*" *Quarterly Review* 108 (1860).

TERMS

Agnosticism—A system of understanding that uses reason to preclude a belief in a God or gods. A person who sees belief in the supernatural as beyond the scope of reason, and who relies solely on reason to determine the scope of belief, would be agnostic. An atheist does not believe in the supernatural, while an agnostic might be said to reserve judgment on this issue. Thomas Huxley coined the term, and Charles Darwin eventually came to think of himself as an agnostic since he could not decide how one might fit the existence of God into an explanation of nature.

Blending Inheritance—Used by naturalists in the nineteenth century and before to understand the appearance of intermediate characters when two individuals with more extreme traits mated. A tall parent and a short parent, for example, often had offspring of intermediate height. Just as black paint mixed with white paint produced grey, intermediate tones could be observed in matings between differently colored parents. By 1900, when Gregor Mendel's conclusions about garden peas became widely known, the conclusions from blending could be dismissed and traits followed more discretely from generation to generation.

Teleology—An argument for the existence of something based on its usefulness or purpose. It is a philosophical stance that assumes the evident purpose of an object provides an adequate explanation for the object's having come into being. Such a philosophy is generally consistent with belief in a Creator or Creators who designed parts of the universe to serve coherent purposes. In evolutionary philosophy, however, teleology often leads to circular arguments. For example, an eye is for seeing, and organisms that need to see must have eyes; therefore, eyes exist so that organisms can see. Natural selection generally avoids circular reasoning by identifying the survival advantages of certain variations and assuming that organisms with such variations will survive; not that they *need* those variations, only that they will not survive without them.

Unity of Plan—Also known as homology refers to the close correspondence of structures in otherwise different organisms. Early anatomists comparing skeletal structures, for example, observed the similarity of arrangement between the wing of a bird and the limb of a horse. Such similarities suggested a relationship among species, but not necessarily in the sense of common ancestry as implied by evolutionary theory. Many naturalists would point to that correspondence as a unity of plan, evidence of a designer's efficiency in using consistently workable forms. As scientists embraced evolution, the plan of nature was supplanted by explanations arising from common ancestry, where descendents would enjoy modifications in response to different environments.

4

EVOLUTION AND ANTI-EVOLUTION IN THE INTERPHASE

From the late nineteenth century through the 1930s, there was considerable disagreement among evolutionary biologists over the proper explanation for evolutionary change. There was not, however, any debate among American biologists over whether or not evolutionary change occurred; there was universal agreement that evolution was a natural phenomenon. Exactly how it happened, though, was not agreed upon. Earlier authors referred to the decades from about 1880 to 1940 as the era of the "eclipse of Darwin," suggesting that the shining light of Darwinism had been obscured, but would surely reappear after the "dark ages" ended. More recently that history has been revised to demonstrate how questions that emerged in Darwin's time, continued throughout the decades that followed, and were slowly resolved throughout the first half of the twentieth century. This more nuanced and continuous historical study has renamed these decades the "interphase." The term, an analogy to a concept in cellular reproduction, refers to a period of great interest. In cell biology as in history, researchers once believed interphase involved little activity, but now realize the vital importance of the phase. The interphase of the history of evolutionary biology emphasizes how the first decades of the twentieth century were an exciting and productive time as evolutionists critically evaluated the state of evolutionary biology, identified shortcomings in evolutionary theory, and sometimes argued about how to best resolve them. The interphase was a productive time in evolutionary theory, but scientists' disagreements had some unintended consequences for the emerging debate over evolution and creation.

There were two principle camps among American biologists, each with its own mechanism to explain precisely how evolution occurred. The first, the Darwinists, included those who accepted Darwin's interpretation that evolution occurred slowly and consistently over long periods of time though the influence of natural selection. In their view, small differences among individuals allowed for a greater or lesser chance of survival to adulthood and reproduction, and those individuals with favorable variations produced more offspring than did those individuals who had variations that made it more difficult for them to survive and reproduce. The greater reproductive success of some meant that succeeding generations would be similar to those individuals who more often survived to adulthood. But the Darwinists had a problem: where did these new variations come from? Experiments in plants suggested that selection did not produce new variations; instead, selection only limited the range of variation. The Darwinists needed an explanation of the source of new variations.

Darwinists found themselves in opposition to a group generally referred to as Mendelian/mutationists. Based on the rediscovery of Mendel's work in 1900 and experiments performed by Hugo DeVries, advocates of Mendelian/mutationism believed that evolution occurred in fits and starts. New variations, they argued, arose through occasional mutations that were passed from one generation to another intact through Mendelian inheritance. Unlike the Darwinists, who posited a slow, continuous line of evolutionary change, Mendelian/mutationists argued that evolutionary change occurred in leaps, as new mutations occurred, proved themselves more successful than earlier organisms, and quickly became the norm via natural selection.

Contention among biologists about the mechanism of evolution sometimes found its way into popular science writing, and there were some who authoritatively told the American public that biologists had lost faith in evolution. This was not at all true; in fact, there was near universal agreement among professional biologists that evolution was a phenomenon of nature. Nonetheless, arguments among working biologists bolstered some antievolutionists' claims. It would take a new generation of biologists, which emerged in the 1930s, to overcome the differences between the Darwinists and the Mendelian/mutationists.

VERNON KELLOGG, *DARWINISM TO-DAY* (1907)

Vernon Lyman Kellogg (1867–1937) was an entomologist, a Darwinist, and a Stanford University professor who wrote and researched extensively on the subject of evolution. Kellogg's work focused on the role of natural selection in the evolution of insects, and he was the author of the single most authoritative book on evolution in the early twentieth century, *Darwinism To-Day*. Kellogg was concerned about the effects of debates among biologists on public perceptions of evolution. After having spent time in Europe, where he witnessed highly contentious debates about the mechanisms of evolution, Kellogg returned to the United States and wrote the book to prevent Americans from thinking that evolution was a debated subject. In *Darwinism To-Day* he explained the nature of the debates while explicitly stating that biologists agreed that evolution took place, but debated precisely how it happened.

Introductory: The Deathbed of Darwinism

"VOM STERBELAGER DES DARWINISMUS!" This is the title of a recent pamphlet lying before me. But ever since there has been Darwinism there have been occasional deathbeds of Darwinism on title pages of pamphlets, addresses, and sermons. Much more worth consideration than any clerical pamphlets or dissertations, under this title, by *frisch-gebackenen* German doctors of philosophy—the title alone proving prejudice or lack of judgment or of knowledge—are the numerous books and papers which, with less sensational headlines but infinitely more important contents, are appearing now in such numbers and from such a variety of reputable sources as to reveal the existence among biologists and philosophers of a widespread belief in the marked weakening, at least, if not serious indisposition, of Darwinism.

A few of these books and papers from scientific sources even suggest that their writers see shadows of a death-bed.

The present extraordinary activity in biology is two-phased; there is going on a most careful re-examination or scrutiny of the theories connected with organic evolution, resulting in much destructive criticism of certain long-cherished and widely held beliefs, and at the same time there are being developed and almost feverishly driven forward certain fascinating and fundamentally important new lines, employing new methods, of biological investigation. Conspicuous among these new kinds of work are the statistical or quantitative study of variations and that most alluring work variously called developmental mechanics, experimental morphology, experimental physiology of development, or, most suitably of all because most comprehensively, experimental biology. This work includes the controlled modification of conditions attending development and behavior, and the pedigreed breeding of pure and hybrid generations. Now this combination of destructive critical activity and active constructive experimental investigation has plainly resulted, or is resulting, in the distinct weakening or modifying of certain familiar and long, entrenched theories concerning the causative factors and the mechanism of organic evolution. Most conspicuous among these theories now in the white light of scientific scrutiny are those established by Darwin, and known, collectively, to biologists, as Darwinism.

> To too many general readers Darwinism is synonymous with organic evolution or the theory of descent. The word is not to be so used or considered. Darwinism, primarily, is a most ingenious, most plausible, and, according to one's belief, most effective or most inadequate, causo-mechanical explanation of adaptation and species-transforming. It is that factor which, ever since its proposal by Darwin in 1859, has been held by a majority of biologists to be the chief working agent in the descent, that is, the origin, of species. However worthy Darwin is of having his name applied directly to the great theory of descent—for it was only by Darwin's aid that this theory, conceived and more or less clearly announced by numerous pre-Darwinian naturalists and philosophers, came to general and nearly immediate acceptance—the fact is that the name Darwinism has been pretty consistently applied by biologists only to those theories practically original with Darwin which offer a mechanical explanation of the accepted fact of descent. Of these Darwinian theories the primary and all-important one is that of natural selection. Included with this in Darwinism are the now nearly wholly discredited theories of sexual selection and of the pangenesis of gemmules. It may also be fairly said that the theory of the descent of man from the lower animals should be included in Darwinism. For Darwin was practically the first naturalist bold enough to admit the logical and obvious consequences of the general acceptance of the theory of descent, and to include man in the general chain of descending, or ascending, organisms. So that the popular notion that Darwinism is in some way the right word to apply to the doctrine that man has come from the monkeys is rather nearer right than wrong. But biologists do not recognize the descent of man as a special phase of Darwinism, but rather of the whole theory of descent, or organic evolution.

Darwinism, then, is not synonymous with organic evolution, nor with the theory of descent (which two phases are used by the biologist practically synonymously). Therefore when one reads of the "death-bed of Darwinism," it is not of the death-bed of organic evolution or of the theory of descent that one is reading. While many reputable biologists today strongly doubt the commonly reputed effectiveness of the Darwinian selection factors to explain descent—some, indeed, holding them to be of absolutely no species-forming value—practically no naturalists of position and recognized attainment doubt the theory of descent. Organic evolution, that is, the descent of species, is looked on by biologists to be as proved a part of their science as gravitation is in the science of physics or chemical affinity in that of chemistry. Doubts of Darwinism are not, then, doubts, of organic evolution. Darwinism might indeed be on its death-bed without shaking in any considerable degree the confidence of biologists and natural philosophers in the theory of descent.

But the educated reader, the scientific layman, the thinker and worker in any line of sociologic, philosophic, or even theologic activity is bound to be disturbed and unsettled by rumors from the camp of professional biologists of any weakness or mortal illness of Darwinism. We have only just got ourselves and our conceptions of nature, of sociology and philosophy, well-oriented and adjusted with regard to Darwinism. And for relentless hands now to come and clutch away our foundations is simply intolerable. *Zum Teufel* with these German professors! For it is precisely the German biologists who are most active in this undermining of the Darwinian theories. But there are others with them; Holland, Russia, Italy, France, and our own country all contribute their quota of disturbing questions and declarations of protest and revolt. The English seem mostly inclined to uphold the glory of their illustrious countryman. But there are rebels even there. Altogether it may be stated with full regard to facts that a major part of the current published output of general biological discussions, theoretical treatises, addresses, and brochures dealing with the great evolutionary problems, is distinctly anti-Darwinian in character. This major part of the public discussion of the status of evolution and its causes, its factors and mechanism, by working biologists and thinking natural philosophers, reveals a lack of belief in the effectiveness or capacity of the natural selection theory to serve as a sufficient causo-mechanical explanation of species-forming and evolution. Nor is this preponderance of anti-Darwinian expression in current biological literature to be wholly or even chiefly attributed to a dignified silence on the part of the believers in selection. Answers and defenses have appeared and are appearing.

But in practically all these defenses two characteristics are to be noted, namely, a tendency to propose supporting hypotheses or theories, and a tendency to make certain distinct concessions to the beleaguering party. The fair truth is that the Darwinian selection theories, considered with regard to their claimed capacity to be an independently sufficient mechanical explanation of descent, stand to-day seriously discredited in the biological world. On the other hand, it is also fair truth to say that no replacing hypothesis or theory of species-forming has been offered by the opponents of selection which has met with any general or even considerable acceptance by naturalists. Mutations seem to be too few and far between; for orthogenesis we can discover no satisfactory mechanism; and the same is true for the Lamarckian theories of modification by the cumulation, through inheritance, of acquired or ontogenic characters. *Kurz* and *gut*, we are immensely unsettled.

Now but little of this philosophic turmoil and wordy strife has found its way as yet into current American literature. Our bookshop windows offer no display, as in Germany, of volumes and pamphlets on the newer evolutionary study; our serious-minded quarterlies, if we have any, and our critical monthlies and weeklies contain no debates or discussions over *"das Sterbelager des Darwinismus."* Our popular magazines keep to the safe and pleasant task of telling sweetly of the joys of making Nature's acquaintance through field-glasses and the attuned ear. But just as certainly as the many material things "made in Germany" have found their way to us so will come soon the echoes and phrases of the present intellectual activity in evolutionary affairs, an activity bound to continue as long as the new lines of biological investigation continue their amazing output of new facts to serve as the bases for new critical attacks on the old notions and for the upbuilding of new hypotheses. If now the first of these echoes to come across the water to us prove to be, as wholly likely, those from the more violent and louder debaters, they may lead to an undue dismay and panic on our part. Things are really in no such desperate way with Darwinism as the polemic vigor of the German and French anti-Darwinians lead them to suggest Says one of them: "Darwinism now belongs to history, like that other curiosity of our century, the Hegelian philosophy; both are variations on the theme: how one manages to lead a whole generation by the nose." The same writer also speaks of the "softening of the brain of the Darwinians." Another one, in similarly relegating Darwinism to the past, takes much pleasure in explaining that "we [anti-Darwinians] are now standing by the death-bed of Darwinism, and making ready to send the friends of the patient a little money to insure a decent burial of the remains." No less intemperate and indecent is Wolff's reference to the "episode of Darwinism" and his suggestion that our attitude toward Darwin should be "as if he never existed." Such absurdity of expression might pass unnoticed in the mouth of a violent non-scientific debater—let us say an indignant theologian of Darwin's own days—but in the mouth of a biologist of recognized achievement, of thorough scientific training and unusually keen mind—for this expression came from just such a man—it can only be referred to as a deplorable example of those things that make the judicious to grieve. Such violence blunts or breaks one's own weapons.

While I have said that the coming across the water of the more vigorous anti-Darwinian utterances might cause some dismay and panic in the ranks of the educated reader—really unnecessary panic, as I hope to point out—it will doubtless occur to some of my readers to say that this fear of panic is unwarranted. If the first phrases to come are as injudicious and intemperate, hence as unconvincing, as those just cited, the whole anti-Darwinian movement will be discredited and get no attention. Which, I hasten to reply, will be as much of a mistake as panic would be. There is something very seriously to be heeded in the chorus of criticism and protest, and wholly to stop one's ears to these criticisms is to refuse enlightenment and to show prejudice. I have thought it, therefore, worth while to try to anticipate the coming of fragmentary and disturbing extracts from the rapidly increasing mass of recent anti-Darwinian literature by presenting in this book a summary account not alone of these modern criticisms, but of the answers to them by the steadfast Darwinians, and of the concessions and supporting hypotheses which the supporters of both sides have been led to offer during the debates. I shall try to give a fair statement of the recent attacks on, and the defense and present scientific standing of, the familiar Darwinian theories, and to give also concise expositions, with some critical comment, of the more important new, or newly remodeled alternative and auxiliary theories of species-forming and descent, such as heterogensis, orthogenesis, isolation, etc., and an estimate of their degree of acceptance by naturalists.

Chapter XII: Darwinism's Present Standing

A river rises from a perennial spring on the mountain side; gravitation compels the water to keep moving, and rock walls, intervening hills, and soft loam banks determine the course of the stream. The living stream of descent finds its never-failing primal source in ever-appearing variations; the eternal flux of Nature, coupled with this inevitable primal variation, compels the stream to keep always in motion, and selection guides it along the ways of least resistance. Although there can be no modification, no evolution, without variation, yet neither can this variation, whatever its character and extent, whether slight and fluctuating, large and mutational, determinate or fortuitous, long compel descent to go contrary to adaptation. And the guardian of the course is natural selection. Selection will inexorably bar the forward movement, will certainly extinguish the direction of any orthogenetic process, Nagelian, Eimerian, or de Vriesian, which is not fit, that is, not adaptive. Darwinism, then, as the natural selection of the fit, the final arbiter in descent control, stands unscathed, clear and high above the obscuring cloud of battle. At least, so it seems to me. But Darwinism, as the all-sufficient or even most important causo-mechanical factor in species-forming and hence as the sufficient explanation of descent, is discredited and cast down. At least, again, so it seems to me. But Darwin himself claimed no *Allmacht* for selection. Darwin may well cry to be saved from his friends!

The selection theories do not satisfy present-day biologists as efficient causal explanations of species-transformation. The fluctuating variations are not sufficient handles for natural selection; the hosts of trivial, indifferent species differences are not the result of an adaptively selecting agent. On the other hand the declarations of Korschinsky, Wolff, Driesch, and others that natural selection is nonexistent, is a vagary, a form of speech, or a negligible influence in descent, are unconvincing; they are unproved.

And these bitter antagonists of selection are especially unconvincing when they come to offer a replacing theory, an alternative explanation of transformation and descent. To my mind every theory of heterogenesis, of orthogenesis, or of modification by the transmission of acquired characters, confesses itself ultimately subordinate to the natural selection theory. However independent of selection and Darwinism may be the beginnings of modification, the incipiency of new species and of new lines of descent; even, indeed, however necessary to natural selection some auxiliary or supporting theory to account for the beginnings of change confessedly is, the working factor or influence postulated by any such auxiliary theory soon finds its independence lost, its influence in evolution dominated and controlled by natural selection. As soon as the new modifications, the new species characters, the new lines of descent, if they may come so far, attain that degree of development where they have to submit to the test of utility, of fitness, just there they are practically delivered over to the tender mercies of selection. No orthogenetic line of descent can persist in a direction not adaptive, that is, not fit, and certainly no present-day biologist is ready to fall back on the long deserted standpoint of teleology and ascribe to heterogenesis or orthogenesis an auto-determination toward adaptiveness and fitness. Modification and development may have been proved to occur along determinate lines without the aid of natural selection. I believe they have. But such development cannot have an aim; it cannot be assumed to be directed toward advance; there is no independent progress upward, i.e., toward higher specialisation. At least, there is no scientific proof of any such capacity in organisms. Natural selection remains the one causo-mechanical explanation of the large and general progress toward fitness; the movement toward specialisation; that is, descent as we know it.

But what Darwinism does not do is to explain the beginnings of change, the modifications in indifferent characters and in indifferent directions. And all this is tremendously important, for there are among animals and plants hosts of existent indifferent characters, and many apparently indifferent directions of specialisation. As to the obvious necessity of beginnings nothing need be said. What is needed, then, is a satisfactory explanation of the pre-useful and pre-hurtful stages in the modifications of organisms: an explanation to relieve Darwinism of its necessity of asking natural selection to find in the fluctuating individual variations a handle for its action; an explanation of how there ever comes to be a handle of advantage or disadvantage of life-and-death determining degree. With such an explanation in our possession—and whether any one or more of the various theories proposed to fill this need, such as Eimerian orthogenesis, de Vriesian heterogenesis, Rouxian battle of the parts, or Weismannian germinal selection, etc., give us this explanation, may be left for the moment undebated—with such a satisfactory explanation, I say, once in our hands, we may depend with confidence on natural selection to do the rest of the work called for by the great theory of descent. Among all the divergent lines of development and change, instituted by this agent of beginnings, natural selection will choose those to persist by saying No to those that may not. And the result is organic evolution. . . .

Then, after the explanation of the why and how of variability, comes the necessity of explaining the cumulation of this variability along certain lines, the first visible issuance of these lines being as species, and later becoming more and more pronounced as courses of descent. This explanation has got to begin lower down in phyletic history than natural selection can begin. Before ever there can be utility and advantage there must have come about a certain degree of heaping up, of cumulating, of intensifying variations. What are these factors? They are possibly only two: (1) orthogenetic or determinate variation as the outcome of plasm preformation or of epigenetic influences, and (2) the segregation of similar variations by physiologic or topographic conditions. Hence, next to the cause or origin of variability the great desideratum is a knowledge of the means of cumulating and directing variability. And both these great fundamental needs of a satisfactory understanding of organic evolution seem to me to be wholly unreferred to in the theory of natural selection. To be sure the control and cumulation of such large differences among organisms and species as are positively sufficient to determine the saving or the loss of life are explicable by selection. And this factor is sooner or later in any phyletic history bound to step in and probably be the dominant one. But a species, or a character, will always have a longer or shorter preselective existence and history, and it is precisely these days before the Inquisition of which we demand information. For of one thing we are now certain, and that is, that evolution and the origin of species have both their beginnings and a certain period of history before the day of the coming of the Grand Inquisitor, selection.

Finally there is still another desideratum and one whose seeking will carry us into dangerous country. For while there may be and are selectionists who might allow us to fumble about in the darkness of preselective time for first causes, there is probably none who will allow us to question his right to explain that other element in evolution besides species transformation, namely, adaptation, or, as the Germans untranslatably put it, *Zweckmässigkeit*. But by no means all biologists find in natural selection a sufficient explanation of adaptation. . . .

If variation is thus simply the wholly natural and unavoidable effect of this inevitable non-identity of vital process and environmental condition, why does not evolution possess in this state of affairs the much sought for, often postulated, all-necessary, automatic modifying principle antedating and preceding selection which must effect change, determinate though not purposeful? Nageli's automatic perfecting principle is an impossibility to the thorough-going evolutionist seeking for a causo-mechanical explanation of change. But an automatic modifying principle which results in determinate or purposive change, that is, in the change needed as the indispensable basis for the upbuilding of the great fabric of species diversity and descent; is not that the very thing provided by the simple physical or mechanical impossibility of perfect identity between process and environment in the case of one individual and process and environment in the case of any other? It seems so to me.

But I do not know. Nor in the present state of our knowledge does any one know, nor will any one know until, as Brooks says of another problem, we find out. We are ignorant; terribly, immensely ignorant. And our work is, to learn. To observe, to experiment, to tabulate, to induce, to deduce. Biology was never a clearer or more inviting field for fascinating, joyful, hopeful work. To question life by new methods, from new angles, on closer terms, under more precise conditions of control; this is the requirement and the opportunity of the biologist of to-day. May his generation hear some whisper from the Sphinx!

SOURCE: Kellogg, Vernon. *Darwinism To-Day*. New York: Henry Holt and Company, 1907.

WILLIAM BATESON, "EVOLUTIONARY FAITH AND MODERN DOUBTS" (1921)

William Bateson (1861–1926) was an English biologist and one of the earliest advocates of mutationism. He popularized Mendelism in Britain and considerably advanced it through his research. He also coined the word "genetics." Bateson delivered the below speech at the 1921 meeting of the American Association for the Advancement of Science and it met with considerable discussion by both the scientists present and by Americans generally. His candid discussion of the shortcomings of modern evolutionary theory empowered some anti-evolutionists to incorrectly claim that biologists were abandoning the concept of evolution.

I visit Canada for the first time in delightful circumstances. After a period of dangerous isolation, intercourse between the centres of scientific development is once more beginning, and I am grateful to the American Association for this splendid opportunity of renewing friendship with my Western colleagues in genetics, and of coming into even a temporary partnership in the great enterprise which they have carried through with such extraordinary success.

In all that relates to the theme which I am about to consider we have been passing through a period of amazing activity and fruitful research. Coming here after a week in close communion with the wonders of Columbia University, I may seem behind the times in asking you to devote an hour to the old topic of evolution. But though that subject is no longer in the forefront of debate, I believe it is never very far from the threshold of our minds, and it was with pleasure that I found it appearing in conspicuous places in several parts of the program of this meeting.

Standing before the American Association, it is not unfit that I should begin with a personal reminiscence. In 1883 I first came to the United States to study the development of Balanoglossus at the Johns Hopkins summer laboratory, then at Hampton, VA. This creature had lately been found there in an easily accessible place. With a magnanimity, that on looking back I realize was superb, Professor W. K. Brooks had given me permission to investigate it, thereby handing over to a young stranger one of the prizes which in this age of more highly developed patriotism, most teachers would keep for themselves and their own students. At that time one morphological laboratory was in purpose and aim very much like another. Morphology was studied because it was the material believed to be most favorable for the elucidation of the problems of evolution, and we all thought that in embryology the quintessence of morphological truth was most palpably presented. Therefore every aspiring zoologist was an embryologist, and the one topic of professional conversation was evolution. It had been so in our Cambridge school, and it was so at Hampton.

I wonder if there is now a single place where the academic problems of morphology which we discussed with such avidity can now arouse a moment's concern. There were of course men who saw a little further, notably Brooks himself. He was at that time writing a book on heredity, and, to me at least, the notion on which he used to expatiate, that there was a special physiology of heredity capable of independent study, came as a new idea. But no organized attack on that problem was begun, nor had any one an inkling of how to set about it. So we went on talking about evolution. That is barely 40 years ago; to-day we feel silence to be the safer course.

Systematists still discuss the limits of specific distinction in a spirit, which I fear is often rather scholastic than progressive, but in the other centers of biological research a score of concrete and immediate problems have replaced evolution.

Discussions of evolution came to an end primarily because it was obvious that no progress was being made. Morphology having been explored in its minutest corners, we turned elsewhere. Variation and heredity, the two components of the evolutionary path, were next tried. The geneticist is the successor of the morphologist. We became geneticists in the conviction that there at least must evolutionary wisdom be found. We got on fast. So soon as a critical study of variation was undertaken, evidence came in as to the way in which varieties do actually arise in descent. The unacceptable doctrine of the secular transformation of masses by the accumulation of impalpable changes became not only unlikely but gratuitous. An examination in the field of the interrelations of pairs of well characterized but closely allied "species" next proved, almost wherever such an inquiry could be instituted, that neither could both have been gradually evolved by natural selection from a common intermediate progenitor, nor either from the other by such a process. Scarcely ever where such pairs co-exist in nature, or occupy conterminous areas do we find an intermediate normal population as the theory demands. The ignorance of common facts bearing on this part of the inquiry which prevailed among evolutionists, was, as one looks back, astonishing and inexplicable. It had been decreed that when varieties of a species co-exist in nature, they must be connected by all intergradations, and it was an article of faith of almost equal validity that the intermediate form must be statistically the majority, and the extremes comparatively rare. The plant breeder might declare that he had varieties of Primula or some other plant, lately constituted, uniform in every varietal character breeding strictly true in those respects, or the entomologist might state that a polymorphic species of a beetle or of a moth fell obviously into definite types, but the evolutionary philosopher knew better. To him such statements merely showed that the reporter was a bad observer, and not improbably a destroyer of inconvenient material. Systematists had sound information but no one consulted them on such matters or cared to hear what they might have to say. The evolutionist of the eighties was perfectly certain that species were a figment of the systematist's mind, not worthy of enlightened attention.

Then came the Mendelian clue. We saw the varieties arising. Segregation maintained their identity. The discontinuity of variation was recognized in abundance. Plenty of the Mendelian combinations would in nature pass the scrutiny of even an exacting systematist and be given "specific rank." In the light of such facts the origin of species was no doubt a similar phenomenon. All was clear ahead.

But soon, though knowledge advanced at a great rate, and though whole ranges of phenomena which had seemed capricious and disorderly fell rapidly into a coordinated system, less and less was heard about evolution in genetical circles, and now the topic is dropped. When students of other sciences ask us what is now currently believed about the origin of species we have no clear answer to give. Faith has given place to agnosticism for reasons which on such an occasion as this we may profitably consider.

Where precisely has the difficulty arisen? Though the reasons for our reticence are many and present themselves in various forms, they are in essence one; that as we have come to know more of living things and their properties, we have become more and more impressed with the inapplicability of the evidence to these questions of origin. There is no apparatus which can be brought to bear on them which promises any immediate solution.

In the period I am thinking of it was in the characteristics and behavior of animals and plants in their more familiar phases, namely, the Zygotic phases that attention centered. Genetical research has revealed the world of gametes from which the zygotes—the products of fertilization are constructed. What has been there witnessed is of such extraordinary novelty and so entirely unexpected that in presence of the new discoveries we would fain desist from speculation for a while. We see long courses of analysis to be traveled through and for some time to come that will be a sufficient occupation. The evolutionary systems of the eighteenth and nineteenth centuries were attempts to elucidate the order seen prevailing in this world of zygotes and to explain it in simpler terms of cause and effect: we now perceive that that order rests on and is determined by another equally significant and equally in need of "explanation." But if we for the present drop evolutionary speculation it is in no spirit of despair. What has been learned about the gametes and their natural history constitutes progress upon which we shall never have to go back. The analysis has gone deeper than the most sanguine could have hoped.

We have turned still another bend in the track and behind the gametes we see the chromosomes. For the doubts—which I trust may be pardoned in one who had never seen the marvels of cytology, save as through a glass darkly—can not as regards the main thesis of the Drosophila workers, be any longer maintained. The arguments of Morgan and his colleagues, and especially the demonstrations of Bridges, must allay all scepticism as to the direct association of particular chromosomes with particular features of the zygote. The transferable characters borne by the gametes have been successfully referred to the visible details of nuclear configuration.

The traces of order in variation and heredity which so lately seemed paradoxical curiosities have led step by step to this beautiful discovery. I come at this Christmas season to lay my respectful homage before the stars that have arisen in the West. What wonder if we hold our breath? When we knew nothing of all this the words came freely. How easy it all used to look. What glorious assumptions went without rebuke. Regardless of the obvious consideration that "modification by descent" must be a chemical process, and that of the principles governing that chemistry science had neither hint, nor surmise, nor even an empirical observation of its working, professed men of science offered very confidently positive opinions on these nebulous topics which would now scarcely pass muster in a newspaper or a sermon. It is a wholesome sign of return to sense that these debates have been suspended.

Biological science has returned to its rightful place, investigation of the structure and properties of the concrete and visible world. We cannot see how the differentiation into species came about. Variation of many kinds, often considerable, we daily witness, but no origin of species. Distinguishing what is known from what may be believed we have absolute certainty that new forms of life, new orders and new species have arisen on the earth. That is proved by the paleontological record. In a spirit of paradox even this has been questioned. It has been asked how do you know for instance that there were no mammals in Paleozoic times? May there not have been mammals somewhere on the earth though no vestige of them has come down to us? We may feel confident there were no mammals then, but are we sure? In very ancient rocks most of the great orders of animals are represented. The absence of the others might by no great stress of imagination be ascribed to accidental circumstances.

Happily however there is one example of which we can be sure. There were no Angiosperms—that is to say "higher plants" with protected seeds—in the carboniferous epoch. Of that age we have abundant remains of a world wide and rich flora. The Angiosperms are cosmopolitan. By their means of dispersal they must immediately have become so. Their remains are very readily preserved. If they had been in existence on the earth in carboniferous times they must have been present with the carboniferous plants, and must have been preserved with them. Hence we may be sure that they did appear on the earth since those times. We are not certain, using certain in the strict sense, that the Angiosperms are the lineal descendants of the carboniferous plants, but it is very much easier to believe that they are than that they are not.

Where is the difficulty? If the Angiosperms came from the carboniferous flora why may we not believe the old comfortable theory in the old way? Well so we may if by belief we mean faith, the substance, the foundation of things hoped for, the evidence of things not seen. In dim outline evolution is evident enough. From the facts it is a conclusion which inevitably follows. But that particular and essential bit of the theory of evolution which is concerned with the origin and nature of species remains utterly mysterious. We no longer feel as we used to do, that the process of variation, now contemporaneously occurring, is the beginning of a work which needs merely the element of time for its completion; for even time can not complete that which has not yet begun. The conclusion in which we were brought up, that species are a product of a summation of variations ignored the chief attribute of species first pointed out by John Ray that the product of their crosses is frequently sterile in greater or less degree. Huxley, very early in the debate pointed out this grave defect in the evidence, but before breeding researches had been made on a large scale no one felt the objection to be serious. Extended work might be trusted to supply the deficiency. It has not done so, and the significance of the negative evidence can no longer be denied.

When Darwin discussed the problem of inter-specific sterility in the "Origin of Species" this aspect of the matter seems to have escaped him. He is at great pains to prove that inter-specific crosses are not always sterile, and he shows that crosses between forms which pass for distinct species may produce hybrids which range from complete fertility to complete sterility. The fertile hybrids he claims in support of his argument. If species arose from a common origin, clearly they should not always give sterile hybrids. So Darwin is concerned to prove that such hybrids are by no means always sterile, which to us is a commonplace of everyday experience. If species have a common origin, where did they pick up the ingredients which produce this sexual incompatibility? Almost certainly it is a variation in which something has been added. We have come to see that variations can very commonly—I do not say always—be distinguished as positive and negative. The validity of this distinction has been doubted, especially by the Drosophila workers. Nevertheless in application to a very large range of characters, I am satisfied that the distinction holds, and that in analysis it is a useful aid. Now we have no difficulty in finding evidence of variation by loss. Examples abound, but variation by addition are rarities, even if there are any which must be so accounted. The variations to which interspecific sterility is due are obviously variations in which something is apparently added to the stock of ingredients. It is one of the common experiences of the breeder that when a hybrid is partially sterile, and from it any fertile offspring can be obtained, the sterility, once lost, disappears. This has been the history of many, perhaps most of our cultivated plants of hybrid origin.

The production of an indubitably sterile hybrid from completely fertile parents which have arisen under critical observation from a single common origin is the event for which we wait. Until this event is witnessed, our knowledge of evolution is incomplete in a vital respect. From time to time a record of such an observation is published, but none has yet survived criticism. Meanwhile, though our faith in evolution stands unshaken, we have no acceptable account of the origin of "species."

Curiously enough, it is at the same point that the validity of the claim of natural selection as the main directing force was most questionable. The survival of the fittest was a plausible account of evolution in broad outline, but failed in application to specific difference. The Darwinian philosophy convinced us that every species must "make good" in nature if it is to survive, but no one could tell how the differences—often very sharply fixed—which we recognize as specific, do in fact enable the species to make good. The claims of natural selection as the chief factor in the determination of species have consequently been discredited.

I pass to another part of the problem, where again, though extraordinary progress in knowledge has been made a new and formidable difficulty has been encountered. Of variations we know a great deal more than we did. Almost all that we have seen are variations in which we recognize that elements have been lost. In addressing the British Association in 1914 I dwelt on evidence of this class. The developments of the last seven years, which are memorable as having provided in regard to one animal, the fly Drosophila, the most comprehensive mass of genetic observation yet collected, serve rather to emphasize than to weaken the considerations which I then referred. Even in Drosophila, where hundreds of genetically distinct factors have been identified, very few new dominants, that is to say positive additions, have been seen, and I am assured that none of them are of a class which could be expected to be viable under natural conditions. I understand even that none are certainly viable in the homozygous state.If we try to trace back the origin of our domesticated animals and plants, we can scarcely ever point to a single wild species as the probable progenitor. Almost every naturalist who has dealt with these questions in recent years has had recourse to theories of multiple origin, because our modern races have positive characteristics which we cannot find in any existing species, and which combination of the existing species seem unable to provide. To produce our domesticated races it seems that ingredients must have been added. To invoke the hypothetical existence of lost species provides a poor escape from this difficulty, and we are left with the conviction that some part of the chain of reasoning is missing. The weight of this objection will be most felt by those who have most experience in practical breeding. I can not, for instance, imagine a round seed being found on a wrinkled variety of pea except by crossing. Such seeds, which look round, sometimes appear, but this is a superficial appearance, and either these seeds are seen to have the starch of wrinkled seeds or can be proved to be the produce of stray pollen. Nor can I imagine a fern-leaved Primula producing a palm-leaf, or a star-shaped flower producing the old type of sinensis flower. And so on through long series of forms which we have watched for twenty years.

Analysis has revealed hosts of transferable characters. Their combinations suffice to supply in abundance series of types which might pass for new species, and certainly would be so classed if they were met with in nature. Yet critically tested, we find that they are not distinct species and we have no reason to suppose that any accumulations of characters of the same order would culminate in the production of distinct species. Specific difference therefore must be regarded as probably attaching to the base upon which these transferables are implanted, of which we know absolutely nothing at all. Nothing that we have witnessed in the contemporary world can colorably be interpreted as providing the sort of evidence required.

Twenty years ago, de Vries made what looked like a promising attempt to supply this so far as Oenothera was concerned. In the light of modern experiments, especially those of Renner, the interest attaching to the polymorphism of Oenothera has greatly developed, but in application to that phenomenon the theory of mutation falls. We see novel forms appearing, but they are no new species of Oenothera, nor are the parents which produce them pure or homozygous forms. Renner's identification of the several complexes allocated to the male and female sides of the several types is a wonderful and significant piece of analysis introducing us to new genetical conceptions. The Oenotheras illustrate in the most striking fashion how crude and inadequate are the suppositions which we entertained before the world of gametes was revealed. The appearance of the plant tells us little or nothing of these things. In Mendelism, we learnt to appreciate the implication of the fact that the organism is a double structure, containing ingredients derived from the mother and from the father respectively. We have now to admit the further conception that between the male and female sides of the same plant these ingredients may be quite differently apportioned, and that the genetical composition of each may be so distinct that the systematist might without extravagance recognize them as distinct specifically. If then our plant may by appropriate treatment be made to give off two distinct forms, why is not that phenomenon a true instance of Darwin's origin of species? In Darwin's time it must have been acclaimed as exactly supplying all and more than he ever hoped to see. We know that that is not the true interpretation. For that which comes out is no new creation.

Only those who are keeping up with these new developments can fully appreciate their vast significance or anticipate the next step. That is the province of the geneticist. Nevertheless, I am convinced that biology would greatly gain by some cooperation among workers in the several branches. I had expected that genetics would provide at once common ground for the systematist and the laboratory worker. This hope has been disappointed. Each still keeps apart. Systematic literature grows precisely as if the genetical discoveries had never been made and the geneticists more and more withdraw each into his special "claim"—a most lamentable result. Both are to blame. If we cannot persuade the systematists to come to us, at least we can go to them. They too have built up a vast edifice of knowledge which they are willing to share with us, and which we greatly need. They too have never lost that longing for the truth about evolution which to men of my date is the salt of biology, the impulse which made us biologists. It is from them that the raw materials for our researches are to be drawn, which alone can give catholicity and breadth to our studies. We and the systematists have to devise a common language.

Both we and the systematists have everything to gain by a closer alliance. Of course we must specialize, but I suggest to educationists that in biology at least specialization begins too early. In England certainly harm is done by a system of examinations discouraging to that taste for field natural history and collecting, spontaneous in so many young people. How it may be on this side, I can not say, but with us attainments of that kind are seldom rewarded, and are too often despised as trivial in comparison with the stereotyped biology which can be learnt from text-books. Nevertheless, given the aptitude, a very wide acquaintance with nature and the diversity of living things may be acquired before the age at which more intensive study must be begun, the best preparation for research in any of the branches of biology.

The separation between the laboratory men and the systematists already imperils the work, I might almost say the sanity, of both. The systematists will feel the ground fall from beneath their feet, when they learn and realize what genetics has accomplished, and we, close students of specially chosen examples, may find our eyes dazzled and blinded when we look up from our work-tables to contemplate the brilliant vision of the natural world in its boundless complexity.

I have put before you very frankly the considerations which have made us agnostic as to the actual mode and processes of evolution. When such confessions are made the enemies of science see their chance. If we cannot declare here and now how species arose, they will obligingly offer us the solutions with which obscurantism is satisfied. Let us then proclaim in precise and unmistakable language that our faith in evolution is unshaken. Every available line of argument converges on this inevitable conclusion. The obscurantist has nothing to suggest which is worth a moment's attention. The difficulties which weigh upon the professional biologist need not trouble the layman. Our doubts are not as to the reality or truth of evolution, but as to the origin of species, a technical, almost domestic, problem. Any day that mystery may be solved. The discoveries of the last twenty-five years enable us for the first time to discuss these questions intelligently and on a basis of fact. That synthesis will follow on an analysis, we do not and cannot doubt.

SOURCE: Bateson, William. "Evolutionary Faith and Modern Doubts." *Science*, New Series, 55:1412 (January 20, 1922), 55–61.

HENRY FAIRFIELD OSBORN, "WILLIAM BATESON ON DARWINISM" (1922)

Henry Fairfield Osborn (1857–1935) was a professor of comparative anatomy at Princeton and later at Columbia University before becoming president of the American Museum of Natural History from 1908 until his death in 1935. An advocate of Darwinism, Osborn published the following response to Bateson shortly after hearing Bateson's AAAS address.

Aside from the fine impression created by the admirable series of papers and addresses in biology, zoology and genetics in Toronto at the Naturalists' meeting, a very regrettable impression was made by a number of passages in the addresses of Professor William Bateson, the distinguished representative of Cambridge University and British biology. On the morning following his principal address the Toronto *Globe* (December 29, 1921) published, in large letters: "Bateson Holds That Former Beliefs Must Be Abandoned. Theory of Darwin Still Remains Unproved and Missing Link between Monkey and Man Has Not Yet Been Discovered by Science. Claims Science Has Outgrown Theory of Origin of Species." In intermediate type it announced: "Distinguished Biologist from Britain Delivers Outstanding Address on Failure of Science to Support Theory That Man Arrived on Earth through Process of Natural Selection and Evolution of Species. Have Traced Man Far Back but Still He Remains Man," and, in smaller type:

> The missing link is still missing, and the Darwinian theory of the origin of species is not proved. This was the verdict of one of the foremost British scientists, Professor William Bateson, director of the John Innes Horticultural Institute, Surrey, England, in the course of an epoch-making address on "Evolutionary Faith and Modem Doubts" at the general session of the American Association for the Advancement of Science, held in Convocation Hall last evening. While declaring that his faith in evolution was unshaken, he frankly admitted that he was "agnostic as to the actual mode and process of evolution." Believing in evolution in "dim outline," he pronounced the cause of origin of species as utterly mysterious.

The speaker then reiterated views expressed in previous addresses. Again quoting the *Globe*:

> Referring to the variations occurring in the different species, Dr. Bateson stated that there was no evidence of any one species acquiring new faculties, but that there were plenty of examples of species losing faculties. Species lose things, but do not add to their possessions. "Biological science has returned to its rightful place," said Dr. Bateson, "namely, the investigation of the structure and properties of the concrete of our visible world. We cannot see how the differentiation into species came about. Variation, of many kinds, often considerable, we daily witness, but no origin of species. Distinguishing what is known from what may be believed, we have absolute certainty that new forms of life, new orders and new species have arisen in the earth, but even this has been questioned. It has been asked, for instance, 'How do you know that there were [no] mammals in palæozoic times? May there not have been mammals somewhere on earth though no vestige of them has come down to us?' We may feel confident there were no mammals then, but are we sure? In very ancient rocks most of the great orders of animals are represented. The absence of the others might by no great stress of the imagination be ascribed to accidental circumstances."

It is not surprising that the next day the *Globe* published a signed letter, under the caption, "The Collapse of Darwinism," of which the following is an abstract:

> To an audience rarely paralleled in Canada for scientific eminence and influence, the famous Professor Bateson, with amazing frankness, removed one by one the props that have been considered the very pillars of Darwinism. A scientist of international repute, one of the leading, if not the leading evolutionist, of the day, he exposed the weakness of many of the leading planks in the "Origin of Species," and ruthlessly tore down one by one the once fondly believed links in the great chain of Darwinian evolution.

These citations cannot be dismissed as mere newspaper talk of no import. They are called forth by the fact that many of the statements in Bateson's address as cited below are inaccurate and misleading, especially those relating to the origin of species, natural selection, and infertility between species.

It is not true that we do not know how species originate. The mode of the origin of species has long been known—in fact, it was very clearly stated by the German paleontologist Waagen in the year 1869, a statement which has been absolutely confirmed beyond a possibility of doubt in the fifty years of subsequent research. It is also true that we know the modes of origin of the human species; our knowledge of human evolution has reached a point not only where a number of links in the chain are thoroughly known but the characters of the missing links can be very clearly predicated. The *cause* of the origin of species is another matter and has been sought in all branches of biology and biological research without an adequate solution having been found. Charles Darwin's theory of selection forms a partial solution of causation and, so far from being discarded, now rests upon much stronger evidence than it did when Darwin enunciated it.

The broad impression conveyed to my mind by the brilliant series of papers in the division of Genetics at Toronto is that genetics is essentially a branch of morphology. It is a running comparison between the morphology of the germ cell and the morphology of the adult. It is in this field, to which Professor Bateson has lent such distinction, that he fails to find either the mode or the cause of the origin of species.

Referring again to the ethical question of the dissemination of scientific truth, I am reminded of the precaution pressed upon me by Huxley from his own experience. He once told me that before delivering any of his popular addresses he very carefully wrote out every word he intended to say, lest in the heat of enthusiasm at the moment he might say something which would give a wrong impression of the truth. We men of science are far too careless in the application of this Huxleyan advice, especially in our popular addresses, which are eagerly read by the public. We must state the truth so clearly that it cannot be misunderstood and when we give voice to our own opinions we should clearly indicate them as our opinions and not as facts. Bateson's attitude towards Darwinism has been patronizing ever since he began his evolutionary studies. When he refers epigrammatically in a previous address to reading his Darwin as he would read his Lucretius he is indirectly stating an untruth which is calculated to do untold harm. In his Toronto address he *does not clearly distinguish between his own personal opinions based on his own field of observation* and the great range of firmly established fact that is now within reach of every student of evolution who surveys the world of life under natural conditions.

Since writing the above there has come to hand a copy of Professor Bateson's address, from which the following excerpts may be made:

> Discussions of evolution came to an end primarily because it was obvious that no progress was being made. Morphology having been explored in its minutest corners, we turned elsewhere.... We became geneticists in the conviction that there at least must evolutionary wisdom be found.... The unacceptable doctrine of the secular transformation of masses by the accumulation of impalpable changes became not only unlikely but gratuitous.... Less and less was heard about evolution in genetical circles, and now the topic is dropped. When students of other sciences ask us what is now currently believed about the origin of species we have no clear answer to give. Faith has given place to agnosticism....
>
> ... But if we for the present drop evolutionary speculation it is in no spirit of despair...
>
> Biological science has returned to its rightful place, investigation of the structure and properties of the concrete and visible world. We can not see how the differentiation into species came about. Variation of many kinds, often considerable, we daily witness, but no origin of species...
>
> ... But that particular and essential bit of the theory of evolution which is concerned with the origin and nature of *species* remains utterly mysterious. We no longer feel as we used to do, that the process of variation, now contemporaneously occurring, is the beginning of a work which needs merely the element of time for its completion; for even time can not complete that which has not yet begun....
>
> ... Meanwhile, though our faith in evolution stands unshaken, we have no acceptable account of the origin of "species..."

... The survival of the fittest was a plausible account of evolution in broad outline, but failed in application to specific difference.... The claims of natural selection as the chief factor in the determination of species have consequently been discredited....

... Even in Drosophila, where hundreds of genetically distinct factors have been identified, very few new dominants, that is to say positive additions, have been seen, and I am assured that none of them are of a class which could be expected to be viable under natural conditions. I understand even that none are certainly viable in the homozygous state....

Analysis has revealed hosts of transferable characters.... Yet critically tested, we find that they are not distinct species and we have no reason to suppose that any accumulations of characters of the same order would culminate in the production of distinct species....

Twenty years ago, de Vries made what looked like a promising attempt to supply this so far as (*Enothera* was concerned.... but in application to that phenomenon the theory of mutation falls. We see novel forms appearing, but they are no new species of (*Enothera*, nor are the parents which produce them pure or homozygous forms... If then our plant may by appropriate treatment be made to give off two distinct forms, why is not that phenomenon a true instance of Darwin's origin of species? In Darwin's time it must have been acclaimed as exactly supplying all and more than he ever hoped to see. We know that that is not the true interpretation. For that which comes out is no new creation....

...If we cannot persuade the systematists to come to us, at least we can go to them. They too have built up a vast edifice of knowledge which they are willing to share with us, and which we greatly need. They too have never lost that longing for the truth about evolution which to men of my date is the salt of biology, the impulse which made us biologists....

The separation between the laboratory men and the systematists already imperils the work, I might almost say the sanity, of both....

I have put before you very frankly the considerations which have made us agnostic as to the actual mode and processes of evolution. When such confessions are made the enemies of science see their chance.... Our doubts are not as to the reality or truth of evolution, but as to the origin of species, a technical, almost domestic, problem. Any day that mystery may be solved.... That synthesis will follow on an analysis, we do not and cannot doubt.

These passages seem to me to do great credit to Professor Bateson in so far as they contain a frank expression of his opinion that up to the present time neither the causes nor the mode of origin of species have been revealed by the older study of Variation, the newer study of Mutation, or the still more modern study of Genetics. If this opinion is generally accepted as a fact or demonstrated truth, the way is open to search the causes of evolution along other lines of inquiry.

SOURCE: Osborn, Henry Fairfield. "William Bateson on Darwinism." *Science*, New Series, 55:1417 (February 24, 1922), 194–197.

J. ARTHUR THOMSON, "WHY WE MUST BE EVOLUTIONISTS" (1928)[1]

J. Arthur Thomson (1861–1933) was an English naturalist and a well-known author of popular science books. He was the author of the best-selling book, *The Outline of Science*, and

---

[1]J. Arthur Thomson, "c" Creation by evolution: A consensus of present-day knowledge as set forth by leading authorities in nontechnical language that all may understand, edited by Frances Baker Mason (New York: The Macmillan Co., 1928), pp. 13–23.

he wrote the below piece in 1928, when the concept of evolution, in particular the teaching of evolution in American public schools, was increasingly under attack.

Evidences of Evolution

We use the familiar phrase "evidences of evolution" with some misgiving, because it does not suggest the right way of looking at the question. Evolution means a way of Becoming. Just as it is certain that all the many races of domesticated pigeon are descended from the wild rock dove, so, it is argued, have all the different kinds of wild animals and wild plants descended from ancestors that were on the whole somewhat simpler, and these from simpler ancestors still, and so back and back until we come to the first living creatures, whose origin is all in the mist. Evolution just means that the present is the child of the past and the parent of the future.

But it is not possible to prove this conclusion in an absolutely rigorous way. We can, indeed, see evolution going on now, but we cannot, so to speak, reverse the world-film and see precisely what took place long ago. The records in the rocks do clearly reveal what happened in the past, even millions of years ago, but not in so clear and so detailed a way as the developing egg of a hen reveals the gradual rise and progress of the chick.

Although we do not know of any competent biologist to-day, however skeptical and inquiring he may be, who has any doubt as to the fact of organic evolution, yet no one would assert that it can be demonstrated as one might demonstrate the law of gravitation, or the conservation of matter and energy, or the development of a chick out of a drop of living matter on the top of the yolk of the egg. But how can a conclusion be accepted without hesitation if it is not rigorously demonstrable? The answer is that the evolution-idea is a master key that opens all locks into which we can fit it, and that we do not know of a single fact that can be said to be in any way contradictory. Like Wisdom, the evolution-idea is justified of its children.

A great zoölogist once said that he was willing to stake the validity of the evolution-idea on the evidence afforded by butterflies, and he was quite right. Any fact about an animal or a plant may be an evidence of evolution when we know enough about it. What makes the general idea of evolution convincing is its satisfactoriness in interpretation. It is always borne out by the facts. We repeat the phrase "the general idea of organic evolution" because this must be distinguished from any particular theory in regard to the factors that have operated in the process. In regard to the *factors* or *causes* of evolution there is, and there may well be, difference of opinion among naturalists, for the inquiry is as young as it is difficult; but it is unfair and confused to use this admission of uncertainty as to *causes* as if it implied any hesitation in regard to the *fact* of an age-long evolutionary process in which many of the highly finished and very perfect types of animals are shown by the rock record to be preceded by a succession of animals in less finished stages.

There is eloquence in the evidence from the rock record. As ages passed there was a gradual emergence of finer and nobler forms of life. Among back-boned animals the first were the fishes. These led to the amphibians, and these were succeeded by reptiles. Later there arose birds and mammals. Throughout the ages, life has been slowly creeping upward. Detailed pedigrees are disclosed in the rocks, some of them with marvelous perfection, as in the evolution of horses and elephants, camel and crocodiles. For some animals, such as fresh-water snails and marine cuttlefishes, there is an almost perfect succession of fossils, forming a chain in which link 10 is very different from link 1, yet just a little different from link 9, as link 2 is a little different from link 1. For such animals we can almost see evolution anciently at work!

The geographical evidences are also endless. If the present state of affairs is not the outcome of a natural process of evolution, why should the fauna of oceanic islands be restricted to those animals which can be accounted for by transport over the sea by currents and by winds, or on the feet of birds? Thus there are no amphibians on oceanic islands, because few amphibians can endure salt water.

The inhospitable Galapagos Islands are said to be the submerged tops of cold volcanoes, which belong to an ancient peninsula that became first an island and then an archipelago. They have a peculiar fauna, which includes the famous giant tortoises. There are ten different kinds of giant tortoise on ten different islands, and those that are on the islands that are farthest apart are most unlike. There are five different kinds in different parts of the largest island, which is called Albemarle. Now if we consider fully these facts what can we find them to mean except that isolated groups of one ancient stock of the original peninsula have varied slightly on one or another island and that the isolation prevented any pooling or blending of the new forms? For these large tortoises cannot swim. On Albe Marle Island the isolation is probably topographic; it is due to barriers formed by the rugged volcanic surface. When Darwin, as a young man, visited these islands during the voyage of the Beagle he was greatly struck by the fact that each island seemed to have its own kind or species of giant tortoise, and he tells us that he felt himself "brought near to the very act of creation." This was one of the experiences that made Darwin an evolutionist.

But think also of the anatomical evidence. It is interesting to compare a number of fore limbs—our own arm, a bat's wing, a whale's flipper, a horse's fore leg, a bird's wing, a turtle's paddle, a frog's small arm and a giant giraffe's at the other extreme. They are very different, and yet when we scrutinize them we find the same fundamental bones and muscles and blood-vessels and nerves. "How inexplicable," Darwin said, "is the similar pattern of the hand of man, the foot of a dog, the wing of a bat, the flipper of a seal, on the doctrine of independent acts of creation! How simply explained on the principle of the natural selection of successive slight variations in the diverging descendants from a single progenitor." Few zoölogists of today would use Darwin's words "how simply explained," for we are aware of factors he did not know of, and some of the factors he believed in very strongly are not unanimously accredited today. But all would agree that the evolution-idea illumines the deep identities, amid great superficial diversities, that are disclosed when we consider, let us say, the classes of backboned animals.

Another anatomical argument is to be found in the frequent occurrence of vestigial structures in animals and in ourselves. Useless dwindled relics of the hind limbs of a whale are found buried deep below the surface. In the inner corner of our eye there is just a trace of what is called the third eyelid, a structure that is strongly developed and readily seen in most mammals, as well as in birds and reptiles. It serves to clean the front of the eye; but although it is big enough to do this in most mammals and birds it is a mere relic in man. Take another example: behind the eye of the skate—a familiar flat fish—there is a large hole called the "spiracle." It serves for the incoming of the "breathing water," which washes the gills and passes out by the five pairs of gill-clefts on the under surface. But if we peer into this very useful breathing-hole or spiracle we see a minute comb-like structure, which is the dwindling useless relic of a gill. The cleft or spiracle is of indispensable use to the skate, but the relic or vestigial gill inside the spiracle is of no use at all. Yet it tells us that a spiracle was evolved from a gill-bearing gill-cleft.

One of the most remarkable sets of facts about living creatures—plants as well as animals—is that old structures become transformed into things very new. The poet Goethe helped to make the great discovery that the parts of a flower—sepals, petals, stamens, and carpels—are just four whorls of transfigured leaves, the stamens and carpels being spore-bearing leaves. We sometimes see the whole flower of a flowering plant that has become too vegetative "go back" and become a tuft of green leaves; and it is an unforgettable lesson to pull the flower of the white water lily to pieces and to find that the green sepals pass gradually into white petals, and these gradually into yellow stamens.

Similar lessons are taught by animals. What is the sting of a bee but a transformed egg-laying organ or ovipositor (therefore never found in drones), and what is an ovipositor but a transformed pair of limbs? The elephant's trunk was a great novelty in its way, but it is just a very long nose it with an additional piece due to a pulling out of the upper lip. This is the evolutionary way!

We live in what has sometimes been called the "age of insects," for of these there are more than a quarter of a million different kinds. Now there must be some meaning in the fact that these can be classified in an orderly way; that one can for many kinds make plausible "genealogical trees." Often one species, with its varieties, seems to grade into another. In many parts of the animal kingdom there are types that link great classes together. Thus the old-fashioned *Peripatus* type, a little creature somewhat like a permanent caterpillar, has some worm characters and some centipede characters. It is to some extent a connecting link. The oldest known bird, a fossil beautifully preserved in lithographic stone of Jurassic age, has numerous reptilian features, such as teeth in both jaws, a long lizard-like tail, a half-made wing, and abdominal ribs. Yet it was a genuine feathered bird! And this fossil is unexplainable unless we recognize the fact that this bird had reptilian ancestors.

Very striking, again, are the embryological facts which show that the development of the individual is like a condensed recapitulation of the probable evolution of the race. An embryo bird is for some days almost indistinguishable from an embryo reptile; they progress along the same high road together; but soon there comes a parting of the ways and each goes off on its own path. The gill-slits of fishes and tadpoles—the slits through which the water used in breathing passes—are persistent in all the embryos of reptiles, birds, and mammals, though in these higher back-boned animals they have nothing to do with respiration. All of them are merely transient passages except the one that becomes the "eustachian tube," which leads from the ear to the back of the mouth. They are straws which show how the evolutionary wind has blown. In a great many ways the individual animal climbs up its own genealogical tree, but we must be careful not to think that an embryo mammal is at an early stage of its development like a little fish, as some writers have carelessly, said. Each living creature, from the very first stage of its development, is itself and no other; and though the tadpole of a frog has for some weeks certain features like that of a fish, especially a larval mudfish, it is an amphibian from first to last. The embryo is the *memory* of a fish or of a reptile-like ancestor. There is no doubt that the hand of the past is upon the present, living and working; and this is evolution.

Many living creatures today are like ever-changing fountains; they are continually giving rise to something new. The beautiful evening primrose (*Oenothera*) and the American fruit-fly (*Drosophila*) are notable examples of changeful types; they are always giving birth to novelties or new forms, technically called "variations" or "mutations"; but the fact of variability is widespread.

In some forms the breeder or the cultivator is able to provoke great changes, for instance, by altering surroundings and food; but he usually has to wait for what the natural fountain of change supplies. This has been our experience with the domesticated animals and cultivated plants that interested Darwin so much. All the domestic pigeons have been derived, under man's care from the blue rock dove; and there is strong evidence that the multitudinous breeds of poultry are all descended from the Indian jungle-fowl. What Darwin said was this: If man can fix and foster this and that novelty and make it the basis of a true-breeding race, and all in a comparatively short time, what may Nature not have accomplished in an unthinkably long time? And when it was objected: But what is there in Nature corresponding to Man the Breeder, his characteristically Darwinian answer was that the Struggle for Existence implied a process of sifting, which he called Natural Selection. Testing all things and holding fast that which is good or fit: that has been the evolutionary method!

These few examples should make plain the nature of the argument for evolution. It is what is called a cumulative argument. All the lines of facts meet in the same conclusion—the present is the child of the past. There is no conflicting evidence; every new discovery points in the same direction. On many sides we find striking facts, which become luminous when we see them in the light of the evolution-idea. But without that light they are worse than puzzling. All the facts conspire toward the conclusion that animate nature has come to be as it is by a continuous natural process, comparable to that which we can study in the history of domesticated animals and cultivated plants. But we do not give a satisfying account of what has taken place until we can state all the factors that have operated, and that is the subject of the much-debated detailed theories of evolution, like Darwinism and Lamarckism. And even if we were agreed about the factors we should still have to inquire into the meaning or significance of the whole. But that is a religious question.

An Enriching Outlook

Another great reason why we must be evolutionists will come as a surprise to some people. The evolutionist outlook is one that lightens the eyes and enriches us. We are impoverishing ourselves if we shut out the light of evolution. Let us consider three points only.

1. The evolution-idea gives the world of animate nature a new unity. All living creatures are part and parcel of a great system that has moved sublimely from less to more. All animals are blood-relations; there is kinship throughout animate nature.

2. It is indeed a sublime picture that the evolutionist discloses—a picture of an advancement of life by continuous natural stages, without haste, yet without rest. No doubt there have been blind alleys, side-tracks, lost races, parasitisms, and retrogressions, but *on the whole* there has been something like what man calls progress. If that word is too "human" we must invent another.

3. One of the greatest facts of organic evolution—a fact so great that it is often not realized at all—is that there has been not merely an increase in complexity but a growing dominance of mind in life. Animals have grown in intelligence, in mastery of their environment, in fine feeling, in kin-sympathy, in freedom, and in what we may call the higher satisfactions.

No evolutionist believes that man sprang from any living kind of ape, yet none can hesitate to believe in his emergence—"a new creation"—from a stock common to the anthropoid apes and to the early "tentative men." Long ago there was a parting of the ways—it could not be less than a million years ago: the anthropoids remained arboreal and the ancestors of the men we know became terrestrial. So far as we can judge from links that are certainly not missing, but always increasing in number, there were for long ages only tentative men like *Pithecanthropus* the Erect, in Java, and *Eoanthropus*, the Piltdown man of the Sussex Weald. Even these were rather collateral offshoots than beings on the main line of man's ancestry. They were Hominids, but not yet Homo. What trials and siftings there seem to have been before there appeared "the man-child glorious!" Doubtless some great brain change led to clearer self-consciousness, to language, to a power of forming general ideas, to greater uprightness of body and mind; and it is very important to realize that a steady advance in brain development, on a line different from that of other mammals, is discernible in the very first monkeyish animals. Man stands apart and is in important ways unique, but he was not an abruptly created novelty. That is not the way in which evolution works. Man, at his best, is a flower on a shoot that has very deep roots. What the evolutionist discloses is man's solidarity, his kinship, with the rest of creation. And the encouragement we find in this disclosure is twofold. In the first place, though we inherit some coarse strands from pre-human pedigree, it is an *ascent*, not a *descent* that we see behind us. In the second place, the evolutionist world is congruent with religious interpretation. It is a world in which the religious man can breathe freely. To take one example: there are great trends discernible in organic evolution, and the greatest of these are toward health and beauty: toward the love of mates, parental care, and family affection; toward self-subordination and kin-sympathy; toward clear-headedness and healthy-mindedness; and *the momentum of these trends is with us at our best*. And evolution, with these great trends, is going on: Who shall set it limits?

SOURCE: Thomson, J. Arthur. "Why We Must Be Evolutionists," in *Creation by Evolution: A Consensus of Present-Day Knowledge as Set Forth by Leading Authorities in Non-Technical Language That All May Understand*. Ed. by Frances Baker Mason. New York: The Macmillan Co., 1928.

RICHARD GOLDSCHMIDT, "SOME ASPECTS OF EVOLUTION" (1933)

Richard Goldschmidt (1878–1958) was a German-born geneticist best known for his theory of evolution through macromutations. Popularly known as the hopeful monster hypothesis, it sought to explain how small mutations led to large evolutionary changes. In 1933, on the eve of the modern evolutionary synthesis, he offered the below paper at the annual meeting of the American Association for the Advancement of Science in 1933, which demonstrated his interest in using genetics as the basis for resolving the ongoing problems within evolutionary theory. It also showed some of the divisions that existed among and between both evolutionary biologists and geneticists over the exact mechanism of evolution.

In his much discussed presidential address at the 1914 meeting of the British Association, the great skeptic William Bateson finished with the following sentence: "Somewhat reluctantly and rather from a sense of duty I have devoted most of this address to the evolutionary aspects of genetic research. We can not keep these things out of our heads, as sometimes we wish we could. The outcome, as you will have seen, is negative, destroying much that till lately passed for gospel." This negative standpoint was certainly justified to a certain extent by the results of early Mendelian work, which led more in the direction of evolutionary skepticism than optimism. Almost twenty years have passed since, which have witnessed an unbelievable increase in the knowledge of genetical facts. And whereas, as Bateson says, we cannot keep these things, namely, the evolutionary aspect of genetics, out of our heads, geneticists from time to time like to leave their bottles, breeding cages and seed pans and to review the advances of experimental work in regard to their bearing on problems of evolution. I must confess to have been repeatedly guilty myself of this sin during the past 15 years, with the result that the curve of my deliberations was oscillating between skepticism and optimism and still is doing so. Let me not be misunderstood: not skepticism in regard to evolution, which I regard as a historic fact, as all biologists do; but skepticism and optimism regarding the insight into the means of evolution on the basis of genetic facts.

You all know that the majority of the geneticists are to-day rather optimistic. Genetic experimentation certainly has shown that the sudden changes of the hereditary units, the genes, called mutations occur with sufficient frequency to furnish material for selection; it has shown that in plants at least considerable changes, amounting to the formation of what might be termed new species, may be brought about by the different types of chromosome-arrangements which play such an important role in present genetical research; and genetics may rightfully claim to have performed experimental changes of forms into other different ones by means which could be conceived as effectual occasionally also in nature; this is at least true for the plant kingdom, but not for animals. In addition, it has been shown that after all Darwin's theory of selection, if properly applied and based upon the present-day knowledge of what Darwin termed generally variation, is still the best guide to an understanding of some of the ways of evolution. This means that, given a certain frequency of mutations, which produce slight changes in a haphazard way and given the selective action of the environment which wipes out certain mutations and lets pass or even favors others, considerable transformations are possible within the time available for evolution. It is not my intention to enlarge here on this topic, which has been treated repeatedly in recent years by leading geneticists. But I have not been satisfied yet that these groups of facts and conclusions, important as they are, tell us the whole story; and I believe that, especially for the animal kingdom, much work has still to be done before we can see clearly how evolution, which we can observe in its great lines as an actual historic fact, has proceeded in detail. I should like then to discuss a few of the fundamental questions regarding the first steps of evolution in nature, which I met in the course of my own experimental work, and then bring to your attention some facts and lines of thought which might assist a deeper insight into our problem.

When Darwin spoke of the origin of species, the Linnean species seemed to be a rather clear-cut unit. Meanwhile we have recognized the existence of microspecies and of subspecies and racial groups, and if we were to define the units which are meant if we are talking about the origin of species, the difficulties would be found insurmountable. In one taxonomic group, what is called a species is hardly distinguishable from the next species, and in another taxonomic group, the species are more different than genera in the first. In my younger days I was working on the minute histology of the nematode worms Ascaris lumbricoides and megalocephala. These species, though well known to every zoologist as very much alike, proved to be different practically in every cell of their body. At that time I could have undertaken to determine the species from a single isolated cell of many organs of these worms. Compare with this the almost complete impossibility of distinguishing a lion's and tiger's skeleton, in order to realize the hopeless situation for a proper definition. As a matter of fact the only case of a taxonomic difference between two forms, which can be properly defined, is the difference between a homozygous strain of an animal or plant and one of its mutations. Then, if we are talking about the formation of species, what we actually mean is the origin of very different forms within a group, without consideration of their taxonomic designation as species, genera or even families, which more or less depends upon the personal judgment of the taxonomist.

The majority of the geneticist's work is done with domestic animals and plants or with such wild forms as have given plenty of mutations under cultivation. The obvious reason is that natural species or still more distant units are either sterile inter se or produce sterile hybrids and therefore do not lend themselves to the methods of genetic analysis by hybridization.

There is only one taxonomic category about which genetic research has given us proper information: This is the so-called Rassenkreis, a conception which in some taxonomic groups, as birds and mollusks, is gradually replacing the species concept. A Rassenkreis is a series of typically different forms or subspecies found at different points within the geographic range of a species and often showing a typical order of their characters if arranged geographically. As the end members of such a group might be rather different, the idea has arisen that the formation of a geographic Rassenkreis is the beginning of speciation. The idea is that distant members of such a group become finally isolated and will come under the influence of new selective agencies, which carry the stream of further mutational changes into new directions towards the formation of new species and genera. Further, whereas it is found that the differential characters of these subspecies may have adaptational value, it is frequently reasoned that the influence of the environment has produced these forms. To quote only one prominent witness: Henry Fairfield Osborn in a recent address has stood up most emphatically in favor of such views. He writes:

> ...the Buffon-St. Hilaire principle of direct environmental action both on body and germ is now universally admitted as one of the great causes of evolution. As shown in the experiments of Sumner it is directly responsible for speciation in animals like Peromyscus (a deer mouse). Sumner has positively demonstrated that modifications in color and form and proportion traceable to the prolonged direct action of environment, are hereditary and therefore true germinal characters. Perhaps the best established zoological generalization of modern times is that subspeciation, and ultimately full speciation is the inevitable result of prolonged change of environment....

I am sorry to say that I can not agree with the eminent paleontologist, either in regard to the evolutionary nature of subspecies or in regard to the origin of their adaptational traits. Simultaneously, with Sumner's work on Peromyscus I have analyzed the case of the geographic variation of the gipsy-moth Lymantria dispar, and owing to the great regularity of behavior of these geographic races in respect to climatic conditions and also to the possibility of working with large numbers, I was able to make what I believe to be the most complete genetic analysis of a Rassenkreis. As a matter of fact, where Sumner's and my work is comparable the results are also identical, as far as facts are concerned. And I would do injustice to Sumner if I would not state that in his last review of his work he expresses himself rather cautiously in regard to the conclusions to which Osborn points, saying, "While admitting the paucity if not the total lack of direct evidence in this field I still lean strongly towards the view that the process of natural selection must be supplemented by adaptive responses of a more direct nature."

My own work, however, permits, I think, of taking a definite stand towards both problems, mentioned in Osborn's sentence which I quoted before, namely, the problem whether the formation of subspecies is the beginning of speciation and whether unknown actions of the environment are responsible for the adaptational features of geographic variation. Regarding the second point, I could prove that certain characters of a more physiological order show within the geographic range of the species a gradient of different heritable conditions which are perfectly parallel to a gradient of certain climatic conditions. For two of these characters, namely, the length of time of hibernation, the so-called diapause and the rate of larval growth, it could be shown in detail that the definite hereditary type found in definite areas constitutes an adaptation of the life-cycle of the animal to the seasonal cycle of nature. To mention only one example, which is typical for all similar cases: In a region with strong winter and short summer the hibernating individuals would be wiped out if they hatched too early; on the other side, the race would be wiped out if they hatched so late that the short summer would not give them enough time to finish their lifecycle. Correspondingly, the genetic constitution of the races inhabiting such a region is such that a certain sum of heat makes the individual hatch within a short time, whereas races inhabiting warmer areas with mild winter require a much larger sum of heat for the same purpose, also on a hereditary basis. And of course all imaginable intermediate conditions are also found in their proper area.

Here, then, we have a series of typical adaptations to the conditions of a series of typically different environments, and these adaptations are caused by different constitutions in regard to Mendelian genes. Changes in the genetic make-up concerning individual genes are known thus far only to occur in the form of mutations, and no geneticist will doubt therefore that also in this case the different genetic constitutions of the races, those with and those without adaptational value, are the result of mutations and their proper recombinations which once must have taken place in the same manner as mutations observed in the laboratory. But how about the adaptational side, in our case the close parallel between the gene-controlled details of the life-cycle which we just mentioned and those of the seasonal cycle in different regions? If I am not mistaken, Davenport and Cuenot were the first to pronounce the principle of preadaptation, which to most, if not all geneticists, seems to furnish the only workable idea in cases like the one here discussed. Preadaptation means that adaptations are not originated in the surroundings in which they are found and also not caused by whatever action of these surroundings; moreover, adaptive characters appear as chance mutations, without any relation to their future adaptational value, as preadaptations. But these changes allow the organism to migrate into new surroundings, into which it will fit on the basis of its preadaptations. Applied to our case, it would mean that among the population in the original environment mutations were found which produced different conditions in regard to adaptational characters, in our example, mutations which prolong or shorten the inherited length of the hibernation period. Such mutated forms were preadapted to another environment. Brought by chance into another environment with a correspondingly different seasonal cycle, they were able to establish themselves. It is needless to say, then, that we must regard such preadaptational mutations as a prerequisite for the spreading of a species into new areas with different conditions, which would be inaccessible to the original form, and therefore also for the formation of geographic races or subspecies; and further that it will be the physiological characters, not the visible traits, which will be of primary importance in this case. In my material, Lymantria, as a matter of fact the diversity of physiological characters is considerably greater within the Rassenkreis than the diversity of forms which the taxonomist could recognize.

May I mention finally two facts which show the principle at work in our material. Every American knows that the few caterpillars of the gipsy-moth which were blown out of Monsieur Trouvelot's window two generations ago established themselves only two well in Massachusetts. In the light of our work their hereditary life-cycle must have been well preadapted to the seasonal cycle in Massachusetts. The same moth has been introduced into England any number of times, but never was established, in my opinion only for lack of preadaptation to the seasonal cycle. The second fact is the following: Some years ago, I had succeeded in producing mutations in Drosophila by the action of high temperature. The Japanese geneticist, Y. Tanaka, informed me then that he succeeded in producing mutations in the silkworm by a similar method applied at a definite stage. I then occasionally treated the gipsy-moth in a similar fashion. One mutation, which was produced, made the young caterpillars hatch without hibernation. Within the present range of distribution of the moth, such a mutation, if occurring in nature, would be absolutely lethal, because in a moderate climate there would be no possibility of finishing a second generation before winter sets in. But introduced into a tropical climate, the same mutation might permit the otherwise unlikely establishment of the form. I do not doubt, then, that the adaptational side of the facts of geographic variation is to be explained on ordinary genetic grounds, namely, chance mutation of preadaptational nature within a population and subsequent migration into and survival in another suitable area. I may add finally that our material is not the only example, but that Brown has since found a parallel case in Daphnids and that also Turesson's work on ecospecies in plants fits perfectly into these lines.

Let us turn now to the other problem stated above and answered in the affirmative by Osborn and probably by most taxonomists: Is the formation of geographic subspecies the beginning of speciation? My own work was started with the idea of proving that it was. As I have already stated at last year's International Congress of Genetics, the results of the analysis led me to the conclusion that it was not. The different subspecies in the different regions occupied by the species are genetically different in many characters. Most of these are found to form quantitative gradients which run parallel to definite features of the climatic conditions. But the series of local changes in regard to one character is not exactly paralleled by those of other characters, so that in a given area one hereditary and differential character might be found over the whole area, another be subdivided into three types and another into more types. But I was unable to find one or a combination of subspecific characters which could be regarded as leading out of the limits of the species or towards another one.

There are found within the same region two other species of the same genus which show practically the same life-cycle and which must be adapted to the same general features of the region. But they are different in practically every detail of their form, structure, larva and even their type of genetic variation. Of course their differences might be also adaptational in a certain sense. But here is the great difference: The different adaptational characters of the subspecies are of a quantitative nature, and show a plus-minus character. For example, we find a longer diapause in warmer and a shorter in colder regions, similarly different rates of development, different sizes, degrees of pigmentation, etc. The adaptation to local conditions then takes place by genetic shifts of a quantitative nature within the typical characters of the species and, as I may now add, running in the same directions as the non-heritable reactions to the environment. The different species, however, may solve one and the same adaptational problem by entirely different methods. For example, the species Lymantria dispar, the gipsy moth, lays her eggs in the shade on wooden or stony surfaces and covers them with a sponge-like mass of hair, the problem being to ensure proper conditions for hibernation, especially regarding moisture. The nearly related species, L. monacha, pastes her eggs without covering into clefts of the bark of trees, and another species, L. mathura, still in the same area, lays below the bark and within a cement-like mass. Of course, within the different genetic systems represented by related species, parallel types of genetic variation, subspeciation, may be found, as is well known. For example, many species of rodents may form pale desert forms, and many species of birds form subspecies with brighter colors in warmer climates. But in other cases even the trend of genetic variation might be different: Lymantria monacha tends towards formation of melanic forms; L. dispar does not. These two species are able to spread all over the moderate, regions by proper adaptive changes, but not into the tropics, the nearly related species L. mathura, however, inhabiting certain regions together with the former, spreads into the tropics but not into cold regions.

I am perfectly aware of the dangers of generalizing from one case, even the best known one. I know also the objections to such conclusions, for example: There are Rassenkreise, the most distant members of which might be so different that in case of isolation they might become the starting point for quite new developments towards another species. Looking closely at the facts concerning the typical differences within a Rassenkreis, I can not see why the isolation of two members of a Rassenkreis could give better chances for new developments than the isolation of individuals within a subspecies: The changes necessary for the formation of a new species are so large that the relatively small differences of the subspecies as a starting point would hardly count. And I can not help confessing that after trying to get acquainted with the taxonomist's material, the skeptical standpoint derived from my own genetic analysis could not be shaken. There is in my opinion no reliable fact known which would force us to assume that geographic variation or formation of subspecies has anything to do with speciation; the results of genetical analysis and of sober evaluation of the other facts are positively in contradiction to such an assumption. We just mentioned the fact that different species and also as a matter of fact members of different families may show a trend towards formation of comparable mutations and parallel series of subspecies, which are, after all, combinations of mutations strained through the sieve of fitness to environment. It is known that especially Vavilov has made such facts the basis of evolutionary considerations. But we also mentioned that nearly related species might show different trends of genetic variation. And this leads us to a point which, I believe, will be considered of paramount importance in future discussions of evolution. The transformation of one species into another is possible only if permanent changes in the genetic make-up occur, and if the changed forms stand the test of selection. Both these points have long been in the foreground of evolutionary discussion. But there is a third point, often neglected, which lies, I think, at the basis of the whole problem, namely, the nature of the developmental system of the organism which is to undergo evolutionary change. The appearance of a genetic form, whether we call it a species or a genus, which is to be considerably different from the ancestral forms, requires that a considerable number of developmental processes between egg and adult have to be changed, in order to lead to a different organization. Development, however, within a species is, we know, considerably one-tracked. The individual developmental processes are so carefully interwoven and arranged so orderly in time and space that the typical result is only possible if the whole process of development is in any single case set in motion and carried out upon the same material basis, the same substratum and under the same control by the germ plasm or the genes. From this it follows that changes in this developmental system leading to new stable forms are only possible as far as they do not destroy or interfere with the orderly progress of developmental processes. Of course, everybody knows that this is the reason why most mutations are lethal. But not everybody keeps in mind that here also is touched one of the basic points of the problem of evolution. The nature and the working of the developmental processes of the individual then should, if known, permit us to form certain notions regarding the possibilities of evolutionary changes.

There are, as far as I can see, two general notions in regard to the causal understanding of individual development which are of importance for the problem under discussion. One is the notion which I have tried to develop from experimental evidence that the action of the genes in controlling development is to be understood as working through the control of reactions of definite velocities, properly in tune with each other and thus guaranteeing the same event always to occur at the same time and at the same place, as worked out in detail in my physiological theory of heredity. The second notion is that derived from the results of experimental embryology. It says that two types of differentiation are closely interwoven in the process of development, namely, independent and dependent differentiation. Independent differentiation means that a once started process of differentiation takes place within an organ or part of the embryo, even if completely isolated from the rest; dependent differentiation, however, requires the presence and influence of other parts of the embryo for orderly differentiation. If, for example, the group of cells which is to be regarded as the primordium of an eye in the embryo of a vertebrate, is removed from its proper place, it will nevertheless be able to develop into an eye. If, however, the part of the skin of the head which is to form the lens of the eye is isolated, no lens is formed because the presence of the eye is necessary for the determination of a lens. Such are the two general notions, which together describe fairly well the essentials of gene-controlled development, namely, the notion which considers development as an orderly interwoven series of developmental reactions of definite velocities, properly in tune with each other, and the notion of dependent and independent differentiation. Both together will allow us to discuss some of the possibilities of evolutionary change as viewed from the standpoint of stable, orderly development.

Let us begin with an experimental fact. It has been known for a long time that it is possible to change the appearance of certain butterflies by proper experimental procedure within a sensitive period of development so that they can not be distinguished from heritable geographic subspecies found in nature in other regions. If, for example, the young pupa of the Central-European swallowtail is treated with extreme temperatures, some individuals will hatch which can not be distinguished from the typical forms inhabiting Palestine. Of course the characteristic features are not heritable in the former case, but strictly heritable in the latter. These and similar facts have since been extended in many ways, also to cases of ordinary gene mutations. I was, for example, able to produce in similar experiments with Drosophila the non-heritable likeness of many well-known mutations. I do not doubt either that it would be possible to perform the same experiment in regard to any known mutations, if the proper method would be found. Speaking generally, this would mean that the more frequently occurring genetic changes, called mutations, are such as change certain developmental processes in a direction which lies within the ordinary range of changes which might occur within the developmental system under purely environmental influences. An explanation is very simple on the basis of the assumption that in the developmental processes in question reaction-velocities are involved; the external influences in question change the rate of some reaction or system of reactions underlying the differentiation of the character in question and the mutation which produces the same phenotypic effect is a change in a gene, which controls the same differentiating reaction, with the effect of a corresponding change of the speed of the reaction. It is perfectly clear, then, that within similar developmental systems, represented by taxonomically related forms, the same types of mutational changes, parallel mutations, will have the greater chance of not being lethal, because in such a system of exactly tuned and interwoven reactions, only few changes of the rate of individual processes will be possible which do not interfere with the others. And there is another consequence: if there are only a few avenues free for the action of mutational changes without knocking out of order the whole properly balanced system of reactions, the probability is exceedingly high that repeated mutations will go in the same direction, will be orthogenetic. Orthogenesis means that evolution, once started, proceeds further in exactly the same direction until sometimes extreme forms are evolved which lead to the ultimate extinction of the whole line. Paleontologists have found the most beautiful examples of this type, facts with which any theory of evolution has to reckon. Many theories have been advocated to explain such facts. We have pointed out a long time ago and still hold that orthogenesis is not the result of the action of selection or of a mystical trend, but a necessary consequence of the way in which the genes control orderly development—a way which makes only a few directions available to mutational changes, directions which if once started and not acted upon by counter-selection, will be continued. I shall not go into the purely genetic details of such a situation. But it might be mentioned that recently some of the younger generation of paleontologists (Beurlen, Schindewolf, Kaufmaiin) have taken up these views. This is indeed very gratifying, because the problem of orthogenesis has always been a stumbling block to an understanding between geneticists and paleontologists.

At this point, we have to think of the second notion, mentioned before, regarding the general control of embryonic differentiation, namely, dependent and independent differentiation. It is obvious that processes of dependent development are so closely linked with the whole of normal development that mutational changes within them can hardly lead to a normal organism. It is therefore to be expected that successful mutations of eventual evolutionary value act upon such developmental processes which themselves are not inductive of further important steps. This means that viable mutations will mostly be concerned in the animal kingdom with end-processes of embryonic differentiation, affecting the organism only after the characteristics of the species have been laid down.

But how about the possibility of occasional successful mutational changes acting upon earlier developmental processes? Would such a change, if possible at all without breaking up the whole system of the orderly sequence of development, not at once have the consequence of changing the whole organization and bridging with one step the gap between taxonomically widely different forms? Let us for a moment dwell upon such an idea, which I pointed out a long time ago as a logical consequence of my views on gene-controlled development and which has repeatedly cropped up since in evolutionary literature (e.g., De Beer, Haldane, Huxley). Again, the most probable mutational change with a chance to lead to a normal organism is a change in the typical rate of certain developmental processes. Of course, in most cases such a shift of a partial process would lead to the production of monstrosities and, as a matter of fact, Stockard has always advocated such a cause for many monstrosities. But we must not forget that what appears to-day as a monster will be to-morrow the origin of a line of special adaptations. The dachshund and the bulldog are monsters. But the first reptiles with rudimentary legs or fish species with bulldog-heads were also monsters. Correspondingly, we certainly know of many cases of mutational shifts of the rate of certain developmental processes leading to non-viable results, for example, caterpillars with pupal antennae, larvae of beetles with wings and similar cases of so-called pro- and opisthotely. But I can not see any objection to the belief that occasionally, though extremely rarely, such a mutation may act on one of the few open avenues of differentiation and actually start a new evolutionary line. Let us assume a mutational change in rate of differentiation of the limb-bud of a vertebrate, to take up the example just mentioned. The consequent rudimentation of the organ would probably not interfere with orderly development of the organism. Here, then, an avenue would be open to considerable evolutionary change with a single basic step, provided that the new form could stand the test of selection, and that a proper environmental niche could be found to which the newly formed monstrosity would be preadapted and where, once occupied, other mutations might improve the new type. And in addition, the possibility for an orthogenetic line of limbrudimentation would be a further consequence in accordance with—what we have heard before. Of course, these are speculations, which we can not help but enjoy occasionally as long as unfortunately there is no way visible of attacking such problems with the methods of genetics. But meanwhile some important insight might already be gathered from purely morphological work, as that of Sewertzoff, or experimental work of the type of Twitty's work on rudimentary eyes.

At the best, such viable mutations concerning rates of earlier developmental processes must be rare, even when processes are involved such as the differentiation of appendages which are not so closely interwoven with the whole of development. Still lower is the chance if we try to imagine changes in differentiation which are of consequence for the whole of development. Let our imagination run wild for a moment and let us consider the possible event of three more and more violent and therefore less and less probable changes of the type under consideration, produced by a viable mutation acting upon earlier embryonic differentiation by changing relative rates of development. D'Arcy Thompson has shown that extremely different forms of organs or of whole organisms may be geometrically transformed into each other by a Cartesian transformation of the system of coordinates: Translated into phylogenetic language, this would mean that immense evolutionary effects could be brought about by changing the differential growth rates of the whole body or organ at an early point in development, with all the necessary secondary effects of such a change. I could imagine, and I have actually pointed out, that a single mutation involving the rate of one of the important reactions connected with growth, acting on the principle underlying Thompson's transformations, could start a perfectly new evolutionary line, leading at once far away from the original form and being able to be completed by orthogenetic development within the once blasted new avenue. Or another example: There are innumerable cases known where no intermediate forms between two extremely different ones are imaginable. Take, for example, the Pleuronectid fishes, the flounders and their kin, lying flat on one side, the eyes being translocated during embryonic development to the other side with all the following asymmetries of skull, fins, muscles. Cuenot expressed his conviction a long time ago that no slow accumulation of variations and selections is needed to explain the origin of such forms. There exist flat symmetrical fishes with the habit of resting lying flat on one side. Given the proper arrangement of the eye muscles and the interorbital septum of the skull, a single step was only necessary to start the migration of the eye, all the rest of the transformations being necessary consequences of the first step. I can not help agreeing with Cuenot and adding that at the proper moment in the evolutionary line a single mutation in regard to the rate of certain embryological processes of the type which ordinarily produce a monster, may have given birth to a monstrous new family with all its essential traits and preadapted to certain modes of living. Of course the further differentiation, the slow evolutionary working out of the details, would be brought about by new mutations of the different types, including as well other large steps, as accumulations of small mutations under the influence of selection.

A third example, which I have repeatedly used to explain the general idea, appears still more fantastic. Let us consider one of the famous lines of transformation which the comparative anatomy of vertebrates has brought to light, for example, the series of transformations of the visceral arches. I believe that these facts constitute one of the most beautiful proofs of evolution; and in addition I believe that their analysis by the methods of comparative anatomy is one of the greatest achievements of biological thinking, though some biologists of today are inclined to prefer the most meaningless experiment to such a piece of masterful morphological analysis. In the case of the visceral skeleton we see, for example, that the so-called hyomandibular bone of fishes loses its function as connective element between jaws and skull, and is transformed into an auditory ossicle situated within the skull and playing an important role in the transmission of sound, a transformation which takes place simultaneously with the appearance of the tympanical membrane as adaptation to terrestrial life. In this transformation two major steps are observed: First, the formation of a new connection between skull and jaw, thus excluding the hyomandibular bone from its former function; second, the appearance of the tympanical membrane in this region and the inclusion of the hyomandibular bone into the ear cavity, with the change of its function to that of an auditory ossicle. The first step is found in the Crossopterygian fishes, the second in Amphibia. In both cases a slow transformation by accumulation of advantageous mutations is hardly imaginable. There are no steps possible between a tympanical membrane and none and also no steps between two types of articulation of the jaw with the skull. But I could not find much difficulty in the idea that the decisive step was taken by a single mutation affecting the relative rate of differentiation of the cranial end of the hyoid arch from which springs the hyomandibular bone, with the effect of forcing these parts, left behind in development, into new surroundings and connections, where future developments could make use of them for quite different purposes. It would certainly be of no use, and sheer speculation, to try to work out such an idea in detail. But I think that we can get hardly around the principle underlying it. Of course, there is no way visible to attack such a problem by the methods of genetical research. But I am not so sure that this means that it can not be attacked at all.

At the beginning of this lecture I said that my mind, like that of many geneticists, is oscillating between skepticism and optimism with regard to the views on the means of evolution as derived from genetical work. I have now presented to you examples of both states of mind: First, a bit of skepticism with regard to the role which the formation of geographic races or subspecies may have played in evolution; and then a bit of optimism in trying to show that the physiological system underlying orderly development, on the basis of the genetic constitution, allows some of the larger steps in evolution to be understood as sudden changes by single mutations concerning the rate of certain embryological processes. But whoever tries to formulate views on the means of evolution on the basis of the actual knowledge of facts must be aware that any day new facts might come to light which could force our ideas into quite different channels. Therefore I wish to return at the end of this lecture again to the results of actual experimentation and to draw your attention to some new lines of experiment which perhaps will finally influence our general conceptions considerably.

A number of years ago I found, as already mentioned, that it is possible to produce gene mutations by the action of extreme temperatures of almost lethal dose. Unfortunately, there is still an unknown element in the technique of these experiments which makes success dependent upon some conditions which have not been isolated as yet. Progress in this line of, research is therefore slow. One of the most startling results of this work was that in a series of experiments a few mutations were always produced again. Jollos, who continued this work, had similar results, but in his experiments other mutations were preponderant and also appeared over again. I then repeated the experiments and in successful cultures had now the same mutations which appeared also in Jollos' cultures.

Thus it seems that there is a relation between stimulus, maybe also material, and the type of genetic response. There was another interesting result. I have mentioned already that in such experiments quite a number of phenotypic changes are produced which resemble well-known mutations, but are of the nature of non-heritable modifications. In a few instances, cases were found where the treated animals themselves showed such a visible change, namely, dark body color, and where the offspring of the same animals showed the same phenotype as mutation. The explanation which had to be given to such a case of so-called parallel induction was that there was simply a chance overlapping of two independent phenomena, namely, the production of a modification and of a mutation of the same phenotype; this would be made possible by the aforementioned assumption that in both cases the same developmental process was changed either by environmental action or by genie action.

But there were still other strange facts. I had observed that the typical non-heritable changes which resembled heritable mutations in appearance, and which always were found in the flies which had been treated with heat during definite larval stages, were different, if the details of treatment were changed. For example, with one type of treatment, a certain peculiarity of the wing-shape was produced; with another type of treatment the majority of changed individuals presented a very different type of wing-form. In recent experiments, Jollos, who had had the same experience, could add some most interesting facts. In the lines with ordinary treatment the most frequent mutations were those of body color, called sooty, and of eye color, called eosin. If the usual treatment was replaced by one with dry heat, the non-heritable variations which appeared in the treated animals were of a different type than usual. Predominant were flies with extended wings, with curly wings, with asymmetrically shortened wings and with scalpelliform wings. Jobs continued treating the normal offspring of these lines with the same method, and during the following generations a number of mutations appeared, some repeatedly; and among these were the mutations, the phenotype of which is identical with the aforementioned non-heritable variations produced in the same line, namely, extended, curly, scalpelloid and asymmetrically shortened wings. Of course, this has nothing to do with an inheritance of acquired characters; the mutations had appeared among the offspring of normal individuals. There are now altogether seven cases in which a mutation has been produced in the same lines in which exactly the same phenotype occurs frequently as a non-heritable modification as a consequence of the same treatment. Among these seven cases, one of which was found by myself and the others by Jollos, is one mutation which before was observed only once in the whole Drosophila work and two which had never been observed.

These certainly are interesting facts, which might lead to strange consequences. I personally am willing to wait for further results before drawing conclusions. Jollos, who has not yet published the results which I quoted, permits me to mention that he is inclined to derive the following interpretation: The genes produce within the protoplasm active stuffs which are of the same constitution as the genes themselves. Both will react in the same way upon external conditions, but those within the protoplasm easier than those protected within the chromosomes. Such a view, of course, would lead to many interesting consequences. We shall, however, dismiss the subject with the mention of the actual facts, which one day may be of great importance not only for problems of special genetics but also for discussions on evolution.

The title of this lecture was: "Some Aspects of Evolution." But as I said at the beginning, it was not meant that the idea of evolution itself, which all biologists consider a historic fact, should be under discussion, but some of the ways and means by which nature makes the transformation of species possible. The three aspects which I chose for representation were, first, an aspect where I had to express skepticism in regard to well-established beliefs. I tried to show on the basis of large experimental evidence that the formation of subspecies or geographic races is not a step towards the formation of species but only a method to allow the spreading of a species to different environments by forming preadaptational mutations and combinations of such, which, however, always remain within the confines of the species. The second aspect which I discussed was one where I felt again optimistic. I tried to emphasize the importance of the methods of normal embryonic development for an understanding of possible evolutionary changes. I tried to show that a directed orthogenetic evolution is a necessary consequence of the embryonic system which allows only certain avenues for transformation. I further emphasized the importance of rare but extremely consequential mutations affecting rates of decisive embryonic processes which might give rise to what one might term hopeful monsters, monsters which would start a new evolutionary line if fitting into some empty environmental niche. Finally, I discussed a third aspect of the problem, this time under the slogan of watchful waiting, namely, new lines of genetic research concerning the problem of mutation and therefore also of evolution. With these discussions we touched certainly only a small fraction of the manifold problems of evolution. But if we would try to visualize all the contributions which the science of genetics has recently made in this direction, we might be entitled to say that our insight into one of the most complex biological problems is constantly increasing. Progress of science follows of course a slowly ascending, wavy curve, with always recurring valleys. But viewed from some distance, the waves disappear and only the upward trend remains visible. Such is also the case with our knowledge of the methods and means of evolution.

SOURCE: Goldschmidt, Richard. "Some Aspects of Evolution." *Science*, New Series, 78: 2033 (December 15, 1933), 539–547.

TERMS

Mendelism—A series of principles that explain the inheritance of traits that are inherited in their entirety from one generation to the next, rather than as a blended combination of the two parents' traits. First described by Gregor Mendel (1822–1884) in 1866 but not generally appreciated until 1900, they explain how certain traits can be passed from one generation to the next intact. Mendel's work was rediscovered at a time when many working biologists questioned Darwinism's ability to explain the source of the new variations that natural selection work upon, and combined with the concept of mutationism served as a competing explanation for evolutionary change.

Mutation—A sudden change in the characteristics of an organism that can be transmitted from parent to offspring. Its influence on the history of evolutionary thought was felt most through the work of Hugo DeVries (1848–1935), who posited that mutations that appeared from one generation to the next served as the source for new variations. Throughout the first third of the twentieth century, so-called Mendelian/mutationists argued against self-described Darwinists about precisely how evolution occurred and what role mutations, or as Darwinists called them, saltations or sports, played in the emergence of new variations.

Orthogenesis—A theory that the emergence of new variations and therefore the path of evolution follow a specific line or head in a specific direction rather than occurring arbitrarily in all directions. The theory supposes some sort of force, either internal to individual organisms or externally guiding the evolution of species of organisms.

5

THE RISE OF FUNDAMENTALISM AND ITS OPPOSITION TO EVOLUTION

In the midst of increasingly contentious debates among American biologists over the proper explanation for evolutionary change, their public statements on the subject encouraged increasing resistance to evolution as a subject in public school science classrooms. Combined with growing animosity toward higher criticism and the effects of World War I, the biologists' uncertainty helped generate an increasingly influential anti-evolution movement in the United States.

American antievolutionism in the 1920s had two principle sources: Protestant fundamentalism and increasing anxieties about the effects of modernism and industrialism. Fundamentalism emerged shortly after the turn of the century and as a reaction to higher criticism, which viewed the Bible as a piece of literature as an historical document, rather than as a piece of scripture.

Fundamentalism and concerns about the increasingly violent nature of Western culture combined in the works of authors like William Jennings Bryan, the three-time Democratic candidate for President and former secretary of state. Funded by the World's Christian Fundamentals Association, Bryan traveled to Dayton, Tennessee, in the summer of 1925, to aid the prosecution in the famous Scopes Trial. Bryan's opposition to the teaching of evolution in public schools began shortly after World War I and synthesized fundamentalist motivations to return to the Bible's teaching and anti-modernist anxieties about the impact of science, technology, and the industrial worldview.

WORLD'S CHRISTIAN FUNDAMENTALS ASSOCIATION DOCTRINAL STATEMENT (1919)

The World's Christian Fundamentals Association [WCFA] was a fundamentalist organization based at Northwestern Bible College in Roseville, Minnesota. The college's president, William Bell Riley, founded the association in 1919 with the goal of promoting traditional faith in the face of modernist challenges to religion. Throughout the 1920s Riley and the WCFA participated in public debates over evolution. In 1925, when the town of Dayton,

Tennessee, decided to accept the American Civil Liberties Union's challenge to prosecute a teacher under the Butler Act, the WCFA paid William Jennings Bryan's expenses to aid the prosecution.

I. We believe in the Scriptures of the Old and New Testaments as verbally inspired of God, and inerrant in the original writings, and that they are of supreme and final authority in faith and life.

II. We believe in one God, eternally existing in three persons, Father, Son and Holy Spirit.

III. We believe that Jesus Christ was begotten by the Holy Spirit, and born of the Virgin Mary and is true God and true man.

IV. We believe that man was created in the image of God, that he sinned and thereby incurred not only physical death, but also that spiritual death which is separation from God; and that all human beings are born with a sinful nature, and, in the case of those who reach moral responsibility, become sinners in thought, word and deed.

V. We believe that the Lord Jesus Christ died for our sins according to the Scriptures as a representative and substitutionary sacrifice; and that all that believe in Him are justified on the ground of His shed blood.

VI. We believe in the resurrection of the crucified body of our Lord, in His ascension into heaven, and in His present life there for us, as High Priest and Advocate.

VII. We believe in "that blessed hope," the personal, premillennial and imminent return of our Lord and Saviour Jesus Christ.

VIII. We believe that all who receive by faith the Lord Jesus Christ are born again of the Holy Spirit and thereby become children of God.

IX. We believe in the bodily resurrection of the just and the unjust, the everlasting blessedness of the saved, and the everlasting, conscious punishment of the lost.

SOURCE: World's Christian Fundamentals Association Doctrinal Statement, 1919.

VERNON KELLOGG, HEADQUARTERS NIGHTS: A RECORD OF CONVERSATIONS AND EXPERIENCES AT THE HEADQUARTERS OF THE GERMAN ARMY IN FRANCE AND BELGIUM (1917)

Vernon Kellogg (1867–1937), the Stanford biologist who wrote *Darwinism To-Day*, resigned his tenured position in 1915 to join his former student, Herbert Hoover, distributing food and clothing to civilians trapped in German-occupied Belgium and Northern France. Through 1916 he led the relief efforts, but when the United States joined the war against Germany in 1917, he was deported and found his way to Washington, DC. An ardent pacifist, Kellogg's humanitarian work demonstrated his belief that cooperation, rather than violence, was the true path to progress. However, once he returned to the States and his nation joined the war effort, he threw his support behind American involvement. His 1917 *Headquarters Nights* was originally published in serial form in the *Atlantic Monthly* and served as a vital propaganda piece supporting President Wilson's decision to go to war. In it, he used his authority as a biologist and as a nationally known advocate of Darwinian evolutionary theory to argue that the U.S. was correct in fighting the Germans because they were inappropriately using

Darwin's work to justify aggressive militarism. While Kellogg had intended his work to merely support American involvement in the war, its impact was felt on the growing antievolution movement. Opponents of the teaching of evolution in public schools, most notably William Jennings Bryan, saw in Kellogg's *Headquarters Nights* the potential of Darwin's work to justify a brutal worldview.

We do not hear much now from the German intellectuals. Some of the professors are writing for the German newspapers, but most of them are keeping silent in public. The famous Ninety-three are not issuing any more proclamations. When your armies are moving swiftly and gloriously forward under the banners of sweetness and light, to carry the proper civilization to an improperly educated and an improperly thinking world, it is easier to make declarations of what is going to happen, and why it is, than when your armies are struggling for life with their backs to the wall—of a French village they have shot and burned to ruin for a reason that does not seem so good a reason now.

But some of the intellectuals still speak in the old strain in private. It has been my peculiar privilege to talk through long evening hours with a few of these men at Headquarters. Not exactly the place, one would think, for meeting these men, but let us say this for them: some of them fight as well as talk. And they fight, not simply because they are forced to, but because, curiously enough, they believe much of their talk. This is one of the dangers from the Germans to which the world is exposed: they really believe much of what they say.

A word of explanation about the Headquarters, and how I happened to be there. It was—it is not longer, and that is why I can speak more freely about it—not only Headquarters but the Great Headquarters—*Grosses Hauptquartier*—of all the German Armies of the West. Here were big Von Schoeler, *General-Intendant*, and the scholarly-looking Von Freytag, *General-Quartiermeister*, with his unscholarly-looking, burly chief of staff, Von Zoellner. Here also were Von Falkenhayn, the Kaiser's Chief of Staff, and sometimes even the All-Highest himself, who never missed the Sunday morning service in the long low corrugated-iron shed which looked all too little like a royal chapel ever to interest a flitting French Bomber.

But not only was this small gray town on the Meuse, just where the water pours out of its beautiful canon course through the Ardennes, the headquarters of the German General Staff—it was also the station, by arrangement with the staff, of the American Relief Commission's humble ununiformed chief representative for the North of France (occupied French territory). For several months I held this position, living with the German officer detached from the General Quartermaster's staff to protect me—and watch me. Later, too, as director of the Commission at Brussels, I had frequent occasion to visit Headquarters for conferences with officers of the General Staff. It was thus that I had opportunity for these Headquarters Nights.

Among the officers and officials of Headquarters there were many strong and keen German militaristic brains—that goes without saying—but there were also a few of the professed intellectuals—men who had exchanged, for the moment, the academic robes of the *Aula* for the field-gray uniforms of the army. The second commandant of the Headquarters town was a professor of jurisprudence at the University of Marburg; and an infantry captain who lived in the house with my guardian officer and men, is the professor of zoology in one of the larger German universities, and one of the most brilliant of present-day biologists. I do not wish to indicate his person more particularly, for I shall say some hard things about him—or about him as representative of many—and we are friends. Indeed, he was *Privat docent* in charge of the laboratory in which I worked years ago at the University of Leipzig, and we have been correspondents and friends ever since. How he came to be at Headquarters, and at precisely the same time that I was there, is a story which has its interest, but cannot be told at present.

Our house was rather a favored centre, for 'my officer,' Graf W—he always called me 'my American,' but he could no more get away from me than I from him—is a generous entertainer, and our dinners were rarely without guests from other headquarters houses. Officers, from veteran generals down to pink-cheeked lieutenants, came to us and asked us to them. The discussions, begun at dinner, lasted long into the night. They sat late, these German officers, over their abundant wine—French vintages conveniently arranged for. And always we talked and tried to understand one another; to get the other man's point of view, his *Weltanschauung.*

Well, I say it dispassionately but with conviction: if I understand theirs, it is a point of view that will never allow any land or people controlled by it to exist peacefully by the side of a people governed by our point of view. For their point of view does not permit of a live-and-let-live kind of carrying on. It is a point of view that justifies itself by a whole-hearted acceptance of the worst of Neo-Darwinism, the *Allmacht* of natural selection applied rigorously to human life and society and *Kultur*.

Professor von Flussen—that is not his name—is a biologist. So am I. So we talked out the biological argument for war, and especially for this war. The captain-professor has a logically constructed argument why, for the goof of the world, there should be this war, and why, for the good of the world, the Germans should win it, win it completely and terribly. Perhaps I can state his argument clearly enough, so that others may see and accept his reasons, too. Unfortunately for the peace of our evenings, I was never convinced. That is, never convinced that for the good of the world the Germans should win this war, completely and terribly. I was convinced, however, that this war, once begun, will determine whether or not Germany's point of view is to rule the world. And this conviction, thus gained, meant the conversion of a pacifist to an ardent supporter, not of War, but of *this* war; of fighting this war to a definitive end—that end to be Germany's conversion to be a good Germany, or not much of any Germany at all. My 'Headquarters Nights' are the confessions of a converted pacifist.

In talking it out biologically, we agreed that the human race is subject to the influence of the fundamental biologic laws of variation, heredity, selection, and so forth, just as are all other animal—and plant—kinds. The factors of organic evolution, generally, are factors in human natural evolution. Man has risen from his primitive bestial stage of glacial time, a hundred or several hundred thousand years ago, when he was animal among animals, to the stage of to-day, always under the influence of these great evolutionary factors, and partly by virtue of them.

But he does not owe all of his progress to these factors, or, least of all, to any one of them, as natural selection, a thesis Professor von Flussen seemed ready to maintain.

Natural selection depends for its working on a rigorous and ruthless struggle for existence. Yet this struggle has its ameliorations, even as regards the lower animals, let alone man.

There are three general phases of this struggle:

1. An inter-specific struggle, or the lethal competition among different animal kinds for food, space, and opportunity to increase;

2. An intra-specific struggle, or lethal competition among the individuals of a single species, resultant on the over-production due to natural multiplication by geometric progression; and,

3. The constant struggle of individuals and species against the rigors of climate, the danger of storm, flood, drought, cold, and heat.

Now any animal kind and its individuals may be continually exposed to all of these phases of the struggle for existence, or, on the other hand, any one or more of these phases may be largely ameliorated or even abolished for a given species and its individuals. This amelioration may come about through a happy accident of time or place, or because the adoption by the species of a habit or mode of life that continually protects it from a certain phase of the struggle.

For example, the voluntary or involuntary migration of representatives of a species hard pressed to exist in its native habitat, may release it from the too severe rigors of a destructive climate, or take it beyond the habitat of its most dangerous enemies, or give it the needed space and food for the support of a numerous progeny. Thus, such a single phenomenon as migration might ameliorate any one or more of the several phases of the struggle for existence.

Again, the adoption by two widely distinct and perhaps antagonistic species of a commensal or symbiotic life, based on the mutual-aid principle—thousands of such cases are familiar to naturalists—would ameliorate or abolish the interspecific struggle between these two species. Even more effective in the modification of the influence due to a bitter struggle for existence, is the adoption by a species of an altruistic or communistic mode of existence so far as its own individuals are concerned. This, of course, would largely ameliorate for that species the intra-specific phase of its struggle for life. Such animal altruism, and the biological success of the species exhibiting it, is familiarly exemplified by the social insects (ants, bees, and wasps).

As a matter of fact, this reliance by animals kinds for success in the world upon a more or less extreme adoption of the mutual-aid principle, is much more widely spread among the lower animals than familiarly recognized, while in the case of man, it has been the greatest single factor in the achievement of his proud biological position as king of living creatures.

Altruism—or mutual aid, as the biologists prefer to call it, to escape the implication of assuming too much consciousness in it—is just as truly a fundamental biologic factor or evolution as is the cruel, strictly self-regarding, exterminating kid of struggle for existence with which the Neo-Darwinists try to fill our eyes and ears, to the exclusion of the recognition of all other factors.

Professor von Flussen is Neo-Darwinian, as are most German biologist and natural philosophers. The creed of the *Allmacht* of a natural selection based on violent and fatal competitive struggle is the gospel of the German intellectuals; all else is illusion and anathema. The mutual-aid principle is recognized only as restricted to its application within limited groups. For instance, it may and does exist, and to positive biological benefit, within single ant communities, but the different ant kinds fight desperately with each other, the stronger destroying or enslaving the weaker. Similarly, it may exist to advantage within the limits of organized human groups—as those which are ethnographically, nationally, or otherwise variously delimited. But as with the different ant species, struggle—bitter, ruthless struggle—is the rule among different human groups.

This struggle not only must go on, for that is the natural law, but it should go on, so that this natural law may work out in its cruel, inevitable way the salvation of the human species. By its salvation is meant its desirable natural evolution. That human group which is in the most advanced evolutionary state as regards internal organization and form of social relationship is best, and should, for the sake of the species, be preserved at the expense of the less advanced, the less effective. It should win in the struggle for existence, and this struggle should occur precisely that the various types may be tested, and the best not only preserved, but put in position to impose its kind of social organization—its *Kultur*—on the others, or, alternatively, to destroy and replace them.

This is the disheartening kind of argument that I faced at Headquarters; argument logically constructed on premises chose by the other fellow. Add to these assumed premises of the *Allmacht* of struggle and selection based on it, and the contemplation of mankind as congeries of different, mutually irreconcilable kinds, like the different ant species, the additional assumption that the Germans are the chose race, and German social and political organization the chosen type of human community life, and you have a wall of logic and conviction that you can break your head against but can never shatter—by headwork. You must long for the muscles of Samson.

SOURCE: Kellogg, Vernon. *Headquarters Nights: A Record of Conversations and Experiences at the Headquarters of the German Army in France and Belgium.* Boston: The Atlantic Monthly Press, 1917.

BUTLER ACT, PUBLIC ACTS OF THE STATE OF TENNESSEE (1925)

In 1925, after reading Darwin's *On the Origin of Species* and *Descent of Man* as well as William Jennings Bryan's essay "Is the Bible True?" and hearing the testimonial of a young woman whose faith in God was destroyed by evolution, John Washington Butler, a Tennessee farmer and occasional school teacher, presented a bill to his state's legislature that forbid the teaching that humans descended from lower order animals. The bill moved quickly through the Committee on Education and passed in the House before joining a similar bill in the Senate. It was begrudgingly signed into law by Governor Austin Peay, who believed that it would not affect the way science was taught and represented little more than a symbolic act. Shortly thereafter the American Civil Liberties Union offered to pay the legal expenses of any Tennessee teacher charged with violating the law, and officials in Dayton seized on the opportunity to put their small but growing town on the map. The Scopes "Monkey" Trial, as quickly came to be known, was based on Butler's law, and John T. Scopes was found guilty of violating the state's ban against teaching the evolutionary origins of humans.

An act prohibiting the teaching of the Evolution theory in all the Universities, Normals and all other public schools of Tennessee, which are supported in whole or in part by the public school funds of the State, and to provide penalties for the violations thereof.

Section 1. Be it enacted by the General Assembly of the State of Tennessee, that it shall be unlawful for any teacher in any the Universities, Normals and all other public schools of the State which are supported in whole or in part by the public school funds of the State, to teach any theory that denies the story of the Divine Creation of man as taught in the Bible, and to teach instead that man has descended from a lower order of animals.

Section 2. Be it further enacted, that any teacher found guilty of the violation of this Act, shall be guilty of a misdemeanor and upon conviction, shall be fined not less than One Hundred Dollars nor more than Five Hundred Dollars for each offense.

Section 3. Be it further enacted, that this Act take effect from and after its passage, the public welfare requiring it.

Passed March 13, 1925, W. F. Barry, Speaker of the House of Representatives, L. D. Hill, Speaker of the Senate, Approved March 21, 1925, Austin Peay, Governor.

SOURCE: Butler Act, Public Acts of the State of Tennessee Passed by the Sixty-Fourth General Assembly, 1925. Chapter no. 27, House Bill No. 185 by Mr. Butler.

JOHN THOMAS SCOPES v. THE STATE, SUPREME COURT OF TENNESSEE (1926)

After being convicted and fined $100 for violating Tennessee's ban against teaching the evolutionary origins of humans in the state's public schools, John T. Scopes appealed his case to the Tennessee Supreme Court. His conviction was overturned on a technicality: the judge had erred in imposing a $100 fine. State law required that fines greater than $50 be levied only by a jury.

Clarence Darrow seated with Judge John F. Raulston [Library of Congress Prints and Photographs Department].

Nashville, December Term, 1926

Opinion filed January 17, 1927

Chief Justice Green delivered majority opinion; Judge Chambliss concurring opinion, and Justice Cook concurred; Judge Colin P. McKinney opinion dissenting, and Judge Swiggart did not participate.

Scopes was convicted of a violation of chapter 27 of the Acts of 1925, for that he did teach in the public schools of Rhea county a certain theory that denied the story of the divine creation of man, as taught in the Bible, and did teach instead thereof that man had descended from a lower order of animals. After a verdict of guilty by the jury, the trial judge imposed a fine of $100, and Scopes brought the case to this court by an appeal in the nature of a writ of error.

The bill of exceptions was not filed within the time fixed by the court below, and, upon motion of the state, at the last term, this bill of exceptions was stricken from the record. Scopes v. State, 152 Tenn. 424.

A motion to quash the indictment was seasonably made in the trial court raising several questions as to the sufficiency thereof and as to the validity and construction of the Statute upon which the indictment rested. These questions appear on the record before us and have been presented and debated in this court with great elaboration. . . .

While the Act was not drafted with as much care as could have been drafted, nevertheless there seems to be no great difficulty in determining its meaning. It is entitled "An Act prohibiting the teaching of the evolution theory in all the Universities, Normals and all other public schools in Tennessee, which are supported in whole or in part by the public school funds of the state, and to provide penalties for the violations thereof."

Evolution, like prohibition, is a broad term. In recent bickering, however, evolution has been understood to mean the theory which holds that man has developed from some pre-existing lower type. This is the popular significance of evolution, just as the popular significance of prohibition is prohibition of the traffic in intoxicating liquors. It was in that sense that evolution was used in this Act. It is that sense that the word will be used in this opinion, unless the context otherwise indicates. It is only to the theory of the evolution of man from a lower type that the Act before us was intended to apply, and much of the discussion we have heard is beside this case. The words of a Statute, if in common use, are to be taken in their natural and ordinary sense. . . .

Thus defining evolution, this Act's title clearly indicates the purpose of the Statute to be the prohibition of teaching in the Schools of the State that man has developed or descended from some lower type or order of animals.

When the draftsman came to express this purpose in the body of the Act, he first forbade the teaching of "any theory that denies the story of the divine creation of man, as taught in the Bible"—his conception evidently being that to forbid the denial of the Bible story would ban the teaching of evolution. To make the purpose more explicit, he added that it should be unlawful to teach "that man had descended from a lower order of animals."

Supplying the ellipsis in section 1 of the act, it reads that it shall be unlawful for any teacher, etc.—"to teach any theory that denies the story of the divine creation of man as taught in the Bible, and to teach instead [of the story on the divine creation of man as taught in the Bible] that man has descended from a lower order of animals."

The language just quoted illustrates what is called in rhetoric exposition by iteration. The different form of the iterated idea serves to expound the first expression of the thought. The undertaking of the Statute was to prevent teaching of the evolution theory. It was considered this purpose could be effected by forbidding the teaching of any theory that denied the Bible story, but to make the purpose clear it was also forbidden to teach that man descended from a lower order of animals.

This manner of expression in written instruments is common, and gives use to the maxim of construction noscitur a sociis. Under this maxim subordinate words and phrases are modified and limited to harmonize with each other and with the leading and controlling purpose or intention of the act. . . .

It thus seems plain that the Legislature in this enactment only intended to forbid teaching that men descended from a lower order of animals. The denunciation of any theory denying the Bible story of creation is restricted by the caption and by the final clause of section 1.

So interpreted, the Statute does not seem to be uncertain in its meaning nor incapable of enforcement for such a reason, notwithstanding the argument to the contrary. The indictment herein follows the language of the Statute. The statute being sufficiently definite in its terms, such an indictment is good. . . . The assignments of error, which challenge the sufficiency of the indictment and the uncertainty of the Act, are accordingly overruled.

It is contended that the Statute violates section 8 of Article 1 of the Tennessee Constitution, and section 1 of the Fourteenth Amendment of the Constitution of the United States—the Law of the Land clause of the state Constitution, and the Due Process of Law clause of the Federal Constitution, which are practically equivalent in meaning.

We think there is little merit in this contention. The plaintiff in error was a teacher in the public schools of Rhea County. He was an employee of the State of Tennessee or of a municipal agency of the State. He was under contract with the State to work in an institution of the State. He had no right or privilege to serve the State except upon such terms as the State prescribed. His liberty, his privilege, his immunity to teach and proclaim the theory of evolution, elsewhere than in the service of the State, was in no wise touched by this law.

The Statute before us is not an exercise of the police power of the State undertaking to regulate the conduct and contracts of individuals in their dealings with each other. On the other hand, it is an Act of the State as a corporation, a proprietor, an employer. It is a declaration of a master as to the character of work the master's servant shall, or rather shall not, perform. In dealing with its own employees engaged upon its own work, the State is not hampered by the limitations of section 8 of Article 1 of the Tennessee Constitution, nor of the Fourteenth Amendment to the Constitution of the United States.

In People v. Crane, 214 N.Y. 154, the validity of a Statute of that State, providing that citizens only should be employed upon public works was sustained. In the course of opinion (page 175), it was said, "The Statute is nothing more, in effect, than a resolve by an employer as to the character of his employees. An individual employer would communicate the resolve to his subordinate by written instructions or by word of mouth. The State, an incorporeal master, speaking through the Legislature, communicates the resolve to its agents by enacting a statute. Either the private employer or the State can revoke the resolve at will. Entire liberty of action in these respects is essential unless the State is to be deprived of a right which has heretofore been deemed a constituent element of the relationship of master and servant, namely, the right of the master to say who his servants shall (and therefore shall not) be."

A case involving the same Statute reached the Supreme Court of the United States, and the integrity of the Statute was sustained by that tribunal. Heim v. McCall, 239 U.S. 175, 60 L.Ed. 207. The Supreme Court referred to People v. Crane, supra, and approvingly quoted a portion of the language of BARRETT, Chief judge, that we have set out above.

At the same term of the Supreme Court of the United States an Arizona Statute, prohibiting individuals and corporations with more than five workers from employing less than 80 percent thereof of qualified electors or native-born citizens of the United States was held invalid. Truax v. Raich, 239 U.S. 33, 60 L.Ed. 131.

These two cases from the Supreme Court make plain the differing tests to be applied to a Statute regulating the State's own affairs and a statute regulating the affairs of private individuals and corporations.

A leading case is *Atkin v. Kansas*, 191 U.S. 207, 48 L.Ed. 148. The court there considered and upheld a Kansas Statute making it a criminal offense for a contractor for a public work to permit or require an employee to perform labor upon that work in excess of eight hours each day. In that case it was laid down:

> …For, whatever may have been the motives controlling the enactment of the statute in question, we can imagine no possible ground to dispute the power of the State to declare that no one undertaking work for it or for one of its municipal agencies, should permit or require an employee on such work to labor in excess of eight hours each day, and to inflict punishment upon those who are embraced by such regulations and yet disregard them.
>
> It cannot be deemed a part of the liberty of any contractor that he be allowed to do public work in any mode he may choose to adopt, without regard to the wishes of the State. On the contrary, it belongs to the State, as the guardian and trustee for its people, and having control of its affairs, to prescribe the conditions upon which it will permit public work to be done on its behalf, or on behalf of its municipalities. No court has authority to review its action in that respect. Regulations on this subject suggest only considerations of public policy. And with such considerations the courts have no concern.

In Ellis v. United States, 206 U.S. 246, 51 L.Ed. 1047, Atkins v. Kansas was followed, and an Act of Congress sustained which prohibited, under penalty of fine or imprisonment, except in case of extraordinary emergency, the requiring or permitting laborers or mechanics employed upon any of the public works of the United States or of the District of Columbia to work more than eight hours each day.

These cases make it obvious that the State or Government, as an incident to its power to authorize and enforce contracts for public services, "may require that they shall be carried out only in a way consistent with its views of public policy, and may punish a departure from that way." Ellis v. United States, supra.

To the same effect is Waugh v. Board of Trustees, 237 U.S. 589, 59 L.Ed. 1131, in which a Mississippi Statute was sanctioned that prohibited the existence of Greek letter fraternities and similar societies in the State's educational institutions, and deprived members of such societies of the right to receive or compete for diploma, class honors, etc.

This court has indicated a like view in Leeper v. State, 103 Tenn. 500, in which the constitutionality of chapter 205 of the Acts of 1899, known as the "Uniform Text Book Law", was sustained. In the opinion in that case judge WILKES observed "If the authority to regulate and control schools is legislative, then it [is] must have an unrestricted right to prescribe methods, and the courts cannot interfere with it unless some scheme is devised which is contrary to other provisions of the Constitution. . . ."

In Marshall &Bruce Co. v. City of Nashville, 109 Tenn., 495, the charter of the City of Nashville required that all contracts for goods and supplies furnished the city, amounting to over $ 50, must be let out at competitive bidding to the lowest responsible bidder. In the face of such a charter provision, an ordinance of the city, which provided that all city printing should bear the union label, was held unauthorized—necessarily so. The lowest bidder, provided he was responsible, was entitled to such a contract, whether he employed union labor, and was empowered to affix the union label to his work or not. Other things said in that case were not necessary to the decision.

Traux v. Raich, supra, Meyer v. Nebraska, 262 U.S. 390, Pierce v. Society of Sisters of the Holy Names of Jesus and Mary, 268 U.S. 510, and other decisions of the Supreme Court of the United States, pressed upon us by counsel for plaintiff in error, deal with Statutes affecting individuals, corporations, and private institutions, and we do not regard these cases as in point.

Since the State may prescribe the character and the hours of labor of the employees on its works, just as freely may it say what kind of work shall be performed in its service, what shall be taught in its schools, so far at least as section 8 of Article 1 of the Tennessee Constitution, and the Fourteenth Amendment to the Constitution of the United States, are concerned.

But it is urged that chapter 27 of the Acts of 1925 conflicts with section 12 of Article 11, the Educational clause, and section 3 of Article 1, the Religious Preference clause, of the Tennessee Constitution. It is to be doubted if the plaintiff in error, before us only as the state's employee, is sufficiently protected by these constitutional provisions to justify him in raising such questions. Nevertheless, as the State appears to concede that these objections are properly here made, the court will consider them.

The relevant portion of section 12 of Article 11 of the Constitution is in these words:

> ...It shall be the duty of the General Assembly in all future periods of this government, to cherish Literature and Science.

The argument is that the theory of the descent of man from a lower order of animals is now established by the preponderance of scientific thought and that the prohibition of the teaching of such theory is a violation of the legislative duty to cherish Science.

While this clause of the Constitution has been mentioned in several of our cases, these references have been casual, and no Act of the Legislature has ever been held inoperative by reason of such provision. In one of the opinions in Green v. Allen, 24 Tenn. (5 Humph.) 170, the provision was said to be directory. Although this court is loath to say that any language of the Constitution is merely directory State v. Burrow, 119 Tenn., 376, Webb v. Carter, 129 Tenn., 182, we are driven to the conclusion that this particular admonition must be so treated. It is too vague to be enforced by any court. To cherish Science means to nourish, to encourage, to foster Science.

In no case can the court directly compel the Legislature to perform its duty. In a plain case the court can prevent the Legislature from transgressing its duty under the Constitution by declaring ineffective such a legislative Act. The case, however, must be plain, and the legislative Act is always given the benefit of any doubt.

If a bequest were made to a private trustee with the avails of which he should cherish Science, and there was nothing more, such a bequest would be void for uncertainty. Green v. Allen, 24 Tenn. (5 Humph.) 170, Ewell v. Sneed, 136 Tenn., 602, and the cases cited. It could not be enforced as a charitable use in the absence of prerogative power in this respect which the courts of Tennessee do not possess. A bequest in such terms would be so indefinite that our courts could not direct a proper application of the trust fund nor prevent its misapplication. The object of such a trust could not be ascertained.

If the courts of Tennessee are without power to direct the administration of such a trust by an individual, how can they supervise the administration of such a trust by the Legislature? It is a matter of far more delicacy to undertake the restriction of a coordinate branch of government to the terms of a trust imposed by the Constitution than to confine an individual trustee to the terms of the instrument under which he functions. If language be so indefinite as to preclude judicial restraint of an individual, such language could not possible excuse judicial restraint of the General Assembly.

If the Legislature thinks that, by reason of popular prejudice, the cause of education and the study of Science generally will be promoted by forbidding the teaching of evolution in the schools of the State, we can conceive of no ground to justify the court's interference. The courts cannot sit in judgment on such Acts of the legislature or its agents and determine whether or not the omission or addition of a particular course of study tends "to cherish Science."

The last serious criticism made of the Act is that it contravenes the provision of section 3 of Article 1 of the Constitution, "that no preference shall ever be given, by law, to any religious establishment or mode of worship."

The language quoted is a part of our Bill of Rights, was contained in our first Constitution of the state adopted in 1796, and has been brought down into the present Constitution.

At the time of the adoption of our first Constitution, this government had recently been established and the recollection of previous conditions was fresh. England and Scotland maintained State churches as did some of the Colonies, and it was intended by this clause of the Constitution to prevent any such undertaking in Tennessee.

We are not able to see how the prohibition of teaching the theory that man has descended from a lower order of animals gives preference to any religious establishment or mode of worship. So far as we know, there is no religious establishment or organized body that has in its creed or confession of faith any article denying or affirming such a theory. So far as we know, the denial or affirmation of such a theory does not enter into any recognized mode of worship. Since this cause has been pending in this court, we have been favored, in addition to briefs of counsel and various amid curiae, with a multitude of resolutions, addresses, and communications from scientific bodies, religious factions, and individuals giving us the benefit of their views upon the theory of evolution. Examination of these contributions indicates that Protestants, Catholics, and Jews are divided among themselves in their beliefs, and that there is no unanimity among the members of any religious establishment as to this subject. Belief or unbelief in the theory of evolution is no more a characteristic of any religious establishment or mode of worship than is belief or unbelief in the wisdom of the prohibition laws. It would appear that members of the same churches quite generally disagree as to these things.

Furthermore, chapter 277 of the Acts of 1925 requires the teaching of nothing. It only forbids the teaching of evolution of man from a lower order of animals. Chapter 102 of the Acts of 1915 requires that ten verses from the Bible be read each day at the opening of every public school, without comment, and provided the teacher does not read the same verses more than twice during any session. It is also provided in this Act that pupils may be excused from the Bible readings upon the written request of their parents.

As the law thus stands, while the theory of evolution of man may not be taught in the schools of the State, nothing contrary to that theory is required to be taught it could scarcely be said that the statutory scriptural reading just mentioned would amount to teaching of a contrary theory.

Our school authorities are therefore quite free to determine how they shall act in this state of the law. Those in charge of the educational affairs of the State are men and women of discernment and culture. If they believe that the teaching of the Science of Biology had been so hampered by chapter 27 of the Acts of 1925 as to render such an effort no longer desirable, this course of study may be entirely omitted from the curriculum of our schools. If this be regarded as a misfortune, it must be charged to the Legislature. It should be repeated that the act of 1925 deals with nothing but the evolution of man from a lower order of animals.

It is not necessary now to determine the exact scope of the Religious Preference clause of the Constitution and other language of that section. The situation does not call for such an attempt. Section 3 of Article 1 is binding alike on the Legislature and the school authorities. So far we are clear that the Legislature has not crossed these constitutional limitations. If hereafter the school authorities should go beyond such limits, a case can then be brought to the courts.

Much has been said in argument about the motives of the Legislature in passing this Act. But the validity of a statute must be determined by its natural and legal effect, rather than proclaimed motives. . . .

Some other questions are made, but in our opinion they do not merit discussion, and the assignments of error raising such questions are overruled.

This record disclosed that the jury found the defendant below guilty, but did not assess the fine. The trial judge himself undertook to impose the minimum fine of $100 authorized by the Statute. This was error. Under section 14 of Article 6 of the Constitution of Tennessee, a fine in excess of $50 must be assessed by a jury. The Statute before us does not permit the imposition of a smaller fine than $100.

Since a jury alone can impose the penalty this Act requires, and as a matter of course no different penalty can be inflicted, the trial judge exceeded his jurisdiction in levying this fine, and we are without power to correct his error. The judgment must accordingly be reversed. UP Church v. State, 153 Tenn., 198.

The Court is informed that the plaintiff in error is no longer in the service of the State. We see nothing to be gained by prolonging the life of this bizarre case. On the contrary, we think the peace and dignity of the State, which all criminal prosecutions are brought to redress, will be better conserved by the entry of a nolle prosequi herein. Such a course is suggested to the Attorney-General.

SOURCE: *John Thomas Scopes v. The State, Supreme Court of Tennessee, 1926.*

WILLIAM JENNINGS BRYAN, "SUMMARY ARGUMENT IN SCOPES TRIAL" (1925)

William Jennings Bryan (1860–1925) was hired by the World's Christian Fundamental Association to aid the prosecution at the Scopes Trial. Although not himself a literalist,

Bryan was a fundamentalist and believed that the Bible, though not literally true, should not be judged as a piece of literature nor should the stories in it be considered mere metaphors. For Bryan, belief in evolution was linked to what he viewed as an increasingly violent world, and his opposition to it stemmed from his sincere belief that to be a Christian, one must embrace pacifism and equality. Evolution, he believed, taught children than there was a natural hierarchy to all living things and it encouraged a "might makes right" worldview. The below address was to have been delivered by Bryan as a closing argument in the Scopes Trial, but the trial ended abruptly, and he was never given the chance to present it. Bryan died in his sleep only days after the trial.

William Jennings Bryan [Library of Congress Prints and Photographs Department, LC-USZ62-95709].

...Let us now separate the issues from the misrepresentations, intentional or unintentional, that have obscured both the letter and the purpose of the law. This is not an interference with freedom of conscience. A teacher can think as he pleases and worship God as he likes, or refuse to worship God at all. He can believe in the Bible or discard it; he can accept Christ or reject Him. This law places no obligations or restraints upon him. And so with freedom of speech; he can, so long as he acts as an individual, say anything he likes on any subject. This law does not violate any rights guaranteed by any constitution to any individual. It deals with the defendant, not as an individual, but as an employee, an official or public servant, paid by the State, and therefore under instructions from the State.

The right of the State to control the public schools is affirmed in the recent decision in the Oregon case, which declares that the State can direct what shall be taught and also forbid the teaching of anything "manifestly inimical to the public welfare." The above decision goes even farther and declares that the parent not only has the right to guard the religious welfare of the child, but is in duty bound to guard it. That decision fits this case exactly. The State had a right to pass this law, and the law represents the determination of the parents to guard the religious welfare of their children.

It need hardly be added that this law did not have its origin in bigotry. It is not trying to force any form of religion on anybody. The majority is not trying to establish a religion or to teach it—it is trying to protect itself from the effort of an insolent minority to force irreligion upon the children under the guise of teaching science. What right has a little irresponsible oligarchy of self-styled "intellectuals" to demand control of the schools of the United States, in which twenty-five millions of children are being educated at an annual expense of nearly two billions of dollars?

Christians must, in every State of the Union, build their own colleges in which to teach Christianity; it is only simple justice that atheists, agnostics and unbelievers should build their own colleges if they want to teach their own religious views or attack the religious views of others.

The statute is brief and free from ambiguity. It prohibits the teaching, in the public schools, of "any theory that denies the story of Divine creation as taught in the Bible," and teaches, "instead, that man descended from a lower order of animals." The first sentence sets forth the purpose of those who passed the law. They forbid the teaching of any evolutionary theory that disputes the Bible record of man's creation and, to make sure that there shall be no misunderstanding, they place their own interpretation on their language and specifically forbid the teaching of any theory that makes man a descendant of any lower form of life.

The evidence shows that defendant taught, in his own language as well as from a book outlining the theory, that man descended from lower forms of life. Howard Morgan's testimony gives us a definition of evolution that will become known throughout the world as this case is discussed. Howard, a fourteen-year-old boy, has translated the words of the teacher and the text-book into language that even a child can understand. As he recollects it, the defendant said, "A little germ of one cell organism was formed in the sea; this kept evolving until it got to be a pretty good-sized animal, then came on to be a land animal, and it kept evolving, and from this was man." There is no room for difference of opinion here, and there is no need of expert testimony. Here are the facts, corroborated by another student, Harry Shelton, and admitted to be true by counsel for defense. Mr. White, Superintendent of Schools, testified to the use of Hunter's Civic Biology, and to the fact that the defendant not only admitted teaching evolution, but declared that he could not teach it without violating the law. Mr. Robinson, the chairman of the School Board, corroborated the testimony of Superintendent White in regard to the defendant's admissions and declaration. These are the facts; they are sufficient and undisputed. A verdict of guilty must follow.

But the importance of this case requires more. The facts and arguments presented to you must not only convince you of the justice of conviction in this case but, while not necessary to a verdict of guilty, they should convince you of the righteousness of the purpose of the people of the State in the enactment of this law. The State must speak through you to the outside world and repel the aspersions cast by the counsel for the defense upon the intelligence and the enlightenment of the citizens of Tennessee. The people of this State have a high appreciation of the value of education. The State Constitution testifies to that in its demand that education shall be fostered and that science and literature shall be cherished. The continuing and increasing appropriations for public instruction furnish abundant proof that Tennessee places a just estimate upon the learning that is secured in its schools.

Religion is not hostile to learning; Christianity has been the greatest patron learning has ever had. But Christians know that "the fear of the Lord is the beginning of wisdom" now just as it has been in the past, and they therefore oppose the teaching of guesses that encourage godlessness among the students.

Neither does Tennessee undervalue the service rendered by science. The Christian men and women of Tennessee know how deeply mankind is indebted to science for benefits conferred by the discovery of the laws of nature and by the designing of machinery for the utilization of these laws. Give science a fact and it is not only invincible, but it is of incalculable service to man. If one is entitled to draw from society in proportion to the service that he renders to society, who is able to estimate the reward earned by those who have given to us the use of steam, the use of electricity, and enabled us to utilize the weight of water that flows down the mountainside? Who will estimate the value of the service rendered by those who invented the phonograph, the telephone, and the radio? Or, to come more closely to our home life, how shall we recompense those who gave us the sewing machine, the harvester, the threshing machine, the tractor, the automobile, and the method now employed in making artificial ice? The department for medicine also opens an unlimited field for invaluable service. Typhoid and yellow fever are not feared as they once were. Diphtheria and pneumonia have been robbed of some of their terrors, and a high place on the scroll of fame still awaits the discoverer of remedies for arthritis, cancer, tuberculosis and other dread diseases to which mankind is heir.

Christianity welcomes truth from whatever source it comes, and is not afraid that any real truth from any source can interfere with the divine truth that comes by inspiration from God Himself. It is not scientific truth to which Christians object, for true science is classified knowledge, and nothing therefore can be scientific unless it is true.

Evolution is not truth; it is merely an hypothesis—it is millions of guesses strung together. It had not been proven in the days of Darwin; he expressed astonishment that with two or three million species it had been impossible to trace any species to any other species. It had not been proven in the days of Huxley, and it has not been proven up to today. It is less than four years ago that Prof. Bateson came all the way from London to Canada to tell the American scientists that every effort to trace one species to another had failed—every one. He said he still had faith in evolution but had doubts about the origin of species. But of what value is evolution if it cannot explain the origin of species? While many scientists accept evolution as if it were a fact, they all admit, when questioned, that no explanation has been found as to how one species developed into another.

Darwin suggested two laws, sexual selection and natural selection. Sexual selection has been laughed out of the class room, and natural selection is being abandoned, and no new explanation is satisfactory even to scientists. Some of the more rash advocates of evolution are wont to say that evolution is as firmly established as the law of gravitation or the Copernican theory. The absurdity of such a claim is apparent when we remember that anyone can prove the law of gravitation by throwing a weight into the air, and that anyone can prove the roundness of the earth by going around it, while no one can prove evolution to be true in any way whatever. . . .

There is no more reason to believe that man descended from some inferior animal than there is to believe that a stately mansion has descended from a small cottage. Resemblances are not proof—they simply put us on inquiry. As one fact, such as the absence of the accused from the scene of the murder, outweighs all the resemblances that a thousand witnesses could swear to, so the inability of science to trace any one of the millions of species to another species, outweighs all the resemblances upon which evolutionists rely to establish man's blood relationship with the brutes.

But while the wisest scientists cannot prove a pushing power, such as evolution is supposed to be, there is a lifting power that any child can understand. The plant lifts the mineral up into a higher world, and the animal lifts the plant up into a world still higher. So, it has been reasoned by analogy, man rises, not by a power within him, but only when drawn upward by a higher power. There is a spiritual gravitation that draws all souls toward heaven, just as surely as there is a physical force that draws all matter on the surface of the earth towards the earth's center. Christ is our drawing power; He said, "I, if I be lifted up from the earth, will draw all men unto me," and His promise is being fulfilled daily all over the world.

It must be remembered that the law under consideration in this case does not prohibit the teaching of evolution up to the line that separates man from the lower forms of animal life. The law might well have gone farther than it does and prohibit the teaching of evolution in lower forms of life; the law is a very conservative statement of the people's opposition to an anti-biblical hypothesis. The defendant was not content to teach what the law permitted; he, for reasons of his own, persisted in teaching that which was forbidden for reasons entirely satisfactory to the law-makers.

Most of the people who believe in evolution do not know what evolution means. One of the science books taught in the Dayton High School has a chapter on "The Evolution of

Machinery." This is a very common misuse of the term. People speak of the evolution of the telephone, the automobile, and the musical instrument. But these are merely illustrations of man's power to deal intelligently with inanimate matter; there is no growth from within in the development of machinery.

Equally improper is the use of the word "evolution" to describe the growth of a plant from a seed, the growth of a chicken from an egg, or the development of any form of animal life from a single cell. All these give us a circle, not a change from one species to another.

Evolution—the evolution involved in this case, and the only evolution that is a matter of controversy anywhere—is the evolution taught by defendant, set forth in the books now prohibited by the new State law, and illustrated in the diagram printed on page 194 of Hunter's *Civic Biology*. The author estimates the number of species in the animal kingdom at five hundred and eighteen thousand, nine hundred. These are divided into eighteen classes, and each class is indicated on the diagram by a circle, proportionate in size to the number of species in each class and attached by a stem to the trunk of the tree. It begins with protozoa and ends with the mammals. Passing over the classes with which the average man is unfamiliar, let me call your attention to a few of the larger and better known groups. The insects are numbered at three hundred and sixty thousand, over two-thirds of the total number of species in the animal world. The fishes are numbered at thirteen thousand, the amphibians at fourteen hundred, the reptiles at thirty-five hundred, and the birds are thirteen thousand, while thirty-five hundred mammals are crowded together in a little circle that is barely higher than the bird circle. No circle is reserved for man alone. He is, according to the diagram, shut up in the little circle entitled "Mammals," with thirty-four hundred and ninety-nine other species of mammals. Does it not seem a little unfair not to distinguish between man and lower forms of life? What shall we say of the intelligence, not to say religion, of those who are so particular to distinguish between fishes and reptiles and birds, but put a man with an immortal soul in the same circle with the wolf, the hyena and the skunk? What must be the impression made upon children by such a degradation of man?

In the preface of this book, the author explains that it is for children, and adds that "the boy or girl of average ability upon admission to the secondary school is not a thinking individual." Whatever may be said in favor of teaching evolution to adults, it surely is not proper to teach it to children who are not yet able to think.

The evolutionist does not undertake to tell us how protozoa, moved by interior and resident forces, sent life up through all the various species, and cannot prove that there was actually any such compelling power at all. And yet, the schoolchildren are asked to accept their guesses and build a philosophy of life upon them. If it were not so serious a matter, one might be tempted to speculate upon the various degrees of relationship that, according to evolutionists, exist between man and other forms of life. It might require some very nice calculation to determine at what degree of relationship the killing of a relative ceases to be murder and the eating of one's kin ceases to be cannibalism.

But it is not a laughing matter when one considers that evolution not only offers no suggestions as to a Creator but tends to put the creative act so far away as to cast doubt upon creation itself. And, while it is shaking faith in God as a beginning, it is also creating doubt as to a heaven at the end of life. Evolutionists do not feel that it is incumbent upon them to show how life began or at what point in their long-drawn-out scheme of changing species man became endowed with hope and promise of immortal life. God may be a matter of indifference to the evolutionists, and a life beyond may have no charm for them, but the mass of mankind will continue to worship their Creator and continue to find comfort in the promise of their Saviour that He has gone to prepare a place for them. Christ has made of death a narrow, star-lit strip between the companionship of yesterday and the reunion of tomorrow; evolution strikes out the stars and deepens the gloom that enshrouds the tomb. . . .

Our first indictment against evolution is that it disputes the truth of the Bible account of man's creation and shakes faith in the Bible as the Word of God. This indictment we prove by comparing the processes described as evolutionary with the text of Genesis. It not only contradicts the Mosaic record as to the beginning of human life, but it disputes the Bible doctrine of reproduction according to kind—the greatest scientific principle known.

Our second indictment is that the evolutionary hypothesis, carried to its logical conclusion, disputes every vital truth of the Bible. Its tendency, natural, if not inevitable, is to lead those who really accept it, first to agnosticism and then to atheism. Evolutionists attack the truth of the Bible, not openly at first, but by using weasel-words like "poetical," "symbolical" and "allegorical" to suck the meaning out the inspired record of man's creation.

We call as our first witness Charles Darwin. He began life a Christian. On page 39, vol. I of the Life and Letters of Charles Darwin, by his son, Francis Darwin, he says, speaking of the period from 1828 to 1831, "I did not then in the least doubt the strict and literal truth of every word in the Bible." On page 412 of vol. II of the same publication, he says, "When I was collecting facts for 'The Origin' my belief in what is called a personal God was as firm as that of Dr. Pusey himself." It may be a surprise to your honor and to you, gentlemen of the jury, as it was to me, to learn that Darwin spent three years at Cambridge studying for the ministry.

This was Darwin as a young man, before he came under the influence of the doctrine that man came from a lower order of animals. The change wrought in his religious views will be found in a letter written to a German youth in 1879, and printed on page 277 of vol. I of the Life and Letters above referred to. The letter begins: "I am much engaged, an old man, and out of health, and I cannot spare time to answer your questions fully,—nor indeed can they be answered. Science has nothing to do with Christ, except in so far as the habit of scientific research makes a man cautious in admitting evidence. For myself, I do not believe that there ever has been any revelation. As for a future life, every man must judge for himself between conflicting vague probabilities."

Note that "science has nothing to do with Christ, except in so far as the habit of scientific research makes a man cautious in admitting evidence." Stated plainly, that simply means that "the habit of scientific research" makes one cautious in accepting the only evidence that we have of Christ's existence, mission, teachings, crucifixion, and resurrection, namely the evidence found in the Bible. To make this interpretation of his words the only possible one, he adds, "For myself, I do not believe that there ever has been any revelation." In rejecting the Bible as a revelation from God, he rejects the Bible's conception of God and he rejects also the supernatural Christ of whom the Bible, and the Bible alone, tells. And, it will be observed, he refuses to express any opinion as to a future life.

Now let us follow with his son's exposition of his father's views as they are given in extracts from a biography written in 1876. Here is Darwin's language as quoted by his son:

During these two years (October, 1838, to January, 1839) I was led to think much about religion. Whilst on board the Beagle I was quite orthodox and I remember being heartily laughed at by several of the officers (though themselves orthodox) for quoting the Bible as an unanswerable authority on some point of morality. When thus reflecting, I felt compelled to look for a First Cause, having an intelligent mind in some degree analogous to man; and I deserved to be called an atheist. This conclusion was strong in my mind about the time, as far as I can remember, when I wrote the *Origin of Species*; it is since that time that it has very gradually, with many fluctuations, become weaker. But then arises the doubt, can the mind of man, which has, as I fully believe, been developed from a mind as low as that possessed by the lowest animals, be trusted when it draws such grand conclusions?

I cannot pretend to throw the least light on such abstruse problems. The mystery of the beginning of all things is insoluble by us; and I for one must be content to remain an Agnostic.

When Darwin entered upon his scientific career he was "quite orthodox and quoted the Bible as an unanswerable authority on some point of morality." Even when he wrote *The Origin of Species*, the thought of "a First Cause, having an intelligent mind in some degree analogous to man" was strong in his mind. It was after that time that "very gradually, with many fluctuations," his belief in God became weaker. He traces this decline for us and concludes by telling us that he cannot pretend to throw the least light on such abstruse problems—the religious problems above referred to. Then comes the flat statement that he "must be content to remain an Agnostic"; and to make clear what he means by the word, agnostic, he says that "the mystery of the beginning of all things is insoluble by us"—not by him alone, but by everybody. Here we have the effect of evolution upon its most distinguished exponent; it led from an orthodox Christian, believing every word of the Bible and in a personal God, down and down and down to helpless and hopeless agnosticism.

But there is one sentence upon which I reserved comment—it throws light upon his downward pathway. "Then arises the doubt, can the mind of man which has, as I fully believe, been developed from a mind as low as that possessed by the lowest animals, be trusted when it draws such grand conclusions?"

Here is the explanation; he drags man down to the brute level, and then, judging man by brute standards, he questions whether man's mind can be trusted to deal with God and immortality!

How can any teacher tell his students that evolution does not tend to destroy his religious faith? How can an honest teacher conceal from his students the effect of evolution upon Darwin himself? And is it not stranger still that preachers who advocate evolution never speak of Darwin's loss of faith, due to his belief in evolution? The parents of Tennessee have reason enough to fear the effect of evolution on the minds of their children. Belief in evolution cannot bring to those who hold such a belief any compensation for the loss of faith in God, trust in the Bible, and belief in the supernatural character of Christ. It is belief in evolution that has caused so many scientists and so many Christians to reject the miracles of the Bible, and then give up, one after another, every vital truth of Christianity. They finally cease to pray and sunder the tie that binds them to their Heavenly Father.

The miracle should not be a stumbling block to any one. It raises but three questions: 1st. Could God perform a miracle? Yes, the God who created the universe can do anything He wants to with it. He can temporarily suspend any law that He has made or He may employ higher laws that we do not understand. 2nd. Would God perform a miracle? To answer that question in the negative one would have to know more about God's plans and purposes than a finite mind can know, and yet some are so wedded to evolution that they deny that God would perform a miracle merely because a miracle is inconsistent with evolution.

If we believe that God can perform a miracle and might desire to do so, we are prepared to consider with open mind the third question, namely, Did God perform the miracles recorded in the Bible? The same evidence that establishes the authority of the Bible establishes the truth of the record of miracles performed. . . .

James H. Leuba, a Professor of Psychology at Bryn Mawr College, Pennsylvania, published a few years ago, a book entitled "Belief in God and Immortality." In this book he relates how he secured the opinions of scientists as to the existence of a personal God and a personal immortality. He used a volume entitled "American Men of Science," which, he says, included the names of "practically every American who may properly be called a scientist." There were fifty-five hundred names in the book. He selected one thousand names as representative of the fifty-five hundred, and addressed them personally. Most of them, he said, were teachers in schools of higher learning. The names were kept confidential. Upon the answers received, he asserts that over half of them doubt or deny the existence of a personal God and a personal immortality, and he asserts that unbelief increases in proportion to prominence, the percentage of unbelief being greatest among the most prominent. Among biologists, believers in a personal God numbered less than thirty-one per cent, while believers in a personal immortality numbered only thirty-seven per cent.

He also questioned the students in nine colleges of high rank and from one thousand answers received, ninety-seven per cent of which were from students between eighteen and twenty, he found that unbelief increased from fifteen per cent in the Freshman class up to forty to forty-five per cent among the men who graduated. On page 280 of this book, we read, "The students' statistics show that young people enter college, possessed of the beliefs still accepted, more or less perfunctorily, in the average home of the land, and gradually abandon the cardinal Christian beliefs." This change from belief to unbelief he attributes to the influence of the persons "of high culture under whom they studied."

The people of Tennessee have been patient enough; they acted none too soon. How can they expect to protect society, and even the church, from the deadening influence of agnosticism and atheism if they permit the teachers employed by taxation to poison the minds of the youth with this destructive doctrine? And remember that the law has not heretofore required the writing of the word "poison" on poisonous doctrines. The bodies of our people are so valuable that druggists and physicians must be careful to properly label all poisons; why not be as careful to protect the spiritual life of our people from the poisons that kill the soul?

There is a test that is sometimes used to ascertain whether one suspected of mental infirmity is really insane. He is put into a tank of water and told to dip the tank dry while a stream of water flows into the tank. If he has not sense enough to turn off the stream, he is adjudged insane. Can parents justify themselves if, knowing the effect of belief in evolution, they permit irreligious teachers to inject skepticism and infidelity into the minds of their children?

Do bad doctrines corrupt the morals of students? We have a case in point. Mr. Darrow, one of the most distinguished criminal lawyers in our land, was engaged about a year ago in defending two rich men's sons who were on trial for as dastardly a murder as was ever committed. The older one, "Babe" Leopold, was a brilliant student, nineteen years old. He was an evolutionist and an atheist. He was also a follower of Nietzsche, whose books he had devoured and whose philosophy he had adopted. Mr. Darrow made a plea for him, based upon the influence—that Nietzsche's–philosophy had exerted upon the boy's mind. Here are extracts from his speech:

> Babe took philosophy. . . . He grew up in this way; he became enamored of the philosophy of Nietzsche. Your honor, I have read almost everything that Nietzsche ever wrote. A man

of wonderful intellect; the most original philosopher of the last century. A man who made a deeper imprint on philosophy than any other man within a hundred years, whether right or wrong. More books have been written about him than probably all the rest of the philosophers in a hundred years. More college professors have talked about him. In a way, he has reached more people, and still he has been a philosopher of what we might call the intellectual cult.

He wrote one book called 'Beyond the Good and Evil,' which was a criticism of all moral precepts, as we understand them, and a treatise that the intelligent man was beyond good and evil, that the laws for good and the laws for evil did not apply to anybody who approached the superman. He wrote on the will to power.

I have just made a few short extracts from Nietzsche that show the things that he (Leopold) has read, and these are short and almost taken at random. It is not how this would affect you. It is not how it would affect me. The question is, how it would affect the impressionable, visionary, dreamy mind of a boy—a boy who should never have seen it—too early for him.

Quotation from Nietzsche: "Why so soft, oh, my brethren? Why so soft, so unresisting and yielding? Why is there so much disavowal and abnegation in your heart? Why is there so little fate in your looks? For all creators are hard and it must seem blessedness unto you to press your hand upon millenniums and upon wax. This new table, oh, my brethren, I put over you: Become hard. To be obsessed by moral consideration presupposes a very low grade of intellect. We should substitute for morality the will to our own end, and consequently to the means to accomplish that. A great man, a man whom nature has built up and invented in a grand style, is colder, harder, less cautious and more free from the fear of public opinion. He does not possess the virtues which are compatible with respectability, with being respected, nor any of those things which are counted among the virtues of the herd.

Mr. Darrow says that the superman, a creation of Nietzsche, has permeated every college and university in the civilized world:

There is not any university in the world where the professor is not familiar with Nietzsche, not one. . . . Some believe it and some do not believe it. Some read it as I do and take it as a theory, a dream, a vision, mixed with good and bad, but not in any way related to human life. Some take it seriously. . . . There is not a university in the world of any high standing where the professors do not tell you about Nietzsche and discuss him, or where the books are not there.

If this boy is to blame for this, where did he get it? Is there any blame attached because somebody took Nietzsche's philosophy seriously and fashioned his life up on it? And there is no question in this case but what that is true. Then who is to blame? The university would be more to blame than he is; the scholars of the world would be more to blame than he is. The publishers of the world—are more to blame than he is. Your honor, it is hardly fair to hang a nineteen-year-old boy for the philosophy that was taught him at the university. It does not meet my ideas of justice and fairness to visit upon his head the philosophy that has been taught by university men for twenty-five years.

In fairness to Mr. Darrow, I think I ought to quote two more paragraphs. After this bold attempt to excuse the student on the ground that he was transformed from a well-meaning youth into a murderer by the philosophy of an atheist, and on the further ground that this philosophy was in the libraries of all the colleges and discussed by the professors-some adopting the philosophy and some rejecting it—on these two grounds, he denies that the boy should be held responsible for the taking of human life. He charges that the scholars in the universities were more responsible than the boy, and that the universities were more responsible than the boy, because they furnished such books to the students, and then he proceeds to exonerate the universities and the scholars, leaving nobody responsible. Here is Mr. Darrow's language:

> Now, I do not want to be misunderstood about this. Even for the sake of saving the lives of my clients, I do not want to be dishonest and tell the court something that I do not honestly think in this case. I do not think that the universities are to blame. I do not think they should be held responsible. I do think, however, that they are too large, and that they should keep a closer watch, if possible, upon the individual.
>
> But you cannot destroy thought because, forsooth, some brain may be deranged by thought. It is the duty of the university, as I conceive it, to be the great storehouse of the wisdom of the ages, and to have its students come there and learn and choose. I have no doubt but what it has meant the death of many; but that we cannot help.

This is a damnable philosophy, and yet it is the flower that blooms on the stalk of evolution. Mr. Darrow thinks the universities are in duty bound to feed out this poisonous stuff to their students, and when the students become stupefied by it and commit murder, neither they nor the universities are to blame. I am sure, your honor and gentlemen of the jury, that you agree with me when I protest against the adoption of any such a philosophy in the state of Tennessee. A criminal is not relieved from responsibility merely because he found Nietzsche's philosophy in a library which ought not to contain it. Neither is the university guiltless if it permits such corrupting nourishment to be fed to the souls that are entrusted to its care. But, go a step farther, would the state be blameless if it permitted the universities under its control to be turned into training schools for murderers? When you get back to the root of this question, you will find that the legislature not only had a right to protect the students from the evolutionary hypothesis but was in duty bound to do so.

While on this subject, let me call your attention to another proposition embodied in Mr. Darrow's speech. He said that Dicky Loeb, the younger boy, had read trashy novels, of the blood and thunder sort. He even went so far as to commend an Illinois statute which forbids minors reading stories of crime. Here is what Mr. Darrow said: "We have a statute in this state, passed only last year, if I recall it, which forbids minors reading stories of crime. Why? There is only one reason; because the legislature in its wisdom thought it would have a tendency to produce these thoughts and this life in the boys who read them."

If Illinois can protect her boys, why cannot this state protect the boys of Tennessee? Are the boys of Illinois any more precious than yours?

But to return to the philosophy of an evolutionist. Mr. Darrow said: "I say to you seriously that the parents of Dicky Loeb are more responsible than he, and yet few boys had better parents. . . ." Again, he says, "I know that one of two things happened to this boy; that this terrible crime was inherent in his organism, and came from some ancestor, or that it came through his education and his training after he was born." He thinks the boy was not responsible for anything; his guilt was due, according to this philosophy, either to heredity or to environment.

But let me complete Mr. Darrow's philosophy based on evolution. He says: "I do not know what remote ancestor may have sent down the seed that corrupted him, and I do not know through how many ancestors it may have passed until it reached Dicky Loeb. All I know is, it is true, and there is not a biologist in the world who will not say I am right."

Psychologists who build upon the evolutionary hypothesis teach that man is nothing but a bundle of characteristics inherited from brute ancestors. That is the philosophy which Mr. Darrow applied in this celebrated criminal case. "Some remote ancestor"—he does not know how "remote"—sent down the seed that corrupted him." You cannot punish the ancestor—he is not only dead but, according to the evolutionists, he was a brute and may have lived a million years ago. And he says that all the biologists agree with him—no wonder so small a percent of the biologists, according to Leuba, believe in a personal God.

This is the quintessence of evolution, distilled for us by one who follows that doctrine to its logical conclusion. Analyze this dogma of darkness and death. Evolutionists say that back in the twilight of life a beast, name and nature unknown, planted a murderous seed and that the impulse that originated in that seed throbs forever in the blood of the brute's descendants, inspiring killings innumerable, for which the murderers are not responsible because coerced by a fate fixed by the laws of heredity! It is an insult to reason and shocks the heart. That doctrine is as deadly as leprosy; it may aid a lawyer in a criminal case, but it would, if generally adopted, destroy all sense of responsibility and menace the morals of the world. A brute, they say, can predestine a man to crime, and yet they deny that God incarnate in the flesh can release a human being from this bondage or save him from ancestral sins. No more repulsive doctrine was ever proclaimed by man; if all the biologists of the world teach this doctrine—as Mr. Darrow says they do—then may heaven defend the youth of our land from their impious babblings.

Our third indictment against evolution is that it diverts attention from pressing problems of great importance to trifling speculation. While one evolutionist is trying to imagine what happened in the dim past, another is trying to pry open the door of the distant future. One recently grew eloquent over ancient worms, and another predicted that seventy-five thousand years hence everyone will be bald and toothless. Both those who endeavor to clothe our remote ancestors with hair and those who endeavor to remove the hair from the heads of our remote descendants ignore the present with its imperative demands. The science of "How to Live" is the most important of all the sciences. It is desirable to know the physical sciences, but it is necessary to know how to live. Christians desire that their children shall be taught all the sciences, but they do not want them to lose sight of the Rock of Ages while they study the age of the rocks; neither do they desire them to become so absorbed in measuring the distance between the stars that they will forget Him who holds the stars in His hand.

While not more than two per cent of our population are college graduates, these, because of enlarged powers, need a "Heavenly Vision" even more than those less learned, both for their own restraint and to assure society that their enlarged powers will be used for the benefit of society and not against the public welfare.

Evolution is deadening the spiritual life of a multitude of students. Christians do not desire less education, but they desire that religion shall be entwined with learning so that our boys and girls will return from college with their hearts aflame with love of God and love of fellowmen, and prepared to lead in the altruistic work that the world so sorely needs. The cry in the business world, in the industrial world, in the professional world, in the political world—even in the religious world—is for consecrated talents—for ability plus a passion for service.

Our fourth indictment against the evolutionary hypothesis is that, by paralyzing the hope of reform, it discourages those who labor for the improvement of man's condition. Every upward-looking man or woman seeks to lift the level upon which mankind stands, and they trust that they will see beneficent changes during the brief span of their own lives. Evolution chills their enthusiasm by substituting aeons for years. It obscures all beginnings in the mists of endless ages. It is represented as a cold and heartless process, beginning with time and ending in eternity, and acting so slowly that even the rocks cannot preserve a record of the imaginary changes through which it is credited with having carried an original germ of life that appeared sometime from somewhere. Its only program for man is scientific breeding, a system under which a few supposedly superior intellects, self-appointed, would direct the mating and the movements of the mass of mankind—an impossible system! Evolution, disputing the miracle, and ignoring the spiritual in life, has no place for the regeneration of the individual. It recognizes no cry of repentance and scoffs at the doctrine that one can be born again.

It is thus the intolerant and unrelenting enemy of the only process that can redeem society through the redemption of the individual. An evolutionist would never write such a story as The Prodigal Son; it contradicts the whole theory of evolution. The two sons inherited from the same parents and, through their parents, from the same ancestors, proximate and remote. And these sons were reared at the same fireside and were surrounded by the same environment during all the days of their youth; and yet they were different. If Mr. Darrow is correct in the theory applied to Loeb, namely, that his crime was due either to inheritance or to environment, how will he explain the difference between the elder brother and the wayward son? The evolutionist may understand from observation, if not by experience, even though he cannot explain, why one of these boys was guilty of every immorality, squandered the money that the father had laboriously earned, and brought disgrace upon the family name; but his theory does not explain why a wicked young man underwent a change of heart, confessed his sin, and begged for forgiveness. And because the evolutionists cannot understand this fact, one of the most important in the human life, he cannot understand the infinite love of the Heavenly Father who stands ready to welcome home any repentant sinner, no matter how far he has wandered, how often he has fallen, or how deep he has sunk in sin.

Your honor has quoted from a wonderful poem written by a great Tennessee poet, Walter Malone. I venture to quote another stanza which puts into exquisite language the new opportunity which a merciful God gives to every one who will turn from sin to righteousness.

> Though deep in mire, wring not your hands and weep;
> I lend my arm to all who say, "I can."
> No shame-faced outcast ever sank so deep
> But he might rise and be again a man.

There are no lines like these in all that evolutionists have ever written. Darwin says that science has nothing to do with the Christ who taught the spirit embodied in the words of Walter Malone, and yet this spirit is the only hope of human progress. A heart can be changed in the twinkling of an eye and a change in the life follows a change in the heart. If one heart can be changed, it is possible that many hearts can be changed, and if many hearts can be changed it is possible that all hearts can be changed—that a world can be born in a day. It is this fact that inspires all who labor for man's betterment. It is because Christians believe in individual regeneration and in the regeneration of society through the regeneration of individuals that they pray, "Thy Kingdom come, Thy Will be done in earth as it is in Heaven." Evolution makes a mockery of the Lord's Prayer!

To interpret the words to mean that the improvement desired must come slowly through unfolding ages,—a process with which each generation could have little to do—is to defer hope, and hope deferred maketh the heart sick.

Our fifth indictment of the evolutionary hypothesis is that, if taken seriously and made the basis of a philosophy of life, it would eliminate love and carry man back to a struggle of tooth and claw. The Christians who have allowed themselves to be deceived into believing that evolution is a beneficent, or even a rational process, have been associating with those who either do not understand its implications or dare not avow their knowledge of these implications. Let me give you some authority on this subject. I will begin with Darwin, the high priest of evolution, to whom all evolutionists bow.

On pages 149 and 150, in *The Descent of Man*, already referred to, he says: "With savages, the weak in body or mind are soon eliminated; and those that survive commonly exhibit a vigorous state of health. We civilized men, on the other hand, do our utmost to check the process of elimination; we build asylums for the imbecile, the maimed, and the sick; we institute poor laws; and our medical men exert their utmost skill to save the life of everyone to the last moment. There is reason to believe that vaccination has preserved thousands who from a weak constitution would formerly have succumbed to smallpox." Thus the weak members of civilized society propagate their kind. No one who has attended to the breeding of domestic animals will doubt that this must be highly injurious to the race of man. It is surprising how soon a want of care, or care wrongly directed, leads to the degeneration of a domestic race; but, excepting in the case of man himself, hardly anyone is so ignorant as to allow his worst animals to breed.

"The aid which we feel impelled to give to the helpless is mainly an incidental result of the instinct of sympathy, which was originally acquired as part of the social instincts, but subsequently rendered, in the manner previously indicated, more tender and more widely diffused. How could we check our sympathy, even at the urging of hard reason, without deterioration in the noblest part of our nature. . . . We must therefore bear the undoubtedly bad effects of the weak surviving and propagating their kind."

Darwin reveals the barbarous sentiment that runs through evolution and dwarfs the moral nature of those who become obsessed with it. Let us analyze the quotation just given. Darwin speaks with approval of the savage custom of eliminating the weak so that only the strong will survive and complains that "we civilized men do our utmost to check the process of elimination." How inhuman such a doctrine as this! He thinks it injurious to "build asylums for the imbecile, the maimed, and the sick," or to care for the poor. Even the medical men come in for criticism because they "exert their utmost skill to save the life of everyone to the last moment." And then note his hostility to vaccination because it has "preserved thousands who, from a weak constitution would, but for vaccination, have succumbed to smallpox"! All of the sympathetic activities of civilized society are condemned because they enable "the weak members to propagate their kind." Then he drags mankind down to the level of the brute and compares the freedom given to man unfavorably with the restraint that we put on barnyard beasts.

The second paragraph of the above quotation shows that his kindly heart rebelled against the cruelty of his own doctrine. He says that we "feel impelled to give to the helpless," although he traces it to a sympathy which he thinks is developed by evolution; he even admits that we could not check this sympathy "even at the urging of hard reason, without deterioration of the noblest part of our nature." "We must therefore bear" what he regards as "the undoubtedly bad effects of the weak surviving and propagating their kind." Could any doctrine be more destructive of civilization? And what a commentary on evolution! He wants us to believe that evolution develops a human sympathy that finally becomes so tender that it repudiates the law that created it and thus invites a return to a level where the extinguishing of pity and sympathy will permit the brutal instincts to again do their progressive (?) work.

Let no one think that this acceptance of barbarism as the basic principle of evolution died with Darwin. Within three years a book has appeared whose author is even more frankly brutal than Darwin. The book is entitled "The New Decalogue of Science" and has attracted wide attention. One of our most reputable magazines has recently printed an article by him defining the religion of a scientist. In his preface he acknowledges indebtedness to twenty-one prominent scientists and educators, nearly all of them "doctors" and "professors." One of them, who has recently been elevated to the head of a great state university, read the manuscript over twice "and made many invaluable suggestions." The author describes Nietzsche who, according to Mr. Darrow, made a murderer out of Babe Leopold, as "the bravest soul since Jesus." He admits that Nietzsche was "gloriously wrong," not certainly, but "perhaps," "in many details of technical knowledge," but he affirms that Nietzsche was "gloriously right in his fearless questioning of the universe and of his own soul."

In another place, the author says, "Most of our morals today are jungle products," and then he affirms that "it would be safer, biologically, if they were more so now." After these two samples of his views, you will not be surprised when I read you the following:

> Evolution is a bloody business, but civilization tries to make it a pink tea. Barbarism is the only process by which man has ever organically progressed, and civilization is the only process by which he has ever organically declined. Civilization is the most dangerous enterprise upon which man ever set out. For when you take man out of the bloody, brutal, but beneficent, hand of natural selection you place him at once in the soft, perfumed, daintily gloved, but far more dangerous, hand of artificial selection. And, unless you call science to your aid and make this artificial selection as efficient as the rude methods of nature, you bungle the whole task.

This aspect of evolution may amaze some of the ministers who have not been admitted to the inner circle of the iconoclasts whose theories menace all the ideals of civilized society. Do these ministers know that "evolution is a bloody business"? Do they know that "barbarism is the only process by which man has ever organically progressed"? And that "civilization is the only process by which he has ever organically declined"? Do they know that "the bloody, brutal hand of natural selection" is "beneficent"? And that the "artificial selection" found in civilization is "dangerous"? What shall we think of the distinguished educators and scientists who read the manuscript before publication and did not protest against this pagan doctrine?

To show that this is a world-wide matter, I now quote from a book issued from the press in 1918, seven years ago. The title of the book is "The Science of Power," and its author, Benjamin Kidd, being an Englishman, could not have any national prejudice against Darwin. On pages 46 and 47, we find Kidd's interpretation of evolution:

> Darwin's presentation of the evolution of the world as the product of natural selection in never-ceasing war—as a product, that is to say, of a struggle in which the individual efficient in the fight for his own interests was always the winning type—touched the profoundest depths of the psychology of the West. The idea seemed to present the whole order of progress in the world as the result of a purely mechanical and materialistic process resting on force. In so doing it was a conception which reached the springs of that heredity born of the unmeasured ages of conquest out of which the Western mind has come. Within half a century the *Origin of Species* had become the Bible of the doctrine of the omnipotence of force.

Kidd goes so far as to charge that "Nietzsche's teaching represented the interpretation of the popular Darwinism delivered with the fury and intensity of genius." And Nietzsche, be it remembered, denounced Christianity as the "doctrine of the degenerate," and democracy as "the refuge of weaklings."

Kidd says that Nietzsche gave Germany the doctrine of Darwin's efficient animal in the voice of his superman, and that Bernhardi and the military textbooks in due time gave Germany the doctrine of the superman translated into the national policy of the super-state aiming at world power (page 67.)

And what else but the spirit of evolution can account for the popularity of the selfish doctrine, "Each one for himself, and the devil take the hindmost," that threatens the very existence of the doctrine of brotherhood.

In 1900—twenty-five years ago—while an International Peace Congress was in session in Paris, the following editorial appeared in L'Univers:

> The spirit of peace has fled the earth because evolution has taken possession of it. The plea for peace in past years has been inspired by faith in the divine nature and the divine origin of man; men were then looked upon as children of one Father, and war, therefore, was fratricide. But now that men are looked upon as children of apes, what matters it whether they are slaughtered or not?

When there is poison in the blood, no one knows on what part of the body it will break out, but we can be sure that it will continue to break out until the blood is purified. One of the leading universities of the South (I love the State too well to mention its name) publishes a monthly magazine entitled "Journal of Social Forces." In the January issue of this year, a contributor has a lengthy article on "Sociology and Ethics," in the course of which he says:

> No attempt will be made to take up the matter of the good or evil of sexual intercourse among humans aside from the matter of conscious procreation, but as an historian, it might be worth while to ask the exponents of the impurity complex to explain the fact that, without exception, the great periods of cultural efflorescence have been those characterized by a large amount of freedom in sex-relations, and that those of the greatest cultural degradation and decline have been accompanied with greater sex repression and purity.

No one charges or suspects that all or any large percentage of the advocates of evolution sympathize with this loathsome application of evolution to social life, but it is worth while to inquire why those in charge of a great institution of learning allow such filth to be poured out for the stirring of the passions of its students.

Just one more quotation: The Southeastern Christian Advocate of June 25, 1925, quotes five eminent college men of Great Britain as joining in an answer to the question, "Will civilization survive?" Their reply is that:

> The greatest danger menacing our civilization is the abuse of the achievements of science. Mastery over the forces of nature has endowed the twentieth century man with a power which he is not fit to exercise. Unless the development of morality catches up with the development of technique, humanity is bound to destroy itself.

Can any Christian remain indifferent? Science needs religion to direct its energies and to inspire with lofty purpose those who employ the forces that are unloosed by science. Evolution is at war with religion because religion is supernatural; it is, therefore, the relentless foe of Christianity, which is a revealed religion.

Let us, then, hear the conclusion of the whole matter. Science is a magnificent material force, but it is not a teacher of morals. It can perfect machinery, but it adds no moral restraints to protect society from the misuse of the machine. It can also build gigantic intellectual ships, but it constructs no moral rudders for the control of storm-tossed human vessels. It not only fails to supply the spiritual element needed but some of its unproven hypotheses rob the ship of its compass and thus endanger its cargo.

In war, science has proven itself an evil genius; it has made war more terrible than it ever was before. Man used to be content to slaughter his fellowmen on a single plain—the earth's surface. Science has taught him to go down into the water and shoot up from below, and to go up into the clouds and shoot down from above, thus making the battlefield three times as bloody as it was before; but science does not teach brotherly love. Science has made war so hellish that civilization was about to commit suicide; and now we are told that newly discovered instruments of destruction will make the cruelties of the late war seem trivial in comparison with the cruelties of wars that may come in the future. If civilization is to be saved from the wreckage threatened by intelligence not consecrated by love, it must be saved by the moral code of the meek and lowly Nazarene. His teachings, and His teachings alone, can solve the problems that vex the heart and perplex the world.

The world needs a Saviour more than it ever did before, and there is only one "Name under heaven given among men whereby we must be saved." It is this Name that evolution degrades, for, carried to its logical conclusion, it robs Christ of the glory of a virgin birth, of the majesty of His deity and mission, and of the triumph of His resurrection. It also disputes the doctrine of the atonement.

It is for the jury to determine whether this attack upon the Christian religion shall be permitted in the public schools of Tennessee by teachers employed by the State and paid out of the public treasury. This case is no longer local; the defendant ceases to play an important part. The case has assumed the proportions of a battle-royal between unbelief that attempts to speak through so-called science and the defenders of the Christian faith, speaking through the Legislators of Tennessee. It is again a choice between God and Baal; it is also a renewal of the issue in Pilate's court. In that historic trial—the greatest in history—force, impersonated by Pilate, occupied the throne. Behind it was the Roman government, mistress of the world, and behind the Roman Government were the legions of Rome. Before Pilate, stood Christ, the Apostle of Love. Force triumphed; they nailed Him to the tree and those who stood around mocked and jeered and said, "He is dead." But from that day the power of Caesar waned and the power of Christ increased. In a few centuries the Roman government was gone and its legions forgotten; while the crucified and risen Lord has become the greatest fact in history and the growing figure of all time. . .

SOURCE: Bryan, William Jennings. "Summary Argument in Scopes Trial," in *The Memoirs of William Jennings Bryan*, Vol. II. Ed. William Jennings Bryan and Mary Baird Bryan. Port Washington: Kennikat Press, 1925.

"A BRIEF REPLY BY CLARENCE DARROW, LIKENS BRYAN SPEECH TO LAWYER'S ARGUMENTATIVE STATEMENT" (1925)[1]

Clarence Darrow (1857–1938) served as the defense attorney, hired by the American Civil Liberties Union, in the Scopes Trial. Darrow was famous for taking on unpopular causes and had defended Leopold and Loeb, two boys charged with murder several years before the Scopes Trial. Both supporters of progressive reforms, Darrow and Bryan were in fact friends,

---

[1]Courtesy Associated Press.

and Darrow had campaigned in support of Bryan's run for President years earlier. By the mid-1920s, though, the issue of evolution had driven the two men apart. Darrow's response to Bryan demonstrates that unlike Bryan, Darrow saw no connection between evolution and violence.

Clarence Darrow at the Scopes Trial, Dayton Tennessee, July 1925 [Library of Congress Prints and Photographs Department, LC-USZ-15589].

LEXINGTON, KY., July 28 (AP) Clarence Darrow, Chicago lawyer who upheld the theory of evolution at the John T. Scopes trial at Dayton, Tenn., tonight answered very briefly the final message of William Jennings Bryan, his chief opponent at the trial.

"I have read what Mr. Bryan intended for his speech at Dayton only hurriedly," Mr. Darrow said, "but it impresses me as only the argumentative statement of a lawyer. He referred again to the Loeb and Leopold case and philosophy of Nietzsche. He indicates that, in his belief such philosophy may have been responsible for their act.

"Loeb knew nothing of evolution or Nietzsche. It is probable he never heard of either. Leopold did, it is true, and had read Nietzsche. But because Leopold had read Nietzsche, does that prove that this philosophy or education was responsible for the act of two crazy boys?

"Isn't it peculiar that of the millions of young men and women who have attended universities and colleges of the country and studied evolution and perhaps Nietzsche, only one of them should commit such a crime as Leopold did?

"If I remember aright, about a week or so after Loeb and Leopold committed their crime a preacher poisoned his wife and a woman her husband that they could be together. Would any one claim that religion had caused this preacher to do the things he did?

"In this world little, if anything, is accomplished without progress. To make Christians of the Chinese you would be forced to kill many of them. The invention of the printing press was frowned upon and even cost some lives, but no one maintains that it has not done good.

"The building of railroads has cost many lives but they aided humanity. Each year automobiles kill more persons than are killed by homicides; but that is no reason they should be abandoned pack and parcel.

"The trial at Dayton has done several things which are significant. Of the jurors who heard the case at Dayton only one of them had ever heard of evolution. Today in Dayton they are selling more books of evolution than any other kind, and the book shops in Chattanooga and other cities of the State are hardly able to supply the demands for works on evolution. The trial has at least started people to thinking."

SOURCE: "A Brief Reply by Clarence Darrow. Likens Bryan Speech to Lawyer's Argumentative Statement." Associated Press, 1925.

WALTER LIPPMANN, AMERICAN INQUISITORS: A COMMENTARY ON DAYTON AND CHICAGO (1928)

Walter Lippmann (1889–1974) was a journalist, public intellectual, one of the founding editors of *The New Republic*, and an advisor to President Wilson. His optimism about American democracy and his strident anti-communist stance made him a powerful figure in American politics throughout the first half of the twentieth century. His 1928 *American Inquisitors* was based on a series of lectures he gave at the University of Virginia and surveyed the challenges American civilization faced as it negotiated competing claims about freedom and equality in the context of an open society. His satirical play included in the below selection highlights some of the reasons why the subject of evolution is a particularly sticky problem for Americans.

Chapter I: New Phases of an Ancient Conflict

1. Ballyhoo

As one whose business it is to write about public affairs, I have often been made to feel like a man at the theatre who forgets where he is and shouts at the hero to beware of the villain. For of late it has been our mood in politics to regard ourselves as the spectators at a show rather than as participants in real events. At a show well bred people do not hiss the villain. They enjoy the perfection of his villainy and recognize that he is necessary to the show.

We have become very sophisticated. We have become so sophisticated that we not only refuse to mistake make-believe for reality, but we even insist upon treating reality as make-believe. We are so completely debunked that we have almost persuaded ourselves that all beer is near-beer and that every battle is a sham battle.

That part of the American people which likes to think of itself as the civilized minority has insisted for some years now that no intelligent man can afford to be caught holding the illusion that any public event really matters very much. For public affairs are the serious occupation only of dunderheads, cowards, trimmers, frauds, cads on the one hand, and of opponents of prohibition, motion picture censorship, and the obscenity laws on the other. They assure us that in the main public affairs are insufferably dull. Taxation is dull. The maintenance of peace is dull. Imperial responsibilities are dull. Everything is dull,—if you treat it responsibly. But if you are a man of wit and discernment you will not treat anything responsibly. You will not expect to be edified. You will manage to be entertained. Having convinced yourself that nothing matters much, having forgotten that it is fully as difficult to govern a state as to write an essay, you will find that the spectacle of democracy in action is a glorious farce full of captivating nonsense.

I do not know whether newspaper writers belong to the civilized minority or not. But I do know that they have never been so thoroughly convinced as they are today that the measure of events is not their importance but their value as entertainment. This is the mood of the people. When my friend Mr. Mencken says "I enjoy democracy immensely. It is incomparably idiotic, and hence incomparably amusing," the democracy replies, or would if it could express itself, "You said it, old man. Everybody ought to have a sense of humor and enjoy himself. We have enjoyed ourselves mightily with half a dozen gorgeous murders, beauty contests, and the inner secrets of a lot of love nests." For the booboisie and the civilized minority are at one in their conviction that the whole world is a vaudeville stage, and that the purveyors of news are impresarios whose business it is to keep the show going at a fast clip. It is still customary to record the conventionally important affairs of state. But they are like the prescribed courses for freshmen, things which you have to pass in order to pass them by.

The real energies of the enterprising members of my profession have recently gone into the selection, the creation, the staging, and the ballyhooing of one great national act after another. Sometimes it is a sordid act. Sometimes, as in the Lindbergh idyll, it is a beautiful act. What matters is that it should never be a dull act. The technical skill which this requires is great. It is no easy thing to keep the excitement going with never a dull moment, and with intermissions just long enough for the audience to go out into the lobby for a breath of air. It is a new and marvelous profession, this business of entertaining a whole nation at breakfast. It is a profession which the older and more sedate editors look upon much as if they were deacons and had been asked to dance the Black Bottom.

2. Dayton and Chicago

Among the events on which the modern art of ballyhoo has been practiced there are two at least which are not likely to be forgotten soon. The world laughed at them, but it has not yet laughed them off. For they are symbols and portents. I refer to the trial of John T. Scopes at Dayton and to the trial of William McAndrew at Chicago. With your permission I propose to discuss these two cases as marking a new phase in the ancient conflict between freedom and authority.

This place is an appropriate one surely to such a discussion. For the University of Virginia is a temple erected by Jefferson to the belief that the conclusions reached by the free use of the human reason should and will prevail over all conclusions guaranteed by custom or revelation or authority. For this boldness Jefferson was, as you know, fiercely attacked as seditious and godless, not only by the Thompsons and the Bryans of his day, but by many of the important leaders of thought. The first appointment to the faculty of this University aroused a storm of protest in the legislature because the Board of Visitors wished to appoint Dr. Cooper, a man who had been prosecuted under the Sedition Law, and was accused as well of being a Unitarian. A century has passed. Legislatures are still ready to be aroused as they were against Dr. Cooper. But Jefferson's theory has become the acknowledged principle of education in all modern communities. There are no longer educated men anywhere who would openly venture to challenge the principle that there is no higher loyalty for the teacher and the scholar than loyalty to the truth.

And yet this principle is under attack today in all sections of the country. The attacks are made by churchmen and by patriots in the name of God and country. The attack of the churchmen is aimed chiefly at the teaching of the biological sciences, the attack of the patriots at the teaching of history. I need hardly tell you that Dayton and Chicago are exceptional only in the amount of attention they have received. They happened to lend themselves to the art of ballyhoo. They are not unique. They are merely episodes of a wide conflict between scholarship and popular faith, between freedom of thought and popular rule, which irritates American politics with deep discords. The spirit of the Tennessee Statute against the teaching of the theory of evolution is not confined to Tennessee. The purpose behind it has been carried into effect in many American communities either by statute, by administrative ruling, or by the self-denying ordinances of frightened educators. The threat of legislation like that in Tennessee is almost as effective as the actual legislation itself, and that such a threat exists as a determining influence on education in many parts of this country, no one, I think, will deny. The same holds true of the patriotic inquisition which is typified by Mayor Thompson's crusade against the text books of history used in the Chicago schools. Mayor Thompson did not start this crusade. He has merely carried on a little more spectacularly the zealous work which others had begun. There are few communities, therefore, in which there has not been some sort of inquisition recently to find out if the teachers are as religious as Dr. John Roach Straton or as patriotic as Mayor Hylan of New York, Mayor Thompson of Chicago, and Mr. William Randolph Hearst.

These assaults upon the freedom of teaching have been supported by the ignorant part of our population, the spokesmen of these new inquisitions have often been mountebanks, and invariably they have been ignoramuses. As a result, educated men have been disposed, partly because they were sincerely contemptuous, partly because they were prudent, to treat the whole matter as a farce which would soon break down through its own inherent absurdity. It is very easy to make light of the Chicago inquisitor who could not recall in the excitement of his patriotism whether it was Nathan Hale or Ethan Allen who regretted that he had only one life to give for his country. It is fairly funny to read that the Mayor of Chicago has drawn up a list of patriots of Polish, German, and Irish descent, who ought to be celebrated in the Chicago schools. But I am not so sure that it is possible to laugh all this off, and I am not so sure but that at the core of all this confusion there is not something of great importance which it behooves us to understand.

I am inclined to think that Dayton and Chicago are landmarks at which it is profitable to pause and ask ourselves whether the theory of liberty which we inherit is adequate. I do not find it adequate. My own experience as a controversial journalist during the last ten years has convinced me that while the intelligence and the wit of the community are opposed to these clerical and patriotic inquisitions, there exists no logically consistent philosophy of liberty with which to combat them. I am thoroughly persuaded that if Mr. Bryan at Dayton had been as acute as his opponents, he would have conquered them in debate. Given his premises, the logic of his position was unassailable. I am no less persuaded that the objects of Mayor Thompson's crusade could be stated in a way which would compel the respectful attention of every thinking man.

I know perfectly well that Mayor Thompson cannot state them in such a fashion. But I see no advantage in winning a cheap victory just because the opposition has a poor lawyer. I propose, therefore, to ignore as irrelevant a the superficial absurdities of the attacks on learning, to ignore the discreditable motive which sometimes confuse the issue, to ignore above all the squalid ignorance which surrounds these controversies, and instead to examine them sympathetically and dispassionately, not in their weakness and folly, but in their strength. I propose, if you please, to be the Devil's Advocate.

Need I remind you that the real title of that official is Promoter of the Faith?

I should like at the outset to invite your attention to a curious coincidence. I have before me a copy of Jefferson's Bill for Establishing Religious Freedom. This bill, as you know, was accepted in 1786 with a few unimportant changes by the General Assembly of Virginia. It has been called the first law ever passed by a popular assembly giving perfect freedom of conscience, and by common consent it is regarded as one of the great charters of human liberty. I have before me also the text of the bill which was passed by the General Assembly of the State of Tennessee on March 13, 1925, entitled An Act Prohibiting the Teaching of the Evolution Theory.

No two laws could be further apart in spirit and in purpose than these two. And yet at one point there is a strange agreement between them. On one vital matter both laws appeal to the same principle although they aim at diametrically opposite ends. The Virginia statute says that "to compel a man to furnish contributions of money for the propagation of opinions which he disbelieves, is sinful and tyrannical." The Tennessee statute prohibits "the teaching of the evolution theory in all the universities, normal and all other public schools of Tennessee, which are supported in whole or in part by the public school funds of the State." You will note that the Tennessee statute does not prohibit the teaching of the evolution theory in Tennessee. It merely prohibits the teaching of that theory in schools to which the people of Tennessee are compelled by law to contribute money. Jefferson had said that it was sinful and tyrannical to compel a man to furnish contributions of money for the propagation of opinions which he disbelieves. The Tennessee legislators representing the people of their state were merely applying this principle. They disbelieved in the evolution theory, and they set out to free their constituents of the sinful and tyrannical compulsion to pay for the propagation of an opinion which they disbelieved. The late Mr. Bryan made this quit clear: "What right," he asked, "has a little irresponsible oligarchy of self-styled intellectuals to demand control of the schools of the United States in which twenty-five millions of children are being educated at an annual expense of ten billions of dollars?"

Some time ago I pointed out this disturbing coincidence to a friend of mine who has devoted many years of his life to the study of Jefferson. After a few remarks about the devil quoting Scripture, he said that the coincidence shows how dangerous it is to use too broad a principle in justifying a practical aim. That of course is true. Jefferson, like other enlightened men of his time, believed in the separation of church and state. He wished to disestablish the church, which was then supported out of public funds, and so he declared that taxation for the propagation of opinions in which a man disbelieved was tyranny. But while he said "opinions," he really meant theological opinions. For ardently as he desired to disestablish the church, he no less ardently desired to establish a system of public education. He thought it quite proper to tax the people to support the public schools. For he believed that "by advancing the minds of our youth with the growing science of the times" the public schools would be elevating them "to the practice of the social duties and functions of self-government."

One hundred and forty years later the political leader who in his generation professed to be Jefferson's most loyal disciple, asked whether, if it is wrong to compel people to support a creed they disbelieve, it is not also wrong to compel them to support teaching which impugns the creed in which they do believe. Jefferson had insisted that the people should not have to pay for the teaching of Anglicanism. Mr. Bryan asked why they should be made to pay for the teaching of agnosticism.

This was, I believe, a momentous question, which we have been too busy to debate. But perhaps by this time, Mr. Jefferson and Mr. Bryan have met on Olympus where there is plenty of time. If they have, let us hope that Socrates is present.

SOCRATES:	I have been reading your tombstone, Mr. Jefferson, and I see that you are the author of the Declaration of Independence, the Statute for Religious Freedom, and that you are the Father of the University of Virginia. You do not mention more worldly honors. It is evident that your passion was for liberty and for learning.
JEFFERSON:	It was. I had, as I once said to Dr. Rush, sworn upon the altar of God eternal hostility against every form of tyranny over the mind of man.
SOCRATES:	And this I believe is Mr. Bryan, three times the chosen leader of the party which you founded.
JEFFERSON:	In a manner of speaking, yes.
SOCRATES:	A disciple of yours?
JEFFERSON:	You, too, had disciples, I believe.
SOCRATES:	Yes, more than I care to remember. They often quarreled. I shall not go further into that.
JEFFERSON:	You were always kind.
SOCRATES:	We shall see. I shall ask you a few questions.
BRYAN:	Mr. Jefferson can answer them all.
JEFFERSON:	I'm not so sure.
BRYAN:	A good conscience can answer any question.
SOCRATES:	I'm afraid then that I never had a good conscience.
BRYAN:	It was good considering that you were a foreigner and a heathen.
SOCRATES:	You, too, were accused of being a heathen. Were you not, Mr. Jefferson, accused of being an enemy of religion?
BRYAN (interrupting):	That is a foolish question. You may not know it, Mr. Socrates, but he was twice President of the United States.
JEFFERSON:	I was denounced as an atheist by many good people.
SOCRATES:	Were you an atheist?
JEFFERSON:	No, but I disestablished the church in Virginia.
SOCRATES:	On what theory?
JEFFERSON:	I reflected that the earth was inhabited by a thousand million of people, that these professed probably a thousand different systems of religion; that ours was but one of that thousand; that if there were but one right, and ours that one, we should wish to see the nine hundred and ninety-nine sects gathered into the fold of truth. But against such a majority we could not effect this by force. I

	said to myself that reason and persuasion are the only practicable instruments. To make way for these, free inquiry must be indulged; and how could we wish others to indulge it while we refused it ourselves?
SOCRATES:	Had not every state in your day established some religion?
JEFFERSON:	That is true. I replied, with some exaggeration I admit, that no two had established the same religion. Was this, I asked, a proof of the infallibility of establishment?
SOCRATES:	So you disestablished the church?
BRYAN:	He did, sir, and thus proved his sterling Americanism.
SOCRATES:	You also, Mr. Bryan, believe in the complete separation of church and state?
BRYAN:	I do, sir, most certainly. It is fundamental.
SOCRATES:	Can it be done? . . . You look surprised. I was merely wondering.
BRYAN:	It has been done in America.
SOCRATES:	I won't argue with you about that. I should like to ask Mr. Jefferson some more questions. For example: the church which you disestablished had a creed as to how the world originated, how it is governed, and what men must do to be saved? Had it not?
JEFFERSON:	It had.
SOCRATES:	And according to the church this creed was a revelation from God. In refusing to pay taxes in support of the teaching of this creed, you asserted, I suppose, that this creed was not revealed by God?
JEFFERSON:	Not exactly. I argued that the validity of this creed was a matter for each individual to determine in accordance with his own conscience.
SOCRATES:	But all these individuals acting as citizens of the state were to assume, I take it, that God had not revealed the nature of the universe to man.
JEFFERSON:	They were free as private individuals to believe what they liked to believe about that.
SOCRATES:	But as citizens they could not believe what they liked?
JEFFERSON:	They could not make their private beliefs the official beliefs of the state.
SOCRATES:	What then were the official beliefs of the state?
JEFFERSON:	There were none. We believed in free inquiry and letting reason prevail.
SOCRATES:	I don't understand you. You say there were many people in your day who believed that God had revealed the truth about the universe. You then tell me that officially your citizens had to believe that human reason and not divine revelation was the source of truth, and yet you say your state had no official beliefs. It seems to me it had a very definite belief, a belief which contradicts utterly the belief of my friend St. Augustine for example. Let us be frank. Did you not overthrow a state religion based on revelation and establish in its place the religion of rationalism?
BRYAN:	It's getting very warm in here. All this talk makes me very uncomfortable. I don't know what it is leading to.
SOCRATES:	I don't either. If I did, I should not be asking questions. What is your answer, Mr. Jefferson?
JEFFERSON:	I'll begin by pointing out to you that there was no coercion of opinion. We had no inquisition.
SOCRATES:	I understand. But you established public schools and a university?

JEFFERSON:	Yes.
SOCRATES:	And taxed the people to support them?
JEFFERSON:	Yes.
SOCRATES:	What was taught in these schools?
JEFFERSON:	The best knowledge of the time.
SOCRATES:	The knowledge revealed by God?
JEFFERSON:	No, the best knowledge acquired by the free use of the human reason.
SOCRATES:	And did your taxpayers believe that the best knowledge could be acquired by the human reason?
JEFFERSON:	Some believed it. Some preferred revelation.
SOCRATES:	And which prevailed?
JEFFERSON:	Those who believed in the human reason.
SOCRATES:	Were they the majority of the citizens?
JEFFERSON:	They must have been. The legislature accepted my plans.
SOCRATES:	You believe, Mr. Jefferson, that the majority should rule?
JEFFERSON:	Yes, providing it does not infringe the natural rights of man.
SOCRATES:	And among the natural rights of man, if I am not mistaken, is, as you once wrote, the right not to be compelled to furnish contributions of money for the propagation of opinions which he disbelieves, and abhors. Mr. Bryan, I think, disbelieves and abhors the opinion that man evolved from a lower form of life.
BRYAN:	I do. It is a theory which undermines religion and morality.
SOCRATES:	And you objected to being taxed for the teaching of such an opinion?
BRYAN:	I most certainly did.
SOCRATES:	And you persuaded the representatives of a majority of the voters in one state to forbid this teaching in the schools they were compelled to support.
BRYAN:	It was an outrageous misuse of public funds.
SOCRATES:	May I ask whether you meant that nobody should be taxed to support the teaching of an opinion which he disbelieves, or whether you meant that the majority shall decide what opinions shall be taught.
BRYAN:	I argued that if a majority of the voters in Tennessee believed that Genesis was the true account of creation, they had every right, since they pay for the schools, not to have the minds of their children poisoned.
SOCRATES:	But the minority in Tennessee, the modernists, the agnostics, and the unbelievers, also have to pay taxes. Do they not?
BRYAN:	The majority must decide.
SOCRATES:	Did you say you believe in the separation of church and state?
BRYAN:	I did. It is a fundamental principle.
SOCRATES:	Is the right of the majority to rule a fundamental principle?
BRYAN:	It is.
SOCRATES:	Is freedom of thought a fundamental principle, Mr. Jefferson?
JEFFERSON:	It is.
SOCRATES:	Well, how would you gentlemen compose your fundamental principles, if a majority, exercising its fundamental right to rule, ordained that only Buddhism should be taught in the public schools?
BRYAN:	I'd move to a Christian country.
JEFFERSON:	I'd exercise the sacred right of revolution. What would you do, Socrates?
SOCRATES:	I'd re-examine my fundamental principles.

3. Who Pays the Piper Calls the Tune

That is what I should like to attempt in these lectures. The greater part of the American people must of necessity be educated in public schools. These schools are supported by taxation and administered by officials who derive their authority from the voters. The question is: Shall those who pay the piper call the tune?

It may be that to many among you these questions will seem speculative and remote. You may feel that I am making too much of the spectacles at Dayton and Chicago, and that I am wrong in taking them as symbols and portents of great significance. May I remind you, then, that the struggles for the control of the schools are among the bitterest political struggles which now divide the nations? Wherever there is a conflict of religious sects, you will find that the public schools are one of the chief bones of contention. It has been so in Canada for generations. It is so now in Mexico. In every country of Europe where there are national minorities, there is bitter dispute over the public schools. It is inevitable that it should be so. Wherever two or more groups within a state differ in religion, or in language and in nationality, the immediate concern of each group is to use the schools to preserve its own faith and tradition. For it is in the school that the child is drawn towards or drawn away from the religion and the patriotism of its parents.

The reason why this kind of conflict is relatively unfamiliar to us is that America has been until recently a fairly homogeneous community. Those who differed in religion or in nationality from the great mass of the people played no important part in American politics. They did the menial work, they had no influence in society, they were not self-conscious, and they had produced no leaders of their own. There were some sectarian differences and some sectional differences within the American nation. But by and large, within the states themselves, the dominant group was like-minded and its dominion was unchallenged.

But in the generation to which we belong a multitude of circumstances have conspired to break up this like-mindedness of the American people. The children and the grandchildren of the new immigration have come of age, have prospered, and have begun to assert a powerful influence in public life. Great cities have been founded which act, as cities always do, to dissolve the customs and beliefs which were nurtured in rural and provincial society. The United States has become an empire and a world power: its thought is fertilized and infected by all the winds of doctrine. There is no longer a well-entrenched community, settled in its customs, homogeneous in its manners, clear in its ultimate beliefs. There is great diversity, and therefore, there are the seeds of great conflict.

It is quite natural, then, that this generation should have witnessed the spectacles at Dayton and Chicago. It is natural too that they should have caused so much excitement. For this is the first generation which has realized that it is divided within itself about religion and about national destiny. A generation ago John T. Scopes would probably not have thought of teaching evolution in Tennessee. Or if he had, no one would have noticed the implications of such teaching. Or if the implications had been noticed, we would have been disciplined as a matter of course, and that would have been the end of it. But today the division of opinion between fundamentalists and modernists has become acute owing to the increasing strength of the modernists. Because both sides were so representative, the struggle at Dayton interested everybody. So it is with the Thompson crusade in Chicago. A generation ago American history was universally taught as an exercise in piety and patriotism. But within our time criticism and skepticism have succeeded in shaking the whole legendary creed of patriotism, and, in the chaos which has followed, a variety of patriotic sects have appeared each contending that it alone expresses the true American patriotism.

If I read the signs rightly, we are at the beginning of a period of intense struggle for the control of public education. There is no longer a sufficient like-mindedness in most American communities to insure an easy harmony between the teachers and the mass of their fellow citizens. I shall not attempt to enumerate all the different groups actually or potentially in conflict. But there is, for example, a conflict between fundamentalists and modernists which has the profoundest bearing on the future of scientific inquiry in many parts of the West and South; there is a latent and unresolved conflict in the North and East between Catholics and Protestants, in which the extremists among the Catholics are demanding a share of the school funds for their parochial schools, and the extremists among the Protestants are demanding a state monopoly of education which would abolish the parochial schools.

There is a kind of war within the schools between the militarists and the pacifists which comes to a head every so often in rows about military training, in inquisitions as to the patriotism of teachers, in pleas that the schools should emphasize the military virtues, or that they should expound the horrors of war and the blessings of peace. Chambers of Commerce also have taken a hand in the conduct of schools, insisting that they be purged of what is usually called Bolshevism; and trades unions have arisen to plead that the schools should give more attention to the struggle of labor for a better life. All the important national groups of which we are composed have their eye on the schools. The Anglophiles wish the schools to teach that George III was only a miserable German King, and not a good Englishman at all. The Anglophobes wish it made very clear that George V still broods and plots at night over the misfortunes of George III. The unreconstructed Irish wish every school child to dwell long and portentously upon the fact that we have had two wars with Great Britain. Others among us like to dwell upon the fact that we have had no war with Great Britain for a hundred years, and shall have none ever again if we care for the future of civilization. The German societies would like a large place in the textbooks for von Steuben who drilled Washington's troops. The Polish societies would like a large place for Kosciusko. The professional Jews want the schools to stop reading *The Merchant of Venice*. And so it goes.

In fact, it almost seems as if there were hardly an organization in America which has not set up a committee to investigate the schools and to rewrite the textbooks. Apparently every organization feels itself eminently qualified to teach the teachers how to conduct the schools. There are I do not know how many schemes on foot for writing the ideal history book. That may surprise you. But in fact it is much easier than you think to write an ideal history. It is difficult to write a true history. But an ideal history is a history which proves what you want it to prove. Almost everybody, therefore, can write an ideal history. And almost everybody is writing one. . . .

6. Servility of Mind

I am not prepared to say that this vast commotion around the schools is a wholly bad thing. It creates excitement, and I should rather see the teaching faculties excited because they are under fire, than have them go comfortably and complacently to sleep. Then, too, the ultimate effect of attack and counter attack is to weaken the defenses of authority. This teacher or that may lose his job, but his opinions are heard far more widely than if they had been ignored. For it is a curious fact that in the conflict between reason and authority, the conflict itself is a victory for reason. Authority is always on the road to defeat when it has to appeal either to force or to reason. It is secure only when it rests upon unquestioned habit. Inquisitions and heresy hunts are therefore invariably the signs that reformation and emancipation are under way.

There is, moreover, a considerable advantage in compelling men to defend their opinions against attack. It is not an unmixed advantage by any means, but unless thought involved a certain personal risk, it would be too tame for the human animal. It does add to the dignity of scholarship to remember that men have died not only for their gods and their flags, but for the freedom of the human mind. I have no personal desire, mind you, to be roasted alive for my opinions, and I know a fair number of martyrs who would not be half so happy if nobody persecuted them, But it does nobody any harm now and then to put his job, his income, his reputation, and even his automobile, in one pot on the table, and gamble them all on his convictions. It is a great protection against premature hardening of the arteries.

However, we need not fear, I think, that thinking will become too safe an occupation in our lifetime. It will remain an adventure for those who can think well. There is more danger in the constant threat of popular raids upon the schools. The bravest men are drawn off into mean squabbling and bickering which take more of their energy than the thing is worth. The less brave become dangerously prudent, not only in public but in their very souls, and a man who has become prudent in his own thinking has really ceased to think. There is no way of measuring what the public schools lose by the refusal of first-rate men to submit to the democratic inquisition and by the withering away of second-rate men who are terrorized by it. But the loss is a big one, we may be sure.

SOURCE: Lippmann, Walter. *American Inquisitors: A Commentary on Dayton and Chicago*. New York: The Macmillan Company, 1928.

TERMS

Biblical Literalism—The belief that statements made in the Bible are to be accepted as literally true. It is most often associated with the *Genesis* account of creation and the flood. According to a literal interpretation of the description of creation from chapter one of *Genesis*, God personally and intentionally created the world from nothing (*de novo*) over a period of six days, each of which lasted twenty-four hours.

Christian Fundamentalism—A movement generally associated with conservative Christian that arose near the turn of the twentieth century as a reaction against historical and scientific criticism of biblical claims and the associated social and political movements. The term "fundamentalism" was coined by William Bell Riley in reference to the five fundamentals: the inerrancy of the Scriptures, the virgin birth and the deity of Jesus, the doctrine of substitutionary atonement through God's grace and human faith, the physical resurrection of Jesus, and the authenticity of Christ's miracles.

Higher Criticism—Also known as biblical criticism, it is the examination of the Bible as an historical and a literary object. It includes analysis of the Bible with the methods of the humanities and the social and natural sciences in order to determine its authorship, creation dates, and the original composition of the text. Religious conservatives often object to higher criticism's use of rationalistic or naturalistic assumptions, arguing that they lead to unacceptable conclusions.

6

THE MODERN
EVOLUTIONARY
SYNTHESIS

Pierre Teilhard de Chardin provided a response to Darwin that reframed the debate in the period after World War II. He saw in evolution more than a biological explanation for evolutionary change. Darwin's proposal that populations change matched a wider worldview that seemed to sweep the late nineteenth century. Rather than defend a view of religion generally, or of Christianity specifically, Teilhard suggested that new conceptions of science and religion must emerge from a clearer understanding of nature and the spiritual realities of people made aware of nature's processes. Teilhard considered his insights as part of an understanding of God as woven into the fabric of nature. Future societies would be better able to accommodate the kind of belief needed to embrace this understanding. In the present, humans as a species had only too recently begun to reach an awareness of the broad outline of what evolution would come to mean. Rather than evolution as religion, or religion tolerating evolutionary explanations, the future held a promise of a single vision of the universe and its meaning for humanity. Such a vision would give greater meaning to the presently separate enterprises or religion and science, a unity the Teilhard hoped for throughout his writings, which inspired evolutionists for decades.

Like Tielhard, Reinhold Niebuhr offered a philosophical basis for understanding how evolution as science and Christianity as religion could be reconciled. Looking back on what religion had attempted to provide humanity in the modern world, he identified a number of areas where traditional alliances with philosophy had misled theologians. He noted that science, in a relentless search for evidence, had moved more meaningfully in the direction of truth. Niebuhr praised scientists for their diligence and honesty, and insisted that theologians follow suit. Most importantly, as the potential for misunderstanding or overstating the applications of Darwin's contributions became apparent during the Nazi regime, and as Communism took hold around the world, he believed Christians and people of other faiths would need to work together to combat the corruption of truth. Using science in that battle, rather than battling science, would transform societies.

One hundred years after the publication of *On the Origin of Species*, there appeared in print several attempts to commemorate Darwin's contribution and summarize the intervening century. Julian Huxley, grandson of Darwin's "Bulldog," Thomas Huxley, figured prominently in this era, helping to describe the culmination of what came to be known as the "modern evolutionary synthesis." That synthesis represented the union of evolutionary thinking with genetics. Biologists in the early twentieth century had taken sides, with either evolution or genetics, generally assuming each provided its own account of change. By the 1950s, Huxley, Ernst Mayr, George Gaylord Simpson, Theodosius Dobzhansky, and others had worked out the process of evolution at various levels, focusing especially on the population level, to involve the genetics of individuals and populations in ways that meaningfully integrated the principles of both evolution and genetics.

In light of mounting evidence, already sufficient for the scientific community, Huxley offered bold predictions of where evolutionary thinking might lead next. Indeed, biologists in general had taken to considering the limitations of the selection process rather than pondering whether it provided adequate explanations in the first place. As such, the potential for understanding the biological world enjoyed constantly expanding horizons in this period. On the same path, studies of behavior, including comparative psychology, were considered open fields of inquiry. Huxley also acknowledged the ways that evolution could intersect with theology, philosophy, and certain aspects of sociology. He noted that certain promises of social engineering had already proven problematic, largely because those engineers had fallen behind the realities of evolutionary research and understanding, resorting instead to ideology that imagined an always-progressive route of change.

Another major contributor to the evolutionary synthesis, Theodosius Dobzhansky, provided a detailed compilation of evidence to suggest that, from a scientific perspective, biology does not make sense without evolution. He intended to demonstrate how clearly and unequivocally evolution lay at the heart of a comprehensive understanding of biology. At the same time, however, he illustrated some of the key tenets of science. In science, it made sense to test hypotheses and adopt theories after considering the facts. It made sense to accept the simplest of possible explanations. And it made sense to believe in a rational Creator, rather than a creator who might deceive or confuse the intelligence of creation. What made sense to Dobzhansky, in some respects, became the essential view of science that creation science challenged most directly.

Henry Morris came on the scene in the early 1970s and introduced scientific creationism. By this time, Tennessee had repealed the law that banned the teaching of evolution in favor of biblical creation, and Arkansas had sent a case on to the U.S. Supreme Court for a decision that removed an anti-evolution law that had promoted biblical teachings from its books. Morris and the Institute for Creation Research began pushing for the teaching of creation science as an alternative to evolution. The move included careful exclusion of biblical teachings in favor of presenting evidence that the Earth was not as old as it appeared, and that species did not change. Morris also demonstrated that evolution necessarily proceeds from assumptions that there was no god, and that such an assumption provided the same kind of religious basis for atheistic evolution as he suggested for creation science. Just as Teilhard had shifted the debate from the context of science versus religion by indicating the common ground where both Christianity and evolution could be accepted as rational products of human understanding, Morris indicated the common ground where both Christianity and evolution could be accepted as products of human belief. In the end, the shift produced new debate over new dichotomies of human understanding and belief.

PIERRE TEILHARD DE CHARDIN, "THE GOD OF EVOLUTION" (1953)[1]

Pierre Teilhard de Teilhard (1881–1955) made unique contributions to evolutionary biology and paleontology, spending years in China and participating in the discovery of "Peking Man." Respected by scientists at major institutions of science, including the American Museum of Natural History and the National Museum of Natural History in Paris, his correspondence and writings found critical readers throughout his lifetime, although they were not published until after his death. Darwinism, to Teilhard, advanced a philosophical understanding of Aristotle's Prime Mover. Rather than add a layer of religion to evolution, or a place for evolution within his religion, Teilhard examined the contributions of each to a broader understanding of the universe. Able to connect the Christian God to science at its most basic explanatory level, he suggested that God existed at the very point where nature holds together. As such, he saw God in nature, rather than above or apart from it. Drawing even more explicitly from Christian theology, he esteemed the resolution of all manner of crises, historical as well as spiritual, as examples of the activity of Christ. In this way, theology had to rethink the role of Christ as Redeemer, considering how varied this figure must play historically. At the same time, worship must include appreciation of the place of humans within the natural universe.

During these last years I have tried, in a series of short memoranda to pin down and define the exact reason why Christianity, in spite of a certain renewal of its grip on backward-looking (or undeveloped) circles in the world, is decidedly and obviously losing its reputation with the most influential and most progressive portion of mankind and ceasing to appeal to it. Not only among the Gentiles or the rank and file of the faithful, but even in the religious orders themselves, Christianity still to some degree provides a *shelter* for the "modern soul", but it no longer *clothes* it, nor *satisfies* it, nor *leads* it. Something has gone wrong—and so something, in the area of faith and religion, must be supplied without delay on this planet. The question is, what is it we are looking for?

It is a question that is asked on all sides, and I shall try once again to answer it by establishing, in a short sequence of linked propositions, the reality of a phenomenon whose manifest existence has been haunting me for what will soon be half a century. I mean the rise (irresistible and yet still unrecognized) over our horizon of what one might call a God (*the* God) of evolution.

I. The 'Evolution' Event

I am becoming more and more convinced that at the fundamental root of the multiple currents and conflicts that are now convulsing the human mass we must place our generation's gradual awakening to consciousness of a movement which is cosmic in breadth and organicity: a movement which, whether we welcome it or not, is drawing us, through the relentless building up in our minds of a common *Weltanschauung*, towards some 'ultra-human' lying ahead in time.

---

[1] Excerpts from *Christianity and Evolution*, copyright © 1969 by Editions du Seuil, English translation by Rene Hague © 1971 by William Collins Sons & Company Limited and Harcourt, Inc. Reprinted by permission of Harcourt, Inc.

A century ago evolution (so-called) could still be regarded as a mere local hypothesis, framed to meet the problem of the origin of species (and, more particularly, that of human origins). Since that time, however, we cannot avoid recognizing that it has included and now dominates the whole of our experience. 'Darwinism' and 'transformism' are words that already have only an historical interest. From the lowest and least stable nuclear elements up to the highest living beings, we now realize, nothing exists, nothing in nature can be an object of scientific thought except as a function of a vast and single combined process of 'corpusculization' and 'complexification', in the course of which can be distinguished the phases of a gradual and irreversible 'interiorization' (development of consciousness) of what we call (without knowing what it is) matter.

(a) First, at the very bottom, and in vast numbers, we have relatively simple particles (corpuscles), which are still (at least apparently) *unconscious*: Pre-life.

(b) Next, following on the emergence of life, and in relatively small numbers, we have beings that are *simply conscious*.

(c) And now (right now!) we have beings that have suddenly become *conscious of becoming every day a little more conscious* as a result of 'co-reflection.'

This is the position we have reached.

As I said before, evolution has in a few years invaded the whole field of our experience; but, what is more, since we can feel ourselves swept up and sucked up in its convergent flood, this evolution is giving new value, as material for our action, to the whole domain of existence: precisely in as much as the appearance of a peak of unification at the higher term of cosmic ferment is now objectively providing human aspirations (for the first time in the course of history) with an absolute direction and an absolute end.

From this arises, *ipso facto*, the general maladjustment we see on all sides in the old moulds in which either morality or religion is contained.

II. The Divine in Evolution

We still hear it said that the fact that we now see the universe not as a cosmos but henceforth as a cosmogenesis in no way affects the idea we used to be able to form of the Author of all things. 'As though it made any difference to God', is a common objection, 'whether he creates *instantaneously* or *evolutively*.'

I shall not try to discuss now the notion (or pseudo-notion) of 'instantaneous creation', nor dwell on the reasons which make me suspect the presence of an ontological contradiction latent in this association of the two words.

On the other hand I must emphasize with all the power at my command the following cardinal point:

> While, in the case of a static world, the creator (the efficient cause) is still, on any theory, *structurally* independent of his work, and in consequence, without any definable basis to his immanence—in the case of a world which is by nature evolutive, the contrary is true: God is not conceivable (either structurally or dynamically) except in so far as he coincides with (as a sort of 'formal' cause), but without being lost in, the centre of convergence of cosmogenesis. I say, advisedly, either structurally or dynamically: because, if God did not appear to us now at this supreme and exact point at which we see that nature is finally held together, our capacity

to love would inevitably gravitate not towards him but (a situation we could not possibly accept) towards some other 'God.'

Ever since Aristotle there have been almost continual attempts to construct 'models' of God on the lines of an outside Prime Mover, acting *a retro*. Since the emergence in our consciousness of the 'sense of evolution' it has become physically impossible for us to conceive or worship anything but an organic Prime-Mover God, *ab ante*.

In future only a God who is functionally and totally "Omega" can satisfy us

Where, then, shall we find such a God? And who will at last give evolution *its own* God?

III. The Christic Advent and Event

As a result, then, of life's very recent passing through a new critical point in the course of its development, no older religious form or formulation can any longer (either factually or logically) satisfy to the full our need and capacity for worship—satisfy, I mean, what has now become permanently their specifically human quality. So true is this, that a 'religion of the future' (definable as a 'religion of evolution') cannot fail to appear before long: a new mysticism, the germ of which (as happens when anything is born) must be recognizable somewhere in our environment, *here and now*.

The more one considers this psycho-biological situation, the more clearly one can distinguish the *universal* meaning and importance of what may legitimately be called the 'Christic advent.' The gospel tells us that Christ once asked his disciples: '*Quem dicunt esse Filium hominis?*' To which Peter impetuously answered: '*Tu es Christus, Filius Dei vivi*'—which was both an answer and no answer, since it still left the question of knowing what exactly is 'the true living God.'

Consider then: from the earliest days of the Church, has not the whole history of Christian thought been one long, slow and persistent exploration of Peter's testimony to the Man-Jesus?

An extraordinary and absolutely unique phenomenon: as the centuries go by, all the great figures of prophets invariably become blurred or are 'mythologized' in human consciousness— Christ, on the other hand, and Christ alone, as time passes, becomes a more and more real being for a particularly vigorous section of mankind; and this as a result of a twofold process which, paradoxically, continually both personalizes and universalizes him more fully as the years go by. For millions and millions of believers (representing the most consciously aware of human beings), Christ has never ceased since his first coming to re-emerge from every crisis of history with more immediacy, more urgency and greater penetrative power than ever before.

If, then, he is to be able to offer himself once again to our new world as the 'new God' for whom we are looking, what does he still lack?

Two things, to my mind, and two only.

The first is this: that in a universe in which we can no longer seriously entertain the idea that thought is an exclusively terrestrial phenomenon, Christ must no longer be *constitutionally* restricted in his operation to a mere 'redemption' of our planet.

And the second: that in a universe in which we can now see that everything is co-reflective along a single axis, Christ must no longer be offered to our worship (in consequence of a subtle and pernicious confusion between 'super-natural' and 'extra-natural') as a peak distinct from, and a rival to, that to which the biologically continued slope of anthropogenesis is leading us.

In the eyes of everyone who is alive to the reality of the cosmic movement of complexity-consciousness which produces us, Christ, as still presented to the world by classical theology, is both too confined (localized) astronomically, and evolutively too extrinsic, to be able to 'cephalize' the universe as we now see it.

And further, there is undoubtedly a most revealing correspondence between the shapes (the pattern) of the two confronting Omegas: that postulated by modem science, and that experienced by Christian mysticism. A correspondence—and one might even say a parity! For Christ would not still be the Consummator so passionately described by St Paul if he did not take on precisely the attributes of the astonishing cosmic pole already potentially (if not as yet explicitly) demanded by our new knowledge of the world: the pole at whose peak the progress of evolution must finally converge.

Prediction and extrapolation, it is true, are always dangerous.

Nevertheless, it is surely impossible in the present circumstances not to believe that Christ's gradual rise in human consciousness cannot continue much longer without there being produced, in our spiritual climate, the revolutionary event of his coincidence with the definitely foreseeable centre of a terrestrial co-reflection (and, more generally, of the assumed focus of all reflection in the universe).

Forced together ever more closely by the progress of hominization, and drawn together even more by a fundamental identity, the two Omegas (let me emphasize again), the Omegas of experience and of faith, are undoubtedly on the point of reacting upon one another in human consciousness, and finally of *being synthesized*: the cosmic being about fantastically to magnify the Christic; and the Christic (astonishing though it may seem) to amorize (which means to energize to the maximum) the entire cosmic.

It is, in truth, an inevitable 'implosive' meeting; and its probable effect will soon be to weld together science and mysticism in a great tide of released evolutive power—centred around a Christ at last, two thousand years after Peter's confession, identified by the work of centuries as the ultimate summit (that is, the only possible God) of an evolution definitively recognized as a movement of convergence.

That is what I foresee.

And that is what I am waiting for.

SOURCE: Teilhard de Chardin, Pierre. "The God of Evolution," in *Christianity and Evolution*. Trans. Rene Hague. New York: Harcourt, Brace, Jovanovich, 1971 (1953).

REINHOLD NIEBUHR, "CHRISTIANITY AND DARWIN'S REVOLUTION" (1958)

Reinhold Niebuhr (1892–1971) explored the role of Aristotle's philosophy in the history of Christianity and revealed how theologians had misdirected the meaning their religion. In this article, he showed how the perceived conflict between evolution and Christianity relied on the rejection by religious thinkers of scientific evidence that was most obviously true, while opposing an interpretation of science that was clearly false. He hoped that better analysis of Aristotle, as well as of Darwin, would resolve the controversy once and for all. Niebuhr believed that, going back to Darwin's day, scientists had taken the high ground of evidence, logic, and rational argument, while opponents of evolution had relied on religious fervor and faulty logic. The consequences of ongoing disagreement in the middle of the twentieth century might compound the tragedies of World War II and the rise of global Communism. He suggested that the route to peace required both sound science and enlightened religion.

Historically, the discovery by Charles Darwin that biological species were subject to mutation was the capstone of a long erosion of Aristotelian science, which assumed the immutability of the forms and structures of both nature and history and which regarded the temporal flux as merely the cycle of "coming to be and passing away" of the individual representatives of the species, the essence or the structure of existence, which their life explicated. The challenge to this Aristotelianism began in the Renaissance and was initially limited to a consideration of the more obvious development of historical structures. The achievement of Darwin was to prove that natural as well as historical structures were subject to temporal development. The concept of "natural selection," while partially validated, probably obscured the mystery of the emergence of novelty in time. Certainly no natural or scientific cause could be given for the radical uniqueness of Homo sapiens, with his endowments of reason and spirit; which enabled him to transcend the temporal flux in which he was undoubtedly involved. The long controversy about the "missing link" is indicative of the surmise of many scientists that, while Darwin's *Origin of Species* had undoubtedly proved that man was chronologically related to the brutes, even as any analysis of his physical structure had long since proved that he was structurally related, nothing in the evolutionary story could give an adequate account of the radical character of the emergence of the novelty of man.

Incidentally, while it is obvious that man's unique capacities are subject to development both individually and collectively, is significant that all accounts of this development which seek ascribe the uniqueness to this development are forced to assume in their argument the distinctively human capacities which they try to explain in evolutionary terms (as, for instance, in George Meade's *Mind and Society*). The rational capacities of man are obviously subject to development, for both children and primitive lack the capacity for conceptual knowledge. There is, nevertheless, no record of an animal herd gradually evolving into a human society, though it is also significant that primitive societies have some similarities with animal herds.

The resistance of the religious community, or more precisely all religious communities, to the Darwinian discoveries in science was so stubborn and so pathetic that it was almost universally regarded as the final rear-guard action of a dying religious faith embattled with an advancing science. The religious attitude was so stubborn because Christianity had for years compounded Aristotelianism with the Biblical doctrine of creation. Of the two it was the Aristotelian science of fixed forms and species which seemed to be the most formidable opponent, particularly since many scientists challenged Darwinism for Aristotelian reasons. But in the minds of the pious the chief reason for challenging Darwin's conclusions were that they compromised both the majesty of the Creator and the dignity of the creature who had been said to have been made "In His Image:" that is the dignity of man.

One reason why the gradual acceptance of the Darwinian thesis proved not to be lethal to religious faith was that the biblical doctrine of creation was not as dependent upon Aristotelian ontology as Christians had traditionally assumed. The two were, in fact, in contradiction to each other; but that was not discovered until Darwin's triumph shattered the relation and also prevented Christian obscurantists from using the doctrine of creation to obviate the necessity or possibility of inquiring into the sequence of causes. For actually it is in precisely the analysis of these sequences that two facts become apparent. One is that every event has a previous cause, as stated in the Latin maxim *ex nihilo nihil sit*. The other is that no previous cause is a sufficient explanation of a previous event. This becomes particularly apparent in the emergences of striking novelties in the evolutionary chain, of which the most notable are the emergence of organic life and the emergence of man. Here we have the most obvious glimpse into the mystery of creation and may be prompted to realize that Aristotelianism and biblical doctrines are not natural allies but contradictory conceptions. The compounding of these contradictory conceptions was one of the consequences of the confluence of Hebraic and Hellenic culture, which reached its height in the noonday of the medieval period of Western culture.

The Hellenic component of our culture sharpened the rational instruments for the advances of all our sciences by its assumption that there were rational elements in nature which the reason of the mind could explore. Mind and nature had affinities insofar as the penetration of the one could explore the consistencies of the other. Nature is rational in terms of its consistent coherences, which is why mathematics and physics are so closely related. The inner consistencies of mind are related to these natural consistencies, which is why logic and mathematics are so closely related.

Nature is nevertheless not completely rational. That is why science must move by induction rather than deduction and wait upon the fact, which can not be deduced from the coherence and consistency of known facts. The first science in which Aristotelian deduction was successfully challenged was astronomy. Ideally, the triumph of Copernicanism should have shattered the partnership of Aristotelianism and Christian piety and made room for the recognition of the "irrationality of the givenness of things" and for the necessity of inductive as well as of deductive procedures in science. Ideally, the biblical doctrine of creation, or the recognition that there is a mystery of creation above and beyond all of Aristotle's four causes, should have made room for genuinely empirical science. Actually the partnership, though challenged, lasted from Copernicus to Darwin. The greatest philosopher of the last generation, Alfred Whitehead, finally clarified the relation between causes and creation in his monumental work, Process and Reality in which he proved that even the most rational account of the temporal processes could not give a picture of a self-explanatory process, but is forced to posit a "primordial God" as the "principle of concretion." For there is no rational explanation of why just this potentiality of all possible potentialities should be realized in concreteness.

All this was unknown in the age of Darwin, and the hosts of piety were embattled against the impiety of the dread Darwinian conception. It is a well-known drama now with Bishop Wilberforce, otherwise irreverently known as "Soapy Sam," and the redoubtable Thomas Huxley, carrying on the debate in the main theatre which was reenacted in almost every village and hamlet. Religious people ought to remember with some embarrassment that the religious arguments were not always honest or logical and that it was Thomas Huxley who insisted on scrupulous honesty, being in perfect conformity to the great virtue of the scientific enterprise, which was and is to "follow the evidence." Huxley was honest enough to challenge the conclusions of those who drew wrong moral and sociological conclusions from biological facts in his Romanes Lecture.

The world of science with its scrupulous honesty in weighing evidence would regard religious piety from that day to this a breeder of dishonesty, zealously "telling a lot of little lies in the interest of a great truth," (Clutton Brock) perhaps of two great truths: the mystery of creation and the unique dignity of man. Science was meanwhile "telling a lot of little truths" about causes, which could be fashioned into a "big lie." That falsehood was that historical processes and natural processes were sufficiently identical to make the same scientific method applicable to both fields. Before we discuss the consequences of this illusion we must delay for a moment to record that pious statesman of the type of William Gladstone, and more belatedly our own William Jennings Bryan, who futilely lent their rhetorical skills for the purpose of arresting the march of the Darwinian "heresy." The Scopes trial, in an obscure Tennessee village, was the last act in the drama of ignorant piety challenging the march of science, which was, among other achievements, to destroy the partnership between piety and obscurantism, and between religious faith and Aristotelian ontology. It must be confessed that the obscurantist temptation to piety is never overcome; because the religious symbols of ultimate meaning are poetic rather than exact and scientific, and the fearfully pious are always tempted to buttress their validity by a frantic adhesion to some outmoded science, against the challenge of a marching science, which always has immediate truth on its side but which always threatens to construct a scientific world picture in which no meaning can be found for man in his grandeur and his misery.

Subsequent developments, after the triumph of Darwin, proved that the religious impulse to defend the unique dignity of man were not as foolish as they seemed, though the methods of defense were both foolish and futile. For the triumph of Darwinism in biology led to false conclusions in the field of morals in particular and to false interpretations of human history in general.

Perhaps the most glaring example of a triumph of truth in the field of the natural sciences leading to error in the field of the social sciences was the emergence of "Social Darwinism." This creed, which tried to transfer the principle of "the survival of the fittest" to historical and moral issues, gave support to the remnants of the laissez faire principles of classical economics, derived from physiocratic illusions of the Enlightenment in France. The illusion was that history was governed by "laws of nature," with which one must not interfere. Social Darwinism served to dull the conscience of the Western world to the injustices of its rising industrialism. It prevented the adoption of the ameliorations of economic inequality, the creation of adequate equilibria of power by which the West was ultimately saved from communism; but the illusions were potent enough to delay action so that the Marxist rebellion could be initiated among the desperate industrial classes of the Continent. Thus a "class struggle" was prompted which brought Western civilization to the very edge of disaster.

Herbert Spencer was not a social Darwinist, but he also regarded the Darwinian triumph as validating his historical fatalism and optimism. He agreed with the social Darwinists at least on the point of obscuring the fact that man has ambiguous place in the historical process; for he is both creature and creator in the process, and he dare not abdicate his responsibilities as creator or forget his importance as creature.

The post-Darwinian era elaborated a confusion of voluntaristic and deterministic ideologies; but even the voluntaristic ideologies, such as that of August Comte, which disputed the determinism of Spencer, also drew inspiration from the basic error introduced by Darwinian biology into historical studies. For Comte based his historical optimism on the hope of an increasing scientific control of historical forces by an elite of scientific creators, who could only manage historical processes as if the human material were as maleable as the forces of nature. The Comtean type of voluntarism was mistaken for the simple reason that no elite of historical managers was godlike and no "stuff" of history to be managed was as "natural" as the theory assumed. The theory, despite its voluntaristic character, was thus derived as clearly from the error of equating history with nature as the Spencerian theory. No one can hold Darwin responsible for these errors, They are worth recording only to illustrate how human history is a curious drama in which truth sometimes is rescued from error and more frequently error is distilled from the truth. The illusions of Comtean voluntarism did not generate immediate perils for civilization because the elite who were to manage history were not sharply defined and there was no political program for endowing them with the omnipotence, which their destiny required. It remained for the apocalyptic creed of communism to designate such an elite, the "proletariate," with precision, and to elaborate a political program which would make their pretensions dangerous by arming them with power to manage the historical forces toward the dreamed of apocalyptic end.

These various forms of deterministic and voluntaristic optimism which the discoveries of Darwin in biology prompted were confined on the whole to secular thought. But it must be recorded that the general historical optimism, whether deterministic or voluntaristic, invaded the religious communities. It is perhaps one of the greatest ironies of history that one part of the Christian church, that part namely which was not in creative relation to modern culture, opposed Darwinism in the field where it was undoubtedly true. But the other part of the Christian community which was in creative contact with the culture, accepted the erroneous conclusions, which seemed inevitably to flow from the discovery of Darwin, and added religious emotion to interpretations of history which were obviously false.

This was particularly true in America, where the indeterminate possibilities of a great nation, expanding on a virgin continent accentuated the mood of historical optimism, which was initiated in the Renaissance and developed through the centuries until the evolutionary theory of Darwin seemed to be the final validation of the mood. The most outspoken and vapid Christian exponent of this optimism was John Fiske, who was equally assiduous in refuting the errors of the religious opponents and in propagating the errors of the secular proponents of "Darwinism." Fiske's *Cosmic Evolution* was a perfect expression of the historical optimism which characterized Western culture at the end of the nineteenth and the beginning of the twentieth centuries. The optimism was so pervasive because both the voluntaristic versions and the deterministic accounts of historical development contributed to it. Progress was assured in the one case by natural forces, history being regarded as merely an extension of nature. Darwin's discoveries did not create this optimistic determinism, but they seemed to support it. In the other case it was "science" and the "scientific method" which were relied upon to put man in gradual control of historical as well as natural forces, thus guaranteeing the progressive elimination of all manner of evil. The purely deterministic theories failed to measure human freedom, which distinguished man from the brutes and history from nature. The voluntaristic theories, whether Comtean and liberal or Marxist and communist, looked forward to a change in the human situation, either by revolution or evolution, which would alter the ambiguity of man's relation to the historical process in which he was both creator and creature; and make him unambiguously the creator of historical destiny. These theories were primarily secular, but they were so dominant in the culture and expressed the mood of the age so accurately that the portion of the Protestant church which was in more organic contact with modern cultural movements completely capitulated to the optimism.

Some violence had to be done to the traditional tenets of the Christian faith to approach conformity between it and the ideas of progress. The idea of Divine Providence was rather easily translated to that progress and would seem to be a more accurate description of what the idea of providence intended. The religious vision of the "Kingdom of God," which had always given the modern mind some difficulty, was interpreted to mean the goal and fulfillment of all historical striving. The biblical recognition of the importance of man could not be easily transmitted or transcribed to fit into the optimistic scheme, but they could be subordinated to the idea that God had called man to be "co-worker" with him. The secular world generally considers the rearguard action of Christian orthodoxy, in vainly trying to refute the undoubted scientific achievements of Darwin, as an undignified and pathetic spectacle. But modern culture is not generally aware that the uncritical appropriation by Christian liberalism of the illusions, propagated by those who drew false conclusions in the realm of history from truths, which were valid in the realm of nature, was just as futile and pathetic. These errors were not, however, as noticeable because they were committed, not in the teeth of opposition to the main currents of modern culture, but in consonance with its mood.

By a curious irony of history the optimism which was so confidently proclaimed at the end of the past century and the beginning of the current century, was cruelly refuted by the weight of historical facts, beginning with the World War of 1914. The dreams of a "parliament of mankind and federation of the world" thus turned into the reality of a conflict on a world scale. The hope that "methods of persuasion" would gradually overcome "methods of force" was disappointed by the realities of more and more total war because modern technical civilization and democratic government were more capable of harnessing the total resources of the community for any end the community intended to achieve, of danger it intended to counter.

The Second World War followed quickly and presented Western civilization with the agonizing choice of allying itself with one despotism in order to overcome what seemed to be a worse one. But the allied despotism of communism proved in the end to be more an enduring threat to the peace of the world. Its apocalyptic vision of a perfect brotherhood of nations and a classless society, once a revolution had eliminated the institution of property, captivated and still captivates the nascent nations of the Colored Continents, while meanwhile the prophets of this new political religion became the priest-kings of despotic utopian states. None of these terrible emergences and emergencies had been anticipated in the "century of hope." Nor was it anticipated that the continued advancement of the natural sciences would gradually result in the discoveries of nuclear physics and that these achievements would be quickly pounced upon by fearful governments so that the scientists became the armorers of the nations in a "nuclear age" in which the world has the possibility of completely destroying civilization by the lethal and destructive efficacy of its nuclear weapons.

Thus history proves in contemporary experience that man's freedom over nature has both destructive and creative possibilities and that these possibilities grow together with the freedom. Our experience also proves that the triumphs of the natural sciences which have created nuclear energy and nuclear weapons cannot be matched by equal triumphs of the "social sciences" or any other wisdom which might bring this awful energy under social control. This would seem to suggest that man is destructive as well as creative in his unique freedom precisely because the freedom to transcend natural finitude is not as absolute as the previous century supposed; and that there is no possibility of making it more absolute.

It would be foolish to hold Darwin responsible for all the foolish illusions which were generated in his name. It would also be idle to celebrate the triumph of science over religious obscurantism without noting the triumph of enlightened religion and the consequent triumph of illusion.

Thus, the imposing achievements of a great scientist in the past century entered into the complex pattern of man's cultural history and prompted both enlightenment and illusion about the human situation.

SOURCE: Niebuhr, Reinhold. "Christianity and Darwin's Revolution," in *A Book That Shook the World.* Ed. Julian Huxley. Pittsburgh: University of Pittsburgh, 1958.

JULIAN HUXLEY, "DARWIN AND THE IDEA OF EVOLUTION" (1958)

When Julian Huxley (1887–1975) wrote the article excerpted below, he brought the reader up to speed on many details of Darwin's initial contributions and described additional work that had clarified the process of natural selection. He suggested that Darwin himself had provided a wealth of evidence that established the fact of evolution, but that it had taken some decades to resolve numerous problems with natural selection. He also noted that scientists had overlooked some of Darwin's ideas about human evolution and about the extension of evolutionary thinking into sociology and psychology, but now those contributions, too, served as the basis for ongoing research. Huxley focused primarily on the cutting-edge of biological thinking in this excerpt. He suggested that natural selection could replace teleological explanations for the features of organisms in a way that exceeded Darwin's expectations and left little room for speculation about why a creator had made such nuisance-species as mosquitoes. This reversal pushed opponents of evolution to restructure their arguments, in many cases.

Charles Darwin is and will always remain one of the preeminent figures in human history. He rendered evolution inescapable as a fact, comprehensible as a process, all-embracing as a concept. After Darwin it became necessary to think of the phenomenal world in terms of process, not merely in terms of mechanisms, and eventually to grasp that the whole of reality is a single process of evolution. . . .

Darwin's essential achievement was to establish the idea of evolution as a natural process. It remains for me to say something of the significance of the evolutionary idea in present-day thought.

To begin with, if evolution is accepted as a fact, much of the *theological* framework of the world's major religions is destroyed, or is conveniently (but to my mind disingenuously) represented as significant myth.

Here Darwin merely extended the effect of Newton's work into the realm of life. Before Newton it appeared necessary to Christian theologians to postulate a Divine Being to guide the planets in their courses: after Newton, this was seen to be unnecessary and indeed impossible. The universe came to be regarded as a gigantic clockwork mechanism, constructed and set going once for all by God, but then continuing automatically on its course. Miracles in the theological sense became a scientific impossibility: when not the product of ignorant credulity, they turned out to be unusual occurrences or unexplained basic properties of nature—miracles in the etymological sense of things to be wondered at, but not due to Divine intervention or interference.

After Darwin, a similar naturalism was introduced into biology. The idea of creation (including the Cuvierian version of it which postulated a number of successive and different creations, separated by a series of cataclysms) had to be given up in favour of the gradual transformation, diversification, and improvement of one (or a few) extremely simple ancestral forms. And eventually it came to be accepted that ancestral life had not been created: it must have originated from non-living matter at some stage in our planet's history.

Nor could it be supposed that any supernatural agency was needed to guide or interfere with the detailed or general course of evolution: that too is determined by simple natural causes. The apparent purposefulness of biological mechanisms (and, we can now say, evolutionary trends) turns out not to demand conscious purpose by a Divine artificer. The purposefulness is only apparent, and has been brought into existence by the blind and automatic forces of natural selection. Darwin himself stressed that if any case occured where a character of one organism was solely of use to another, he would have to abandon the idea of natural selection; and G. G. Simpson, in *The Meaning of Evolution*, points out that natural selection can never envisage or anticipate future consequences; so that evolution proceeds by a series of improvisations, and the plans of organs (e.g., the eye) are often far from embodying an ideal design.

Indeed, Paley's argument from design now works in reverse. The more remarkable an adaptation is (like the woodpecker's tongue or the bees' communication system), the better it demonstrates the extraordinary efficiency of natural selection. We can (or at least we should) no longer ask what is the use of a mosquito or a tapeworm. It is there because it can survive in certain ecological conditions.

Then, as astronomy has expanded our space-scale, evolution has expanded our time-scale. In place of the recurrent cycles of Hinduism or the few thousand years of Judaeo-Christian theology, and in spite of the grudging estimates of nineteenth century physics, the past of life has been steadily increased by science until it now exceeds the staggering figure of two and one-half billion years. And in place of an imminent last Judgment, life on this planet (barring some improbable cosmic catastrophe) can envisage at least an equal span of evolutionary time in the future.

Our new knowledge of the mechanism of heredity and variation is enlarging our ideas of the power of artificial selection to extend the work of natural selection. By radiation, we are now artificially producing mutations in crop-plants where the range of variation is low, and then selecting and recombining the few favourable ones to make new breeds. By these and other methods we are doing in a few decades what it took natural selection millions of years to effect-extending the range of species into previously prohibited habitats. Artificial insemination could do something similar for animals, and is opening up the prospect of a practicable system of Eugenics, as H. J. Muller stressed in his book, *Out of the Night*.

The example of industrial melanism in moths, to which I referred earlier, deserves fuller treatment as showing how biologists are tackling the problems of selection on neo-Darwinian lines, with a Mendelian basis for heredity and variation. Within the past eighty years moths of many different species have turned black in industrial areas, but not in the open country. Research has already shown that this is due, not to any direct effect of smoke or chemicals, but to the natural selection of black types. Black types crop up as rare dominant mutants in all the moths, and are hardier and more resistant than the normals (recessive blacks also appear, but do not show increased resistance). But the normals resemble the bark on which they rest by day, and the advantage conferred by this protective resemblance outweighs their lesser hardiness. However, in industrial areas the trees were darker coloured, and there were poisonous chemicals on the caterpillars' food, so now hardiness had the advantage; the black types increased in numbers in each generation and in the course of seven or eight decades replaced the non-blacks and became the "normal" type. Experiments have shown that birds are effective in eliminating moths that do not harm and that noxious chemicals in the food cause an increased death rate in non-black caterpillars. Furthermore, these experiments are beginning to give information on the actual selection-pressure that is operating—the quantitative advantage enjoyed by one type over the other similar experiments, on the appearance of DDT-resistant mosquitos, or on the acquired tolerance of bacteria to antibiotic drugs, are yielding important practical results. Biologists are now more and more turning to the detailed study of populations, either in nature or experimentally in special cages or enclosures.

On a more general level, increasing attention is being devoted to evolution as a phenomenon, to its course and its results. We are even beginning to be able to measure its speed: thus Simpson finds that the rate of evolutionary change is nearly three times as high in horses as in early ammonites.

Then there is evolutionary philosophy. It is becoming urgent to clarify certain evolutionary concepts. One of the basic facts of evolution as a process is the succession of types. A previously established group gives birth to a new type whose success, as shown by its rapid radiation into many sub-types, and often by the reduction of the parent type, demonstrates that it is superior or "higher" in its organization. We must attempt to give a scientific meaning to level of organization, to clarify what we mean by "higher" and "lower" types, and by biological "progress."

Darwin himself, in characteristic terms, rightly affirmed that natural selection inevitably caused the improvement of most organisms in relation to their conditions of life. We need to define biological improvement more closely, and to find out what type of improvement occurs in what conditions. Note Darwin's caveat: not all organisms are being "improved." Some types (and indeed many more than Darwin imagined) become stabilized and persist over long periods. This applies not only to "living fossils" like the Coelacanth fish Latimeria and reduced groups like the Reptilia, but also to highly successful terminal types, like higher spiders, modern birds, or ants, all of which have persisted for tens of millions of years with only minor change. Meanwhile, during the early stages of a group's adaptive radiation, numerous types appear which do not persist but become extinct, presumably because their organizational plan is less well integrated.

We need to discover what confers persistence and stability on a type, of whatever taxonomic rank. Is it genetic homostasis; is it efficiency of organizational pattern? Equally we want to discover what are the factors that restrict the progressive change of a group and set a limit to its further evolution, except by a rare break-through to a new and "higher" organizational level.

This links up with a rather radical change in approach. Nineteenth century biologists were mainly interested in origins. Twentieth century biologists are becoming increasingly concerned with possibilities. The new idea of evolution that is emerging is of a dialectic process, tending to the realization of new possibilities, but constantly checked, in one trend after another, by limitations which it cannot transcend. Patterned colour-vision and temperature-regulation are examples of new possibilities realizable only at certain stages in evolutionary history. But there are limits to the acuity of vision and to the accuracy of homothermic regulation. The limits, of course, apply to biological evolution operated by natural selection.

Thus the artificial (exosomatic) sense-organs manufactured by man (e.g. telescope, electron microscope) have enormously enlarged the scope of vision.

Some types, we are finding, possess potentialities which are normally unrealized, and are revealed only when new conditions are provided. In the laboratory, jackdaws are as good as human beings at non-verbal counting. Chimpanzees will create designs when given paper and paints, and will rival human performance when provided with rollerskates. Behaviour is becoming a focal point of evolutionary study. From one angle, it is being clarified by the application of information theory and the ideas of Cybernetics. From another, it is throwing light on the mind-body puzzle, by demonstrating the emergence, the diversification, and the steady intensification of awareness during evolution, and exploring the relation of different types of aware behaviour to the evolution of brain structure. The complexity of the behaviour of higher insects (e.g., of bees) suggests that their tiny brains operate in different ways from those of vertebrates: administration of drugs like lysergic acid are revealing wholly unexpected possibilities of behaviour and subjective experience in mammals: electrical stimulation is mapping the human cortex and showing us the material basis for memory.

Indeed, I would prophecy that the study of organisms as behaviour-systems is likely to be crucial for a better understanding of the problem of organization. Perhaps level of organization is best evaluated not merely by the number of differentiated functional and structural elements in a behaviour-system, but by the intensity of their interactions and the degree to which these interactions are integrated in a patterned whole. This, I think, is what is implied in Teilhard de Chardin's idea of progressive *enroulement* during evolution, which he develops in his remarkable book *Le Phenoméne Humain*: the quality and level of awareness is correlated with the degree of "tension" generated by the central interaction of "information" from different elements of the whole system, not merely with brain structure. However, much work will need to be done before such ideas on psycho-physical correlation become scientifically profitable.

Finally we come to the application of evolutionary ideas to man. Darwin, with typical modesty, concluded the *Origin* with the remark that with the acceptance of the idea of evolution, "light will be thrown on man and his history." At first attention was focused on the animal ancestry of man, and much progress has been made in its elucidation. But to-day the generalization of the idea of evolution is illuminating the entire human problem.

To start with, we now realize that evolution operates in the whole of nature, and that it. can best be defined as a one-way process of change in time which in its course increases diversification, creates novelty, and raises the upper level of organization. Thus, in a certain sense, all phenomenal reality is a single process of evolution.

But this general process is divisible into distinct sectors, separated by critical points, each with its own characteristic tempo and mechanism of operation, its own type of product. The three sectors we can now distinguish are the inorganic, the biological, and the human or psycho-social, the second arising out of the first, the third out of the second. To take only the two last-named, the main mechanism of biological evolution is natural selection, and its products are discrete organic species: while psycho-social or cultural evolution is based on the mechanism of the cumulative transmission of experience, and its results and its products are social groups not rigidly separated but capable of cultural interpenetration.

There have been many attempts to apply ideas derived directly from biological evolution to human affairs—notably to justify individualist laisse faire on the basis of the biological struggle for existence, or the principle of a master race from the succession of dominant types in palaeontology. But all such attempts are bound to be misleading since in man intra-specific competition is much less important than co-operative participation, especially when consciously embarked upon, and since succession in human history is of cultures, not of genetic (racial) types.

Sometimes, again, sociologists continue to think in evolutionary terms which have long been rejected by biologists, notably the assumption that evolution is always progressive, and is confined to a single line or trend. The Victorian idea of universal and inevitable progress, or Comte's procrustean framework of cultural stages, are examples.

Conversely, some historians and anthropologists who rightly reject such naive notions, throw the baby out with the bath-water and deny the possibility of genuine advance, either reverting to the idea of recurrent cycles, or emphasizing only the relativity of all cultural phenomena, such as social structure or morality.

To begin to comprehend cultural evolution, we must first of all make a thorough analysis of its underlying mechanisms, and then survey it on the largest scale: it is useless to confine attention to civilizations, like Toynbee, or to primitive societies, like many anthropologists.

The first major difference between the biological and the psycho-social phase is that man, though a new dominant type, consists of only a single species. The incipient biological divergence which gave rise to the primary races of man was soon complemented by a process of convergence by migration and interbreeding. Of course marked cultural divergence has occurred, leading to the appearance of distinct cultures and types of society. But this trend too was succeeded by one of convergence: this process of cultural diffusion is always tending to spread more and more elements of culture over larger and larger areas. Though marked cultural differentiation remains within cultures, in respect of the basic mechanisms of communication and control there is a clear tendency towards global unification.

The second major difference is the immensely quicker rate of change seen in cultural evolution, and its tendency to show acceleration. This has now reached alarming proportions. It will be one of the tasks of the future to stabilize change at a manageable rate.

The third is that major advance is always dependent on new organizations of knowledge, either in the form of practical applications or of ideas and general approach.

Cultural, like biological evolution, proceeds by steps or stages. I will conclude with two relevant examples from the present. The fact of rapidly increasing population is obtruding itself forcibly on human attention; and it is becoming clear that this phase must tend towards stabilization if many difficulties and possible disasters are to be avoided. This will involve substituting the idea of human quality for mere quantitative increase.

The second is more radical. The process of evolution, as represented by man, is now, for the first time in its long history, becoming conscious of itself and of its nature. Man is the latest dominant type to be produced in evolution and the only one capable of further major advance. I would prophecy that one of the major scientific enterprises of the moderately near future will be a study of human possibilities and the evolutionary implications of attempts to realize them. If so, the idea of evolution, which became scientifically respectable a bare century ago, will find its most important application in the central problem of human destiny.

SOURCE: Huxley, Julian. "Darwin and the Idea of Evolution," in *A Book That Shook the World*. Ed. Julian Huxley. Pittsburgh: University of Pittsburgh, 1958.

THEODOSIUS DOBZHANSKY, "NOTHING IN BIOLOGY MAKES SENSE EXCEPT IN THE LIGHT OF EVOLUTION" (1973)[2]

Population geneticist Theodosius Dobzhansky (1900–1975) contributed enormously to his field through his studies of the fruit fly, Drosophila. He combined detailed collecting excursions with elaborate laboratory tests of the flies from different regions to test the effect of environment and genetics on adaptation. In the debates over evolution and creationism, he contributed a line, which became the title of this article, and which advocates of evolution have embraced and opponents have reviled. Dobzhansky made clear that for working evolutionary biologists, certain components of the theory and evidence from nature "make sense" when interpreted through the mechanism of natural selection. The same issues, interpreted as the result of divine creation, do not make sense to scientists. As he noted repeatedly, evidence contrary to belief in a creator did not make sense to him, and he wondered how it could make sense to believers. Dobzhansky offered examples of the unity and diversity of life, allowing that a creator might have played a role over evolutionary history in laying out the processes, but he could not believe that a creator would take short cuts and then leave evidence of a longer process. He took some pains to point out how use of words like theory and fact can differ between public discourse to scientific communication. He noted that disagreement within science, based on skepticism and an ongoing search for better answers, could not be equated with total uncertainty over basic scientific concepts. Appearing as it did in a journal for biology teachers, the article catalyzed education of evolution as new challenges arose from religious groups.

As recently as 1966, Sheik Abd el Aziz bin Baz asked the king of Saudi Arabia to suppress a heresy that was spreading in his land. Wrote the Sheik:

> The Holy Koran, the Prophet's teachings, the majority of Islamic scientists, and the actual facts all prove that the sun is running in its orbit...and that the earth is fixed and stable, spread out by God for his mankind....Anyone who professed otherwise would utter a charge of falsehood toward God, the Koran, and the Prophet.

The good Sheik evidently holds the Copernican theory to be a "mere theory," not a "fact." In this he is technically correct. A theory can be verified by a mass of facts, but it becomes a proven theory, not a fact. The Sheik was perhaps unaware that the Space Age had begun before he asked the king to suppress the Copernican heresy. The sphericity of the earth has been seen by astronauts, and even by many earth-bound people on their television screens. Perhaps the Sheik could retort that those who venture beyond the confines of God's earth suffer hallucinations, and that the earth is really flat.

Parts of the Copernican world model, such as the contention that the earth rotates around the sun, and not vice versa, have not been verified by direct observations even to the extent the sphericity of the earth has been. Yet scientists accept the model as an accurate representation of reality. Why? Because it makes sense of a multitude of facts which are otherwise meaningless or extravagant. To nonspecialists most of these facts are unfamiliar. Why then do we accept the "mere theory" that the earth is a sphere revolving around a spherical sun? Are we simply

---

[2]Theodosius Dobzhansky, "Nothing in Biology Makes Sense Except in the Light of Evolution." *American Biology Teacher*, 1973.

submitting to authority? Not quite: we know that those who took the time to study the evidence found it convincing.

The good Sheik is probably ignorant of the evidence. Even more likely, he is so hopelessly biased that no amount of evidence would impress him. Anyway, it would be a sheer waste of time to attempt to convince him. The Koran and the Bible do not contradict Copernicus, nor does Copernicus contradict them. It is ludicrous to mistake the Bible and the Koran for primers of natural science. They treat of matters even more important: the meaning of man and his relations to God. They are written in poetic symbols that were understandable to people of the age when they were written, as well as to peoples of all other ages. The king of Arabia did not comply with the Sheik's demand. He knew that some people fear enlightenment, because enlightenment threatens their vested interests. Education is not to be used to promote obscurantism.

The earth is not the geometric center of the universe, although it may be its spiritual center. It is a mere speck of dust in the cosmic spaces. Contrary to Bishop Ussher's calculations, the world did not appear in approximately its present state in 4004 BC. The estimates of the age of the universe given by modern cosmologists are still only rough approximations, which are revised (usually upward) as the methods of estimation are refined. Some cosmologists take the universe to be about 10 billion years old; others suppose that it may have existed, and will continue to exist, eternally. The origin of life on earth is dated tentatively between 3 and 5 billion years ago; manlike beings appeared relatively quite recently, between 2 and 4 million years ago. The estimates of the age of the earth, of the duration of the geologic and paleontologic eras, and of the antiquity of man's ancestors are now based mainly on radiometric evidence—the proportions of isotopes of certain chemical elements in rocks suitable for such studies.

Shiek bin Baz and his like refuse to accept the radiometric evidence, because it is a "mere theory." What is the alternative? One can suppose that the Creator saw fit to play deceitful tricks on geologists and biologists. He carefully arranged to have various rocks provided with isotope ratios just right to mislead us into thinking that certain rocks are 2 billion years old, others 2 million, while in fact they are only some 6000 years old. This kind of pseudo-explanation is not very new. One of the early antievolutionists, P. H. Gosse, published a book entitled *Omphalos* ("the Navel"). The gist of this amazing book is that Adam, though he had no mother, was created with a navel, and that fossils were placed by the Creator where we find them now—a deliberate act on His part, to give the appearance of great antiquity and geologic upheavals. It is easy to see the fatal flaw in all such notions. They are blasphemies, accusing God of absurd deceitfulness. This is as revolting as it is uncalled for.

Diversity of Living Beings

The diversity and the unity of life are equally striking and meaningful aspects of the living world. Between 1.5 and 2 million species of animals and plants have been described and studied; the number yet to be described is probably as great. The diversity of sizes, structures, and ways of life is staggering but fascinating. Here are just a few examples.

The foot-and-mouth disease virus is a sphere 8–12 [micrometers] in diameter. The blue whale reaches 30 [meters] in length and 135 [tons] in weight. The simplest viruses are parasites in cells of other organisms, reduced to barest essentials minute amounts of DNA or RNA, which subvert the biochemical machinery of the host cells to replicate their genetic information, rather than that of the host.

It is a matter of opinion, or of definition, whether viruses are considered living organisms or peculiar chemical substances. The fact that such differences of opinion can exist is in itself highly significant. It means that the borderline between living and inanimate matter is obliterated. At the opposite end of the simplicity-complexity spectrum you have vertebrate animals, including man. The human brain has some 12 billion neurons; the synapses between the neurons are perhaps a thousand times more numerous.

Some organisms live in a great variety of environments. Man is at the top of the scale in this respect. He is not only a truly cosmopolitan species but, owing to his technologic achievements, can survive for at least a limited time on the surface of the moon and in cosmic spaces. By contrast, some organisms are amazingly specialized. Perhaps the narrowest ecologic niche of all is that of a species of the fungus family Laboulbeniaceae, which grows exclusively on the rear portion of the elytra of the beetle *Aphenops cronei*, which is found only in some limestone caves in southern France. Larvae of the fly *Psilopa petrolei* develop in seepages of crude oil in California oilfields; as far as is known they occur nowhere else. This is the only insect able to live and feed in oil, and its adult can walk on the surface of the oil only as long as no body part other than the tarsi are in contact with the oil. Larvae of the fly *Drosophila carciniphila* develop only in the nephric grooves beneath the flaps of the third maxilliped of the land crab *Geocarcinus ruricola*, which is restricted to certain islands in the Caribbean.

Is there an explanation, to make intelligible to reason this colossal diversity of living beings? Whence came these extraordinary, seemingly whimsical and superfluous creatures, like the fungus *Laboulbenia*, the beetle *Aphenops cronei*, the flies *Psilopa petrolei* and *Drosophila carciniphila*, and many, many more apparent biologic curiosities? The only explanation that makes sense is that the organic diversity has evolved in response to the diversity of environment on the planet earth. No single species, however perfect and however versatile, could exploit all the opportunities for living. Every one of the millions of species has its own way of living and of getting sustenance from the environment. There are doubtless many other possible ways of living as yet unexploited by any existing species; but one thing is clear: with less organic diversity, some opportunities for living would remain unexploited. The evolutionary process tends to fill up the available ecologic niches. It does not do so consciously or deliberately; the relations between evolution and environment are more subtle and more interesting than that. The environment does not impose evolutionary changes on its inhabitants, as postulated by the now abandoned neo-Lamarckian theories. The best way to envisage the situation is as follows: the environment presents challenges to living species, to which the latter may respond by adaptive genetic changes.

An unoccupied ecologic niche, an unexploited opportunity for living, is a challenge. So is an environmental change, such as the Ice Age climate giving place to a warmer climate. Natural selection may cause a living species to respond to the challenge by adaptive genetic changes. These changes may enable the species to occupy the formerly empty ecologic niche as a new opportunity for living, or to resist the environmental change if it is unfavorable. But the response may or may not be successful. This depends on many factors, the chief of which is the genetic composition of the responding species at the time the response is called for. Lack of successful response may cause the species to become extinct. The evidence of fossils shows clearly that the eventual end of most evolutionary lines is extinction. Organisms now living are successful descendants of only a minority of the species that lived in the past and of smaller and smaller minorities the farther back you look. Nevertheless, the number of living species has not dwindled; indeed, it has probably grown with time. All this is understandable in the light of evolution theory; but what a senseless operation it would have been, on God's part, to fabricate a multitude of species ex nihilo and then let most of them die out!

There is, of course, nothing conscious or intentional in the action of natural selection. A biologic species does not say to itself, "Let me try tomorrow (or a million years from now) to grow in a different soil, or use a different food, or subsist on a different body part of a different crab." Only a human being could make such conscious decisions. This is why the species *Homo sapiens* is the apex of evolution. Natural selection is at one and the same time a blind and creative process. Only a creative and blind process could produce, on the one hand, the tremendous biologic success that is the human species and, on the other, forms of adaptedness as narrow and as constraining as those of the overspecialized fungus, beetle, and flies mentioned above.

Antievolutionists fail to understand how natural selection operates. They fancy that all existing species were generated by supernatural fiat a few thousand years ago, pretty much as we find them today. But what is the sense of having as many as 2 or 3 million species living on earth? If natural selection is the main factor that brings evolution about, any number of species is understandable: natural selection does not work according to a foreordained plan, and species are produced not because they are needed for some purpose but simply because there is an environmental opportunity and genetic wherewithal to make them possible. Was the Creator in a jocular mood when he made *Psilopa petrolei* for California oil fields and species of *Drosophila* to live exclusively on some body-parts of certain land crabs on only certain islands in the Caribbean? The organic diversity becomes, however, reasonable and understandable if the Creator has created the living world not by caprice but by evolution propelled by natural selection. It is wrong to hold creation and evolution as mutually exclusive alternatives. I am a creationist *and* an evolutionist. Evolution is God's, or Nature's method of creation. Creation is not an event that happened in 4004 BC; it is a process that began some 10 billion years ago and is still under way.

Unity of Life

The unity of life is no less remarkable than its diversity. Most forms of life are similar in many respects. The universal biologic similarities are particularly striking in the biochemical dimension. From viruses to man, heredity is coded in just two, chemically related substances: DNA and RNA. The genetic code is as simple as it is universal. There are only four genetic "letters" in DNA: adenine, guanine, thymine, and cytosine. Uracil replaces thymine in RNA. The entire evolutionary development of the living world has taken place not by invention of new "letters" in the genetic "alphabet" but by elaboration of ever-new combinations of these letters.

Not only is the DNA-RNA genetic code universal, but so is the method of translation of the sequences of the "letters" in DNA-RNA into sequences of amino acids in proteins. The same 20 amino acids compose countless different proteins in all, or at least in most, organisms. Different amino acids are coded by one to six nucleotide triplets in DNA and RNA. And the biochemical universals extend beyond the genetic code and its translation into proteins: striking uniformities prevail in the cellular metabolism of the most diverse living beings. Adenosine triphosphate, biotin, riboflavin, hemes, pyridoxin, vitamins K and B_{12}, and folic acid implement metabolic processes everywhere.

What do these biochemical or biologic universals mean? They suggest that life arose from inanimate matter only once and that all organisms, no matter now diverse, in other respects, conserve the basic features of the primordial life. (It is also possible that there were several, or even many, origins of life; if so, the progeny of only one of them has survived and inherited the earth.) But what if there was no evolution and every one of the millions of species were created by separate fiat? However offensive the notion may be to religious feeling and to reason, the antievolutionists must again accuse the Creator of cheating. They must insist that He deliberately arranged things exactly as if his method of creation was evolution, intentionally to mislead sincere seekers of truth.

The remarkable advances of molecular biology in recent years have made it possible to understand how it is that diverse organisms are constructed from such monotonously similar materials: proteins composed of only 20 kinds of amino acids and coded only by DNA and RNA, each with only four kinds of nucleotides. The method is astonishingly simple. All English words, sentences, chapters, and books are made up of sequences of 26 letters of the alphabet. (They can be represented also by only three signs of the Morse code: dot, dash, and gap.) The meaning of a word or a sentence is defined not so much by what letters it contains as by the sequences of these letters. It is the same with heredity: it is coded by the sequences of the genetic "letters" the nucleotides in the DNA. They are translated into the sequences of amino acids in the proteins.

Molecular studies have made possible an approach to exact measurements of degrees of biochemical similarities and differences among organisms. Some kinds of enzymes and other proteins are quasiuniversal, or at any rate widespread, in the living world. They are functionally similar in different living beings, in that they catalyze similar chemical reactions. But when such proteins are isolated and their structures determined chemically, they are often found to contain more or less different sequences of amino acids in different organisms. For example, the so-called alpha chains of hemoglobin have identical sequences of amino acids in man and the chimpanzee, but they differ in a single amino acid (out of 141) in the gorilla. Alpha chains of human hemoglobin differ from cattle hemoglobin in 17 amino acid substitutions, 18 from horse, 20 from donkey, 25 from rabbit, and 71 from fish (carp).

Cytochrome C is an enzyme that plays an important role in the metabolism of aerobic cells. It is found in the most diverse organisms, from man to molds. E. Margoliash, W. M. Fitch, and others have compared the amino acid sequences in cytochrome C in different branches of the living world. Most significant similarities as well as differences have been brought to light. The cytochrome C of different orders of mammals and birds differ in 2 to 17 amino acids, classes of vertebrates in 7 to 38, and vertebrates and insects in 23 to 41; and animals differ from yeasts and molds in 56 to 72 amino acids. Fitch and Margoliash prefer to express their findings in what are called "minimal mutational distances." It has been mentioned above that different amino acids are coded by different triplets of nucleotides in DNA of the genes; this code is now known. Most mutations involve substitutions of single nucleotides somewhere in the DNA chain coding for a given protein. Therefore, one can calculate the minimum numbers of single mutations needed to change the cytochrome C of one organism into that of another. Minimal mutational distances between human cytochrome C and the cytochrome C of other living beings are as follows:

Monkey	1	Chicken	18
Dog	13	Penguin	18
Horse	17	Turtle	19
Donkey	16	Rattlesnake	20
Pig	13	Fish (tuna)	31
Rabbit	12	Fly	33
Kangaroo	12	Moth	36
Duck	17	Mold	63
Pigeon	16	Yeast	56

It is important to note that amino acid sequences in a given kind of protein vary within a species as well as from species to species. It is evident that the differences among proteins at the level of species, genus, family, order, class, and phylum are compounded of elements that vary also among individuals within a species. Individual and group differences are only quantitatively, not qualitatively, different. Evidence supporting the above propositions is ample and is growing rapidly. Much work has been done in recent years on individual variations in amino acid sequences of hemoglobin of human blood. More that 100 variants have been detected. Most of them involve substitutions of single amino acids—substitutions that have arisen by genetic mutations in the persons in whom they are discovered or in their ancestors. As expected, some of these mutations are deleterious to their carriers, but others apparently are neutral or even favorable in certain environments. Some mutant hemoglobins have been found only in one person or in one family; others are discovered repeatedly among inhabitants of different parts of the world. I submit that all these remarkable findings make sense in the light of evolution: they are nonsense otherwise.

Comparative Anatomy and Embryology

The biochemical universals are the most impressive and the most recently discovered, but certainly they are not the only vestiges of creation by means of evolution. Comparative anatomy and embryology proclaim the evolutionary origins of the present inhabitants of the world. In 1555 Pierre Belon established the presence of homologous bones in the superficially very different skeletons of man and bird. Later anatomists traced the homologies in the skeletons, as well as in other organs, of all vertebrates. Homologies are also traceable in the external skeletons of arthropods as seemingly unlike as a lobster, a fly, and a butterfly. Examples of homologies can be multiplied indefinitely.

Embryos of apparently quite diverse animals often exhibit striking similarities. A century ago these similarities led some biologists (notably the German zoologist Ernst Haeckel) to be carried by their enthusiasm as far as to interpret the embryonic similarities as meaning that the embryo repeats in its development the evolutionary history of its species: it was said to pass through stages in which it resembles its remote ancestors. In other words, early-day biologists supposed that by studying embryonic development one can, as it were, read off the stages through which the evolutionary development had passed. This so-called biogenetic law is no longer credited in its original form. And yet embryonic similarities are undeniably impressive and significant.

Probably everybody knows the sedentary barnacles which seem to have no similarity to free-swimming crustaceans, such as the copepods. How remarkable that barnacles pass through a free-swimming larval stage, the nauplius! At that stage of its development a barnacle and a *Cyclops* look unmistakably similar. They are evidently relatives. The presence of gill slits in human embryos and in embryos of other terrestrial vertebrates is another famous example. Of course, at no stage of its development is a human embryo a fish, nor does it ever have functioning gills. But why should it have unmistakable gill slits unless its remote ancestors did respire with the aid of gills? It is the Creator again playing practical jokes?

Adaptive Radiation: Hawaii's Flies

There are about 2,000 species of drosophilid flies in the world as a whole. About a quarter of them occur in Hawaii, although the total area of the archipelago is only about that of the state of New Jersey. All but 17 of the species in Hawaii are endemic (found nowhere else). Furthermore, a great majority of the Hawaiian endemics do not occur throughout the archipelago: they are restricted to single islands or even to a part of an island. What is the explanation of this extraordinary proliferation of drosophilid species in so small a territory? Recent work of H. L. Carson, H. T. Spieth, D. E. Hardy, and others makes the situation understandable.

The Hawaiian Islands are of volcanic origin; they were never parts of any continent. Their ages are between 5.6 and 0.7 million years. Before man came there inhabitants were descendants of immigrants that had been transported across the ocean by air currents and other accidental means. A single drosophilid species, which arrived in Hawaii first, before there were numerous competitors, faced the challenge of an abundance of many unoccupied ecologic niches. Its descendants responded to this challenge by evolutionary adaptive radiation, the products of which are the remarkable Hawaiian drosophilids of today. To forestall a possible misunderstanding, let it be made clear that the Hawaiian endemics are by no means so similar to each other that they could be mistaken for variants of the same species; if anything, they are more diversified than are drosophilids elsewhere. The largest and the smallest drosophilid species are both Hawaiian. They exhibit an astonishing variety of behavior patterns. Some of them have become adapted to ways of life quite extraordinary for a drosophilid fly, such as being parasites in egg cocoons of spiders.

Oceanic islands other than Hawaii, scattered over the wide Pacific Ocean, are not conspicuously rich in endemic species of drosophilids. The most probable explanation of this fact is that these other islands were colonized by drosophilids after most ecologic niches had already been filled by earlier arrivals. This surely is a hypothesis, but it is a reasonable one. Antievolutionists might perhaps suggest an alternative hypothesis: in a fit of absentmindedness, the Creator went on manufacturing more and more drosophilid species for Hawaii, until there was an extravagant surfeit of them in this archipelago. I leave it up to you to decide which hypothesis makes sense.

Strength and Acceptance of the Theory

Seen in the light of evolution, biology is, perhaps, intellectually the most satisfying and inspiring science. Without that light it becomes a pile of sundry facts some of them interesting or curious but making no meaningful picture as a whole.

This is not to imply that we know everything that can and should be known about biology and about evolution. Any competent biologist is aware of a multitude of problems yet unresolved and of questions yet unanswered. After all, biologic research shows no sign of approaching completion; quite the opposite is true. Disagreements and clashes of opinion are rife among biologists, as they should be in a living and growing science. Antievolutionists mistake, or pretend to mistake, these disagreements as indications of dubiousness of the entire doctrine of evolution. Their favorite sport is stringing together quotations, carefully and sometimes expertly taken out of context, to show that nothing is really established or agreed upon among evolutionists. Some of my colleagues and myself have been amused and amazed to read ourselves quoted in a way showing that we are really antievolutionists under the skin.

Let me try to make crystal clear what is established beyond reasonable doubt, and what needs further study, about evolution. Evolution as a process that has always gone on in the history of the earth can be doubted only by those who are ignorant of the evidence or are resistant to evidence, owing to emotional blocks or to plain bigotry. By contrast, the mechanisms that bring evolution about certainly need study and clarification. There are no alternatives to evolution as history that can withstand critical examination. Yet we are constantly learning new and important facts about evolutionary mechanisms.

It is remarkable that more than a century ago Darwin was able to discern so much about evolution without having available to him the key facts discovered since. The development of genetics after 1900 especially of molecular genetics, in the last two decades has provided information essential to the understanding of evolutionary mechanisms. But much is in doubt and much remains to be learned. This is heartening and inspiring for any scientist worth his salt. Imagine that everything is completely known and that science has nothing more to discover: what a nightmare!

Does the evolutionary doctrine clash with religious faith? It does not. It is a blunder to mistake the Holy Scriptures for elementary textbooks of astronomy, geology, biology, and anthropology. Only if symbols are construed to mean what they are not intended to mean can there arise imaginary, insoluble conflicts. As pointed out above, the blunder leads to blasphemy: the Creator is accused of systematic deceitfulness.

One of the great thinkers of our age, Pierre Teilhard de Chardin, wrote the following: "Is evolution a theory, a system, or a hypothesis? It is much more—it is a general postulate to which all theories, all hypotheses, all systems must henceforward bow and which they must satisfy in order to be thinkable and true. Evolution is a light which illuminates all facts, a trajectory which all lines of thought must follow—This is what evolution is." Of course, some scientists, as well as some philosophers and theologians, disagree with some parts of Teilhard's teachings; the acceptance of his worldview falls short of universal. But there is no doubt at all that Teilhard was a truly and deeply religious man and that Christianity was the cornerstone of his worldview. Moreover, in his worldview science and faith were not segregated in watertight compartments, as they are with so many people. They were harmoniously fitting parts of his worldview. Teilhard was a creationist, but one who understood that the Creation is realized in this world by means of evolution.

SOURCE: Dobzhansky, Theodosius. "Nothing in Biology Makes Sense Except in the Light of Evolution." *American Biology Teacher*, 1973.

REPEAL OF THE BUTLER ACT (1967)

The Scopes Trial took place in 1925, in Dayton, Tennessee. Although the defendant, John Scopes, was convicted for teaching evolution in violation of the state's recently passed law, science was often seen as the victor. William Jennings Bryan died of a heart attack shortly after the trial concluded, probably in part as a result of the exertions of prosecuting Scopes. The defense, led by Clarence Darrow, provided such a reasonable and vivid portrayal of the place of science in modern society that most reports of the trial provided a boost to public acceptance of evolution. Since Scopes' conviction was later overturned due to a technicality in the sentencing, and the law was never enforced or challenged again, people typically assume it was dropped from the books shortly thereafter. In fact, the law remained in effect until the late 1960s, during a time when science had established evolution more firmly than ever, and when religious groups were mounting a new basis for challenging it. The heavily religious and biblical wording of the 1925 Tennessee law conflicted with the new challenge. It was formally repealed in 1967.

Public Acts of the State of Tennessee Passed by the Eighty-Fifth General Assembly 1967

AN ACT to repeal Section 498—1922, Tennessee Code Annotated, prohibiting the teaching of evolution.

Be it enacted by the General Assembly of the State of Tennessee:

Section 1. Section 49—1922, Tennessee Code Annotated, is repealed.

Section 2. This Act shall take effect September 1, 1967.

Passed: May 13, 1967

James H. Cummings, *Speaker of the House of Representatives*

Frank C. Gorrell, *Speaker of the Senate*

Approved: May 17, 1967.

Buford Ellington, Governor.

SOURCE: Repeal of the Butler Act, Public Acts of the State of Tennessee. 1967.

EPPERSON v. ARKANSAS (1968)

In this case, a forty-year-old law on the books in the State of Arkansas came before the U. S. Supreme Court. They would have to decide whether the law forbidding the teaching of Darwinian evolution as an explanation for human origins could be enforced, after the Arkansas State Supreme Court upheld the law without clarifying the legal questions surrounding it. In the decision handed down by the U. S. Court, the status of evolution as acceptable scientific theory carried some weight, although the primary concern was whether the Arkansas law violated the Fourteenth Amendment of the Constitution, or the First Amendment. At this point in American history, few serious objections to evolution remained, and the law itself was seen as an anachronism. Opponents of the law hoped to remove this relic of early twentieth century legislation, but like the repeal of the Butler Act in Tennessee, eliminating biblical creationism paved the way for creation science in the Constitutional battles that ensued.

Mr. Justice Fortas delivered the opinion of the Court.

I.

This appeal challenges the constitutionality of the "anti-evolution" statute which the State of Arkansas adopted in 1928 to prohibit the teaching in its public schools and universities of the theory that man evolved from other species of life. The statute was a product of the upsurge of "fundamentalist" religious fervor of the twenties. The Arkansas statute was an adaptation of the famous Tennessee "monkey law" which that State adopted in 1925. The constitutionality of the Tennessee law was upheld by the Tennessee Supreme Court in the celebrated Scopes case in 1927.

The Arkansas law makes it unlawful for a teacher in any state-supported school or university "to teach the theory or doctrine that mankind ascended or descended from a lower order of animals," or "to adopt or use in any such institution a textbook that teaches" this theory. Violation is a misdemeanor and subjects the violator to dismissal from his position.

The present case concerns the teaching of biology in a high school in Little Rock. According to the testimony, until the events here in litigation, the official textbook furnished for the high school biology course did not have a section on the Darwinian Theory. Then, for the academic year 1965–1966, the school administration, on recommendation of the teachers of biology in the school system, adopted and prescribed a textbook which contained a chapter setting forth "the theory about the origin. . . of man from a lower form of animal."

Susan Epperson, a young woman who graduated from Arkansas' school system and then obtained her master's degree in zoology at the University of Illinois, was employed by the Little Rock school system in the fall of 1964 to teach 10th grade biology at Central High School. At the start of the next academic year, 1965, she was confronted by the new textbook (which one surmises from the record was not unwelcome to her). She faced at least a literal dilemma because she was supposed to use the new textbook for classroom instruction and presumably to teach the statutorily condemned chapter; but to do so would be a criminal offense and subject her to dismissal.

She instituted the present action in the Chancery Court of the State, seeking a declaration that the Arkansas statute is void and enjoining the State and the defendant officials of the Little Rock school system from dismissing her for violation of the statute's provisions. H. H. Blanchard, a parent of children attending the public schools, intervened in support of the action.

The Chancery Court, in an opinion by Chancellor Murray O. Reed, held that the statute violated the Fourteenth Amendment to the United States Constitution. The court noted that this Amendment encompasses the prohibitions upon state interference with freedom of speech and thought which are contained in the First Amendment. Accordingly, it held that the challenged statute is unconstitutional because, in violation of the First Amendment, it "tends to hinder the quest for knowledge, restrict the freedom to learn, and restrain the freedom to teach." In this perspective, the Act, it held, was an unconstitutional and void restraint upon the freedom of speech guaranteed by the Constitution.

On appeal, the Supreme Court of Arkansas reversed. Its two-sentence opinion is set forth in the margin. It sustained the statute as an exercise of the State's power to specify the curriculum in public schools. It did not address itself to the competing constitutional considerations.

Appeal was duly prosecuted to this Court under 28 U. S. C. 1257 (2). Only Arkansas and Mississippi have such "anti-evolution" or "monkey" laws on their books. There is no record of any prosecutions in Arkansas under its statute. It is possible that the statute is presently more of a curiosity than a vital fact of life in these States. Nevertheless, the present case was brought, the appeal as of right is properly here, and it is our duty to decide the issues presented.

II.

At the outset, it is urged upon us that the challenged statute is vague and uncertain and therefore within the condemnation of the Due Process Clause of the Fourteenth Amendment. The contention that the Act is vague and uncertain is supported by language in the brief opinion of Arkansas' Supreme Court. That court, perhaps reflecting the discomfort which the statute's quixotic prohibition necessarily engenders in the modern mind, stated that it "expresses no opinion" as to whether the Act prohibits "explanation" of the theory of evolution or merely forbids "teaching that the theory is true." Regardless of this uncertainty, the court held that the statute is constitutional.

On the other hand, counsel for the State, in oral argument in this Court, candidly stated that, despite the State Supreme Court's equivocation, Arkansas would interpret the statute "to mean that to make a student aware of the theory. . . just to teach that there was such a theory" would be grounds for dismissal and for prosecution under the statute; and he said "that the Supreme Court of Arkansas' opinion should be interpreted in that manner." He said: "If Mrs. Epperson would tell her students that 'Here is Darwin's theory, that man ascended or descended from a lower form of being,' then I think she would be under this statute liable for prosecution."

In any event, we do not rest our decision upon the asserted vagueness of the statute. On either interpretation of its language, Arkansas' statute cannot stand. It is of no moment whether the law is deemed to prohibit mention of Darwin's theory, or to forbid any or all of the infinite varieties of communication embraced within the term "teaching." Under either interpretation, the law must be stricken because of its conflict with the constitutional prohibition of state laws respecting an establishment of religion or prohibiting the free exercise thereof. The overriding fact is that Arkansas' law selects from the body of knowledge a particular segment which it proscribes for the sole reason that it is deemed to conflict with a particular religious doctrine; that is, with a particular interpretation of the Book of Genesis by a particular religious group.

III.

The antecedents of today's decision are many and unmistakable. They are rooted in the foundation soil of our Nation. They are fundamental to freedom.

Government in our democracy, state and national, must be neutral in matters of religious theory, doctrine, and practice. It may not be hostile to any religion or to the advocacy of no-religion; and it may not aid, foster, or promote one religion or religious theory against another or even against the militant opposite. The First Amendment mandates governmental neutrality between religion and religion, and between religion and nonreligion.

As early as 1872, this Court said: "The law knows no heresy, and is committed to the support of no dogma, the establishment of no sect." *Watson v. Jones*, 13 Wall. 679, 728. This has been the interpretation of the great First Amendment which this Court has applied in the many and subtle problems which the ferment of our national life has presented for decision within the Amendment's broad command.

Judicial interposition in the operation of the public school system of the Nation raises problems requiring care and restraint. Our courts, however, have not failed to apply the First Amendment's mandate in our educational system where essential to safeguard the fundamental values of freedom of speech and inquiry and of belief. By and large, public education in our Nation is committed to the control of state and local authorities. Courts do not and cannot intervene in the resolution of conflicts which arise in the daily operation of school systems and which do not directly and sharply implicate basic constitutional values. On the other hand, "[t]he vigilant protection of constitutional freedoms is nowhere more vital than in the community of American schools," *Shelton v. Tucker*, 364 U. S. 479, 487

(1960). As this Court said in *Keyishian v. Board of Regents*, the First Amendment "does not tolerate laws that cast a pall of orthodoxy over the classroom." 385 U. S. 589, 603 (1967).

The earliest cases in this Court on the subject of the impact of constitutional guarantees upon the classroom were decided before the Court expressly applied the specific prohibitions of the First Amendment to the States. But as early as 1923, the Court did not hesitate to condemn under the Due Process Clause "arbitrary" restrictions upon the freedom of teachers to teach and of students to learn. In that year, the Court, in an opinion by Justice McReynolds, held unconstitutional an Act of the State of Nebraska making it a crime to teach any subject in any language other than English to pupils who had not passed the eighth grade. The State's purpose in enacting the law was to promote civic cohesiveness by encouraging the learning of English and to combat the "baneful effect" of permitting foreigners to rear and educate their children in the language of the parents' native land. The Court recognized these purposes, and it acknowledged the State's power to prescribe the school curriculum, but it held that these were not adequate to support the restriction upon the liberty of teacher and pupil. The challenged statute, it held, unconstitutionally interfered with the right of the individual, guaranteed by the Due Process Clause, to engage in any of the common occupations of life and to acquire useful knowledge. *Meyer v. Nebraska*, 262 U. S. 390 (1923). See also *Bartels v. Iowa*, 262 U. S. 404 (1923).

For purposes of the present case, we need not re-enter the difficult terrain which the Court, in 1923, traversed without apparent misgivings. We need not take advantage of the broad premise which the Court's decision in Meyer furnishes, nor need we explore the implications of that decision in terms of the justiciability of the multitude of controversies that beset our campuses today. Today's problem is capable of resolution in the narrower terms of the First Amendment's prohibition of laws respecting an establishment of religion or prohibiting the free exercise thereof.

There is and can be no doubt that the First Amendment does not permit the State to require that teaching and learning must be tailored to the principles or prohibitions of any religious sect or dogma. In *Everson v. Board of Education*, this Court, in upholding a state law to provide free bus service to school children, including those attending parochial schools, said: "Neither [a State nor the Federal Government] can pass laws which aid one religion, aid all religions, or prefer one religion over another." 330 U. S. 1, 15 (1947).

At the following Term of Court, in *McCollum v. Board of Education*, 333 U. S. 203 (1948), the Court held that Illinois could not release pupils from class to attend classes of instruction in the school buildings in the religion of their choice. This, it said, would involve the State in using tax-supported property for religious purposes, thereby breaching the "wall of separation" which, according to Jefferson, the First Amendment was intended to erect between church and state. Id., at 211. See also *Engel v. Vitale*, 370 U. S. 421 (1962); *Abington School District v. Schempp*, 374 U. S. 203 (1963). While study of religions and of the Bible from a literary and historic viewpoint, presented objectively as part of a secular program of education, need not collide with the First Amendment's prohibition, the State may not adopt programs or practices in its public schools or colleges which "aid or oppose" any religion. Id., at 225. This prohibition is absolute. It forbids alike the preference of a religious doctrine or the prohibition of theory which is deemed antagonistic to a particular dogma. As Mr. Justice Clark stated in *Joseph Burstyn, Inc. v. Wilson*, "the state has no legitimate interest in protecting any or all religions from views distasteful to them. . . ." 343 U. S. 495, 505 (1952). The test was stated as follows in *Abington School District v. Schempp*, supra, at 222: "[W]hat are the purpose and the primary effect of the enactment? If either is the advancement or inhibition of religion then the enactment exceeds the scope of legislative power as circumscribed by the Constitution."

These precedents inevitably determine the result in the present case. The State's undoubted right to prescribe the curriculum for its public schools does not carry with it the right to prohibit, on pain of criminal penalty, the teaching of a scientific theory or doctrine where that prohibition is based upon reasons that violate the First Amendment. It is much too late to argue that the State may impose upon the teachers in its schools any conditions that it chooses, however restrictive they may be of constitutional guarantees. *Keyishian v. Board of Regents*, 385 U. S. 589, 605–606 (1967).

In the present case, there can be no doubt that Arkansas has sought to prevent its teachers from discussing the theory of evolution because it is contrary to the belief of some that the Book of Genesis must be the exclusive source of doctrine as to the origin of man. No suggestion has been made that Arkansas' law may be justified by considerations of state policy other than the religious views of some of its citizens. It is clear that fundamentalist sectarian conviction was and is the law's reason for existence. Its antecedent, Tennessee's "monkey law," candidly stated its purpose: to make it unlawful "to teach any theory that denies the story of the Divine Creation of man as taught in the Bible, and to teach instead that man has descended from a lower order of animals." Perhaps the sensational publicity attendant upon the Scopes trial induced Arkansas to adopt less explicit language. It eliminated Tennessee's reference to "the story of the Divine Creation of man" as taught in the Bible, but there is no doubt that the motivation for the law was the same: to suppress the teaching of a theory which, it was thought, "denied" the divine creation of man.

Arkansas' law cannot be defended as an act of religious neutrality. Arkansas did not seek to excise from the curricula of its schools and universities all discussion of the origin of man. The law's effort was confined to an attempt to blot out a particular theory because of its supposed conflict with the biblical account, literally read. Plainly, the law is contrary to the mandate of the First, and in violation of the Fourteenth, Amendment to the Constitution.

The judgment of the Supreme Court of Arkansas is reversed.

Mr. Justice Black, concurring.

I am by no means sure that this case presents a genuinely justiciable case or controversy. Although Arkansas Initiated Act No. 1, the statute alleged to be unconstitutional, was passed by the voters of Arkansas in 1928, we are informed that there has never been even a single attempt by the State to enforce it. And the pallid, unenthusiastic, even apologetic defense of the Act presented by the State in this Court indicates that the State would make no attempt to enforce the law should it remain on the books for the next century. Now, nearly 40 years after the law has slumbered on the books as though dead, a teacher alleging fear that the State might arouse from its lethargy and try to punish her has asked for a declaratory judgment holding the law unconstitutional. She was subsequently joined by a parent who alleged his interest in seeing that his two then school-age sons "be informed of all scientific theories and hypotheses. . . ." But whether this Arkansas teacher is still a teacher, fearful of punishment under the Act, we do not know. It may be, as has been published in the daily press, that she has long since given up her job as a teacher and moved to a distant city, thereby escaping the dangers she had imagined might befall her under this lifeless Arkansas Act. And there is not one iota of concrete evidence to show that the parent-intervenor's sons have not been or will not be taught about evolution. The textbook adopted for use in biology classes in Little Rock includes an entire chapter dealing with evolution. There is no evidence that this chapter is not being freely taught in the schools that use the textbook and no evidence that the intervenor's sons, who were 15 and 17 years old when this suit was brought three years ago, are still in high school or yet to take biology. Unfortunately, however, the State's languid interest in the case has not prompted it to keep this Court informed concerning facts that might easily justify dismissal of this alleged lawsuit as moot or as lacking the qualities of a genuine case or controversy.

Notwithstanding my own doubts as to whether the case presents a justiciable controversy, the Court brushes aside these doubts and leaps headlong into the middle of the very broad problems involved in federal intrusion into state powers to decide what subjects and school-books it may wish to use in teaching state pupils. While I hesitate to enter into the consideration and decision of such sensitive state-federal relationships, I reluctantly acquiesce. But, agreeing to consider this as a genuine case or controversy, I cannot agree to thrust the Federal Government's long arm the least bit further into state school curriculums than decision of this particular case requires. And the Court, in order to invalidate the Arkansas law as a violation of the First Amendment, has been compelled to give the State's law a broader meaning than the State Supreme Court was willing to give it. The Arkansas Supreme Court's opinion, in its entirety, stated that:

> Upon the principal issue, that of constitutionality, the court holds that Initiated Measure No. 1 of 1928, Ark. Stat. Ann. 80–1627 and 80–1628 (Repl. 1960), is a valid exercise of the state's power to specify the curriculum in its public schools. The court expresses no opinion on the question whether the Act prohibits any explanation of the theory of evolution or merely prohibits teaching that the theory is true; the answer not being necessary to a decision in the case, and the issue not having been raised.

It is plain that a state law prohibiting all teaching of human development or biology is constitutionally quite different from a law that compels a teacher to teach as true only one theory of a given doctrine. It would be difficult to make a First Amendment case out of a state law eliminating the subject of higher mathematics, or astronomy, or biology from its curriculum. And, for all the Supreme Court of Arkansas has said, this particular Act may prohibit that and nothing else. This Court, however, treats the Arkansas Act as though it made it a misdemeanor to teach or to use a book that teaches that evolution is true. But it is not for this Court to arrogate to itself the power to determine the scope of Arkansas statutes. Since the highest court of Arkansas has deliberately refused to give its statute that meaning, we should not presume to do so.

It seems to me that in this situation the statute is too vague for us to strike it down on any ground but that: vagueness. Under this statute as construed by the Arkansas Supreme Court, a teacher cannot know whether he is forbidden to mention Darwin's theory at all or only free to discuss it as long as he refrains from contending that it is true. It is an established rule that a statute which leaves an ordinary man so doubtful about its meaning that he cannot know when he has violated it denies him the first essential of due process. See, e.g., *Connally v. General Construction Co.*, 269 U. S. 385, 391 (1926). Holding the statute too vague to enforce would not only follow long-standing constitutional precedents but it would avoid having this Court take unto itself the duty of a State's highest court to interpret and mark the boundaries of the State's laws. And, more important, it would not place this Court in the unenviable position of violating the principle of leaving the States absolutely free to choose their own curriculums for their own schools so long as their action does not palpably conflict with a clear constitutional command.

The Court, not content to strike down this Arkansas Act on the unchallengeable ground of its plain vagueness, chooses rather to invalidate it as a violation of the Establishment of Religion Clause of the First Amendment. I would not decide this case on such a sweeping ground for the following reasons, among others.

1. In the first place I find it difficult to agree with the Court's statement that "there can be no doubt that Arkansas has sought to prevent its teachers from discussing the theory of evolution because it is contrary to the belief of some that the Book of Genesis must be the exclusive source of doctrine as to the origin of man." It may be instead that the people's motive was merely that it would be best to remove this controversial subject from its schools; there is no reason I can imagine why a State is without power to withdraw from its curriculum any subject deemed too emotional and controversial for its public schools. And this Court has consistently held that it is not for us to invalidate a statute because of our views that the "motives" behind its passage were improper; it is simply too difficult to determine what those motives were. See, e.g., *United States v. O'Brien*, 391 U. S. 367, 382–383 (1968).

2. A second question that arises for me is whether this Court's decision forbidding a State to exclude the subject of evolution from its schools infringes the religious freedom of those who consider evolution an anti-religious doctrine. If the theory is considered anti-religious, as the Court indicates, how can the State be bound by the Federal Constitution to permit its teachers to advocate such an "anti-religious" doctrine to schoolchildren? The very cases cited by the Court as supporting its conclusion hold that the State must be neutral, not favoring one religious or anti-religious view over another. The Darwinian theory is said to challenge the Bible's story of creation; so too have some of those who believe in the Bible, along with many others, challenged the Darwinian theory. Since there is no indication that the literal Biblical doctrine of the origin of man is included in the curriculum of Arkansas schools, does not the removal of the subject of evolution leave the State in a neutral position toward these supposedly competing religious and anti-religious doctrines? Unless this Court is prepared simply to write off as pure nonsense the views of those who consider evolution an anti-religious doctrine, then this issue presents problems under the Establishment Clause far more troublesome than are discussed in the Court's opinion.

3. I am also not ready to hold that a person hired to teach school children takes with him into the classroom a constitutional right to teach sociological, economic, political, or religious subjects that the school's managers do not want discussed. This Court has said that the rights of free speech "while fundamental in our democratic society, still do not mean that everyone with opinions or beliefs to express may address a group at any public place and at any time." *Cox v. Louisiana*, 379 U. S. 536, 554; *Cox v. Louisiana*, 379 U. S. 559, 574. I question whether it is absolutely certain, as the Court's opinion indicates, that "academic freedom" permits a teacher to breach his contractual agreement to teach only the subjects designated by the school authorities who hired him.

Certainly the Darwinian theory, precisely like the Genesis story of the creation of man, is not above challenge. In fact the Darwinian theory has not merely been criticized by religionists but by scientists, and perhaps no scientist would be willing to take an oath and swear that everything announced in the Darwinian theory is unquestionably true. The Court, it seems to me, makes a serious mistake in bypassing the plain, unconstitutional vagueness of this statute in order to reach out and decide this troublesome, to me, First Amendment question. However wise this Court may be or may become hereafter, it is doubtful that, sitting in Washington, it can successfully supervise and censor the curriculum of every public school in every hamlet and city in the United States. I doubt that our wisdom is so nearly infallible.

I would either strike down the Arkansas Act as too vague to enforce, or remand to the State Supreme Court for clarification of its holding and opinion.

Mr. Justice Harlan, concurring.

I think it deplorable that this case should have come to us with such an opaque opinion by the State's highest court. With all respect, that court's handling of the case savors of a studied effort to avoid coming to grips with this anachronistic statute and to "pass the buck" to this Court. This sort of temporizing does not make for healthy operations between the state and federal judiciaries. Despite these observations, I am in agreement with this Court's opinion that, the constitutional claims having been properly raised and necessarily decided below, resolution of the matter by us cannot properly be avoided. See, e.g., *Chicago Life Insurance Co. v. Needles*, 113 U.S. 574, 579 (1885).

I concur in so much of the Court's opinion as holds that the Arkansas statute constitutes an "establishment of religion" forbidden to the States by the Fourteenth Amendment. I do not understand, however, why the Court finds it necessary to explore at length appellants' contentions that the statute is unconstitutionally vague and that it interferes with free speech, only to conclude that these issues need not be decided in this case. In the process of not deciding them, the Court obscures its otherwise straightforward holding, and opens its opinion to possible implications from which I am constrained to disassociate myself.

SOURCE: Susan Epperson et al. v. Arkansas. 393 U.S. 97; 89 S. Ct. 266; 21 L. Ed. 2 d 228; 1968 U.S. LEXIS 328.

TERMS

Creation Science—provides an alternative to atheistic evolution. The foundation of Creation Science suggests that a creator established the universe and all living forms according to processes that are no longer in operation. The processes that currently govern the operation of the universe can be studied by science and may reveal evidence of creation. The true account given in the Bible coincides perfectly with the "functioning completeness" of the universe, even if some evidence appears contrary to that completeness. Scientists who embrace creation challenge the evidence for evolution often insist that evolutionists are blinded by a commitment to atheism and cannot see objectively the evidence of a young Earth and the stasis of species.

Evolutionary Synthesis—suggests a time when scientists pursuing disparate biological enterprises came together in aligning the concepts of genetics and evolution. After Gregor Mendel's work on garden peas became widely known in 1900, many biologists accepted genetic change and individual inheritance as the basis for understanding variation among individuals and among species. Since Darwin's explanation for change and variation had existed for over four decades without such a basis, evolutionary biologists were reluctant to adopt a Mendelian view, and most saw the two as conflicting. The principles of population genetics laid down by the so-called architects of the evolutionary synthesis provided common ground for what became a more coherent account of change. Whether these views were significantly far apart and really required synthesis has become an ongoing historical debate, but for opponents of evolution, the split provided evidence of theoretical uncertainty that was crucial to the evolution-creation debates.

Fact—stands as the goal most observers would like to hold all of science to achieve. Scientists and the general public together seem to agree that facts are those observable, irrefutable, agreed-upon bits of evidence that represent a clear understanding of nature. Scientists are, in some respects, less inclined to ascribe to their ideas the status of fact. That is, while the repeated observation that pushing a book off of a table will establish the fact that it falls to the floor, the cause of it falling, gravity, is considered a theory. Gravity is a scientific theory, where competing explanations may exist and can be tested.

Hypothesis—represents the initial thinking of scientists attempting to explain a phenomenon. They may base their hypothesis on observations, facts about nature, and previous experience or inferences. The test their hypothesis repeated, hoping that further evidence will support it, but aware that a test that does not support it may force them to abandon or at least revise the hypothesis.

Naturalism—follows the belief that the universe can be explained from basic principles and examined through direct observation and inference. It denies a role for supernatural explanation. Intervention by an intelligent being or divine creator is unnecessary to explain and understand the natural world. The work of Isaac Newton, in formulating a mathematical and mechanical explanation for gravity, represented a major step in the development of naturalism. He explained motion on Earth and in the universe more generally according to the same principles. Darwin's explanation of evolution provided an important biological basis for advancing naturalism.

Theory—provides for scientists a working explanation for phenomena that relate to a particular set of causes. Unlike hypothesis, theory is well-established by repeated tests and usually by evidence from multiple fields of science. A theory may provoke new questions that need further testing, and occasionally one theory may replace a previous theory, based on new evidence, better data, and a change in the broad understanding of a scientific community. Such revolutions in the history of science are relatively rare and do not happen suddenly, but rather require extensive review of existing evidence and reconceptualization of previous interpretations. In everyday language, the phrase "just a theory" would not match scientists' use of the term. Scientists would more accurately use the word hypothesis in that context. in that context.

7

FROM EQUAL TIME TO THE DE-EMPHASIS OF EVOLUTION IN AMERICAN SCHOOLS

Following the modern evolutionary synthesis, biologists in the second half of the twentieth century boldly outlined the place of evolution within the broader field of biology. They identified connections from well-charted studies in embryology, genetics, and physiology to emerging subfields, such as ecology, molecular biology, and biochemistry. Everywhere they turned their attention, the implications of evolution appeared with unmistakable clarity. The convergence of evidence from diverse fields reinforced for them the conclusion that evolution was fact, and that natural selection as a mechanism for evolution served as the most potent explanation for change.

Evolution took an increasingly central position in the teaching of biology as well. Students being trained to contribute to science and to compete in a global scientific community needed analytical skills and the ability to invoke powerful explanatory processes. Professional scientific organizations partnered with government funding agencies to provide systematic curricula to meet those needs. Associations devoted to science education in particular joined with federal, state, and local education officials to develop criteria and review programs. These movements gained momentum during the cold war, when dominance in science was perceived to equal dominance in the global political arena.

Opponents to evolution, generally fundamentalist Christians, took the movement to institutionalize evolution in American education as a threat to their traditional worldview. They often equated evolution with secular humanism, and with atheism. The response gained momentum just as the last antievolution acts were stricken from state laws. Creationists chose to tackle the issue more directly. Rather than seek to outlaw the teaching of evolution, they demanded that creationism be taught side-by-side with evolution in the form of creation science. They provided arguments against the validity of evidence for evolution, and they offered alternative explanations for additional facts from nature. The arguments made their way again into the courtroom, with particular challenges attracting national attention just as the Scopes Trial had in the 1920s. Rather than a debate between science and religion, however, these challenges drew expert witnesses from philosophy, and hinged on legal interpretations that had far-reaching Constitutional implications. Each victory, and each defeat, was seen as only a small part in a larger war that would inevitably play out for decades to come.

GEORGE GAYLORD SIMPSON, COLIN S. PITTENDRIGH, AND LEWIS H. TIFFANY, *LIFE* (1958)[1]

George Gaylord Simpson (1902–1984) was one of the architects of the modern synthesis in evolutionary biology, as well as being an author of one of the most widely used textbooks in biology education at the advanced high school and introductory college levels, *Life*. The book went through numerous editions, but the fundamental attention given to evolution throughout the book reflects Simpson's commitment and illustrates the centrality of evolution in biology throughout the middle of the twentieth century. He indicated the importance of evolution as a means of approaching science empirically and historically in order to provide students with the tools they would need to become successful scientists in biology or any other field. When the book first appeared, little opposition could be found to this approach. Textbook authors became more circumspect in later years.

Biology and the Scientific Conceptual Scheme

Origins of Science

Men have surely been asking questions about the world and proposing answers ever since the human species began. What questions are asked and how answers are sought always depend on a conceptual scheme, an attitude toward the world and a set of postulates or beliefs about it. A conceptual scheme that sees the world as capricious and chaotic gives rise to no questions about order and natural law. A conceptual scheme that embraces magic and invisible spirits as causes of phenomena does not evoke answers testable by the phenomena themselves. Some such primitive conceptual schemes still have a lingering influence, but in the light of present knowledge, the result of science, we consider them superstitions. At a more sophisticated level are conceptual schemes that seek answers to questions about the material world not from that world but from dogmatic authority or by deduction from subjective philosophical premises. Such answers, not even regarded as subject to observational test, cannot be scientific.

The ancient world of Babylonia, Egypt, and Greece made great advances in knowledge of the world and laid the foundations for science and technology, but the ancients never developed a fully scientific conceptual scheme. Europe of the Middle Ages inherited from the ancient world a conceptual scheme that pictured the world as orderly but that sought answers about its orderliness in authority and in philosophical deduction more than in the world itself. It is an extraordinary fact that the scientific conceptual scheme arose so late in human history, within a single culture, that of Western Europe, and over a comparatively brief span of time, roughly definable as from Nicolaus Copernicus (1473–1543) to Charles Darwin (1809–1882). Beyond the already general concept of order in the world, its whole basis was the strict relation of questions and answers to the observation of the world, the seeking of natural explanations for natural phenomena, the proposal of testable hypotheses, and the testing of them.

---

[1] Excerpts from *Life: An Introduction to Biology*, copyright © 1965 by Harcourt, Inc. and renewed in 1993 by William S. Beck, Elizabeth Simpson Wurr, Helen S. Vishniac, and Joan S. Burns, reprinted by permission of Harcourt, Inc.

It seems quite simple and obvious to us that the only logical means of investigating the material world is by observing it. But we have grown up in a civilization in which the scientific conceptual scheme already existed and was generally (although not exclusively) accepted as the effective way of acquiring material knowledge. Copernicus' scientific theory that the earth circles the sun was firmly grounded in observation, but it was violently rejected by those whose conceptual scheme was still based on authority and philosophical deduction.

Early, Incompletely Scientific Biology

The scientific conceptual scheme arose first in the physical sciences. It brought about a revolution in human thought. Its insistence that natural phenomena obey natural, impersonal laws was a bitter and, at first, a deeply resented blow at age-old superstitions embedded in nonscientific conceptual schemes. Nevertheless, the scientific scheme responded to a refined concept of common sense, and it worked. As regards the physical aspects of the world, its acceptance was soon general, if not quite universal. Yet well into the nineteenth century the great majority of people—even among the most intelligent and most learned—clung to a conceptual scheme in which essential phenomena of life, and most particularly of human life, were believed to transcend physical laws and not to be amenable to strictly scientific explanation. Biology is as old a science as any. It had its roots in antiquity, and in its physical or plainly material aspects it became a true science along with physics, chemistry, astronomy, and the rest when the scientific conceptual scheme was developed in the sixteenth and seventeenth centuries. Until 1859, however, it was impeded by the common view that some of its subject matter did not fit into that scheme.

Biology Fully Enters the Scheme

It was in 1859 that Darwin's book *The Origin of Species* was published. This book accomplished two main objectives. First, it established the theory of evolution, the broadest generalization ever made about the interrelationships of living things. This theory (which in common speech we are now justified in calling a fact) states that all organisms have arisen from common ancestors by a natural, historical process of change and diversification. Second, the book propounded a theory to explain the causes and results of evolution. The most important point that had to be explained was the apparent purposiveness of life, the observation that organisms seem to be designed precisely for the functions they carry out. It was this, more than anything else, that had supported the claim that vital structures and processes could not have entirely natural causes and hence did not fit the scientific conceptual scheme.

Darwin's complex explanation was only partially successful, but its most essential element, natural selection, has stood the test of time and is accepted today in somewhat modified form. We shall return to that subject, and also to the problem of purpose in nature, in Chapters 15 and 16, after sufficient basis for comprehension has been laid. What is significant here is that Darwin sufficiently demonstrated that natural explanations for all of the material phenomena of life should be sought and can be found. Thus *The Origin of Species* actually accomplished a third objective, most important of all: it finally brought biology as a whole, in all its aspects, within the conceptual scheme of science.

SOURCE: Simpson, George Gaylord, Colin S. Pittendrigh, and Lewis H. Tiffany, *Life: An Introduction to Biology*. New York: Harcourt Brace, 1958.

HENRY MORRIS, "EVOLUTION, CREATION, AND THE PUBLIC SCHOOLS" (1973)

As opposition to the teaching of evolution grew, largely from the efforts of the Institute for Creation Research, Henry Morris (1918–2006) published further arguments against the validity of evolution as science. He demonstrated repeatedly that evolution required the adoption of certain beliefs, and even leaps of faith, in order to account for gaps in evidence and logic. Evolution came to stand in for liberal agendas of all sorts, and the significance of providing a balanced curriculum for students or parents who did not share the liberal worldview became paramount. Morris felt the pressure of maintaining a constitutionally valid position, separating church and state by not insisting on biblical creationism, and in the process charted a careful course whereby students, parents, and teachers could consistently challenge evolution in the classroom by presenting facts that highlighted the gaps scientists themselves would acknowledge.

One of the most amazing phenomena in the history of education is that a speculative philosophy based on no true scientific evidence could have been universally adopted and taught as scientific fact, in all the public schools. This philosophy has been made the very framework of modern education and the underlying premise in all textbooks. It constitutes the present world-view of liberal intellectuals in every field.

This is the philosophy of evolution. Although widely promoted as a scientific fact, evolution has never been proved scientifically. Some writers still call it the theory of evolution, but even this is too generous. A scientific hypothesis should be capable of being tested in some way, to determine whether or not it is true, but evolution cannot be tested. No laboratory experiment can either confirm or falsify a process which, by its very nature, requires millions of years to accomplish significant results.

Evolution is, therefore, neither fact, theory, nor hypothesis. It is a belief—and nothing more.

When creationists propose, however, that creation be taught in the schools along with evolution, evolutionists commonly react emotionally, rather than scientifically. Their "religion" of naturalism and humanism has been in effect the established religion of the state for a hundred years, and they fear competition.

In the present world, neither evolution nor creation is taking place, so far as can be observed (and science is supposed to be based on observation!). Cats beget cats and fruit-flies beget fruit-flies. Life comes only from life. There is nothing new under the sun.

Neither evolution nor creation is accessible to the scientific method, since they deal with origins and history, not with presently observable and repeatable events. They can, however, be formulated as scientific models, or frameworks, within which to predict and correlate observed facts. Neither can be proved; neither can be tested. They can only be compared in terms of the relative ease with which they can explain data which exist in the real world.

There are, therefore, sound scientific and pedagogical reasons why both models should be taught, as objectively as possible, in public classrooms, giving arguments pro and con for each. Some students and their parents believe in creation, some in evolution, and some are undecided. If creationists desire only the creation model to be taught, they should send their children to private schools which do this; if evolutionists want only evolution to be taught, they should provide private schools for that purpose. The public schools should be neutral and either teach both or teach neither.

This is clearly the most equitable and constitutional approach. Many people have been led to believe, however, that court decisions restricting "religious" teaching in the public schools apply to "creation" teaching and not to "evolution" teaching. Nevertheless, creationism is actually a far more effective scientific model than evolutionism, and evolution requires a far more credulous religious faith in the illogical and unproveable than does creation. An abundance of sound scientific literature is available today to document this statement, but few evolutionists have bothered to read any of it. Many of those who have read it have become creationists!

What can creationists do to help bring about a more equitable treatment of this vital issue in the public schools? How can they help their own children in the meantime? The following suggestions are in order of recommended priority. All involve effort and expense, but the stakes are high and the need is urgent.

1. Most basic is the necessity for each concerned creationist himself to become informed on the issue and the scientific facts involved. He does not need to be a scientist to do this, but merely to read several of the scholarly creationist books that are now available. He should also study creationist literature that demonstrates the fallacious nature of the various compromising positions (e.g., theistic evolution, day-age theory, gap theory, local flood theory, etc.) in order to be on solid ground in his own convictions.

2. He should then see that his own children and young people, as well as others for whom he is concerned, have access to similar literature on their own level. He also should be aware of the teachings they are currently receiving in school and help them find answers to the problems they are encountering. He should encourage them always to be gracious and respectful to the teacher, but also to look for opportunities (in speeches, term papers, quizzes, etc.) to show that, although they understand the arguments for evolution, the creationist model can also be held and presented scientifically.

3. If he learns of teachers who are obviously bigoted and unfair toward students of creationist convictions, it would be well for him to talk with the teacher himself, as graciously as possible, pointing out the true nature of the issue and requesting the teacher to present both points of view to the students. Under some circumstances, this might be followed up by similar talks with the principal and superintendent.

4. Many teachers and administrators are quite willing to present both viewpoints, but have been unaware that there does exist a solid scientific case for creation, and, therefore, they don't know how to do this. There is thus a great need for teachers, room libraries, and school libraries to be supplied with sound creationist literature. Perhaps some schools, or even districts, will be willing to provide such literature themselves. If not, the other alternative is for parental associations, churches, or individuals to take on such a project as a public service. If sound creationist books are conveniently available, many teachers (not all, unfortunately, but far more than at present) would be willing to use them and to encourage their students to use them.

5. Creationist parents, teachers, pastors, and others can join forces to sponsor meetings, seminars, teaching institutes, etc., in their localities. Qualified creationist scientists can be invited to speak at such meetings, and if adequate publicity (especially on a person-to-person basis) is given, a real community-wide impact can be made in this way. Especially valuable, when such invitations can be arranged, are opportunities for creationist scientists to speak at meetings of scientists or educators. Also such men can be invited to speak in churches or in other large gatherings of interested laymen.

6. Discussions can be held with officials at high levels (state education boards, district boards, superintendents, etc.) to acquaint them with the evidences supporting creation and the importance of the issue. They can be requested to inform the teachers of their state or district that the equal teaching of evolution and creation, not on a religious basis, but as scientific models, is both permitted and encouraged. Cases of unfair discrimination against creationist minorities in classrooms can be reported, and most officials at such levels are sufficiently concerned with the needs of all their constituents that, if they can first be shown there is a valid scientific case for creation and that evolution has at least as much religious character as does creation, they will quite probably favor such a request.

7. Public response can be made (always of a scientific, rather than emotional flavor) to newspaper stories, television programs, etc., which favor evolution. Those responses may be in the form of letters-to-the-editor, protest letters to sponsors, news releases, and other means.

8. Financial support should be provided for those organizations attempting in a systematic way to do scientific research, produce creationist textbooks and other literature, and to provide formal instruction from qualified scientists in the field of creationism. This can be done both through individual gifts and bequests and through budgeted giving by churches and other organizations.

It will be noted that no recommendation is made for political or legal pressure to force the teaching of creationism in the schools. Some well-meaning people have tried this, and it may serve the purpose of generating publicity for the creationist movement. In general, however, such pressures are self-defeating. "A man convinced against his will is of the same opinion still."

Force generates reaction, and this is especially true in such a sensitive and vital area as this. The hatchet job accomplished on the fundamentalists by the news media and the educational establishment following the Scopes trial in 1925 is an example of what could happen, in the unlikely event that favorable legislation or court decisions could be obtained by this route.

Reasonable persuasion is the better route. "The servant of the Lord must not strive; but be gentle unto all men, apt to teach, patient, in meekness instructing those that oppose themselves" (II Timothy 2:24, 25).

SOURCE: Morris, Henry. "Evolution, Creation, and the Public Schools," (1973) in *Creation: Acts, Facts, Impacts.* San Diego, CA: ICR Pub. Co., 1974.

HENRY MORRIS, "THE MATHEMATICAL IMPOSSIBILITY OF EVOLUTION" (1972)

Without insisting on the truth of biblical creation, and without even mentioning divine intervention, Morris consistently provided arguments against the validity of evolution. While scientists often conceded that there were gaps in the evidence, their opponents went farther in presenting how gaps in evidence and in logic proved fatal to any argument in favor of evolution. Relying on conservative estimates for how often mutations might arise, and how they might persist in living forms, Morris illustrated in straightforward fashion how unlikely evolution would be to create, even once, a simplified organism. By extension of that logic, the vast diversity of life, appearing even over vast stretches of apparent geological time, would be unfathomable. From there, readers could easily conclude that instead of evolution, the explanation for life on Earth required a designer and a creator.

According to the accepted theory of evolution today, the sole mechanism for producing evolution is that of random mutation and natural selection. Mutations are *random* changes in genetic systems. Natural selection is considered by evolutionists to be a sort of sieve which retains the "good" mutations and allows the others to pass away.

Now random changes in ordered systems almost always will decrease the amount of order in those systems, and therefore, nearly all mutations are harmful to the organisms which experience them. Nevertheless, the evolutionist insists that each complex organism in the world today has arisen by a long string of gradually accumulated, good mutations preserved by natural selection.

No one has ever actually *observed* a genuine mutation occurring in the natural environment which was beneficial, and therefore, retained by the selection process. For some reason, however, the idea has a certain persuasive quality about it and seems eminently reasonable to many people—until it is examined *quantitatively*, that is!

For example, consider a very simple organism composed of only 200 integrated and functioning parts, and the problem of deriving that organism by this type of process. Obviously, the organism must have started with only one part and then gradually built itself up over many generations into its 200-part organization. The developing organism, at each successive stage, must itself be integrated and functioning in its environment in order to survive until the next stage. Each successive stage, of course, becomes less likely than the preceding one, since it is far easier for a complex system to break down then [*sic*] to build itself up. A four-component integrated system can more easily mutate into a three-component system (or even a four-component non-functioning system) than into a five-component integrated system. If, at any step in the chain, the system mutates backward or downward, then it is either destroyed altogether or else moves backward.

Therefore, the successful production of a 200-component functioning organism requires, *at least*, 200 successive, successful mutations, each of which is highly unlikely. Even evolutionists recognize that true mutations are very rare, and beneficial mutations are *extremely* rare—not more than one out of a thousand mutations are beneficial, at the very most.

But let us give the evolutionist the benefit of every consideration. Assume that, at each mutational step, there is equally as much chance for it to be good as bad. Thus, the probability for the success of each mutation is assumed to be one out of two, or one-half. Elementary statistical theory shows that the probability of 200 successive mutations being successful is then $(1/2)^{200}$, or, one chance out of 10^{60}. The number 10^{60} if written out, would be "one" followed by sixty "zeros." In other words, the chance that a 200-component organism could be formed by mutation and natural selection is less than one chance out of a trillion, trillion, trillion, trillion, trillion! Lest anyone think that a 200-part system is unreasonably complex, it should be noted that even a one-celled plant or animal may have millions of molecular "parts."

The evolutionist might react by saying that even though any one such mutating organism might not be successful, surely some around the world would be, especially in the 10 billion years (or 10^{18} seconds) of assumed earth history. Therefore, let us imagine that every one of the earth's 10^{14} square feet of surface harbors a billion (i.e., 10^9) mutating systems and that each mutation requires one-half second (actually it would take far more time than this). Each system can thus go through its 200 mutations in 100 seconds and then, if it is unsuccessful, start over for a new try. In 10^{18} seconds, there can, therefore, be $10^{18}/10^2$, or 10^{16}, trials by each mutating system. Multiplying all these numbers together, there would be a total possible number of attempts to develop a 200-component system equal to 10^{14} (10^9) (10^{16}), or 10^{39} attempts. Since the probability against the success of any one of them is 10^{60}, it is obvious that the probability just one of these 10^{39} attempts might be successful is only one out of $10^{60}/10^{39}$ or 10^{21}.

All of this means that the chance that any kind of a 200-component integrated functioning organism could be developed by mutation and natural selection just once, anywhere in the world, in all the assumed expanse of geologic time, is less than one chance out of a billion trillion. What possible conclusion, therefore, can we derive from such considerations as this except that evolution by mutation and natural selection is mathematically and logically indefensible!

SOURCE: Morris, Henry. "The Mathematical Impossibility of Evolution," (1972) in *Creation: Acts, Facts, Impacts.* San Diego, CA: ICR Pub. Co., 1974.

WILLIAM R. OVERTON, MCLEAN *v.* ARKANSAS (1982)

When the State of Arkansas established a law requiring that equal time be spent in science classrooms on the theories of evolution and creation science, opponents of the new law wasted little time in bringing suit against the state Board of Education. Those opponents included religious leaders and organizations as well as parents, education societies, and civil liberties groups. Witnesses for the defense included well-known creationists, while witnesses for the plaintiffs included scientists and philosophers of science. The focus of the case was not to determine the truth of either evolution or creation, but to determine whether creation science could properly be taught as science. The plaintiffs held that creation science was, in fact, religion, and thus violated the U.S. Constitution's First Amendment. Creationists called to testify generally confirmed the strong connection between their scientific views and religious views, while scientists and philosophers demonstrated that creation science was not science according to the criteria generally adopted by the scientific community. As a result, the ruling came down, less than ten months after the law was enacted, that creation science did not meet the standards of science that could be taught in public schools because it violated the First Amendment in advancing a particular religious belief.

McLean v. Arkansas Board of Education Decision by U.S. District Court Judge William R. Overton Dated this January 5, 1982

Introduction

On March 19, 1981, the Governor of Arkansas signed into law Act 590 of 1981, entitled "Balanced Treatment for Creation-Science and Evolution-Science Act." The Act is codified as Ark. Stat. Ann. & 80-1663, *et seq.*, (1981 Supp.). Its essential mandate is stated in its first sentence: "Public schools within this State shall give balanced treatment to creation-science and to evolution-science." On May 27, 1981, this suit was filed challenging the constitutional validity of Act 590 on three distinct grounds.

Evolutionists and humanists who participated in the debate over teaching creation science in public schools often poked fun at the extent to which creationists explained the short geological and biological history of the earth. [Dorothy Sigler, Cover Illustration. *Creation/Evolution* (1985) 5:1. Reprinted by permission of the illustrator and the National Center for Science Education].

First, it is contended that Act 590 constitutes an establishment of religion prohibited by the First Amendment to the Constitution, which is made applicable to the states by the Fourteenth Amendment. Second, the plaintiffs argue the Act violates a right to academic freedom which they say is guaranteed to students and teachers by the Free Speech Clause of the First Amendment. Third, plaintiffs allege the Act is impermissibly vague and thereby violates the Due Process Clause of the Fourteenth Amendment.

The individual plaintiffs include the resident Arkansas Bishops of the United Methodist, Episcopal, Roman Catholic, and African Methodist Episcopal Churches, the principal official of the Presbyterian Churches in Arkansas, other United Methodist, Southern Baptist, and Presbyterian clergy, as well as several persons who sue as parents and next friends of minor children attending Arkansas public schools. One plaintiff is a high school biology teacher. All are also Arkansas taxpayers. Among the organizational plaintiffs are the American Jewish Congress, the Union of American Hebrew Congregations, the American Jewish Committee, the Arkansas Education Association, the National Association of Biology Teachers, and the National Coalition for Public Education and Religious Liberty, all of which sue on behalf of members living in Arkansas.

The defendants include the Arkansas Board of Education and its members, the Director of the Department of Education, and the State Textbooks and Instructional Materials Selecting Committee. The Pulaski County Special School District and its Directors and Superintendent were voluntarily dismissed by the plaintiffs at the pre-trial conference held October 1, 1981.

The trial commenced December 7, 1981, and continued through December 17, 1981. This Memorandum Opinion constitutes the Court's findings of fact and conclusions of law. Further orders and judgments will be in conformity with this opinion.

I

There is no controversy over the legal standards under which the Establishment Clause portion of this case must be judged. The Supreme Court has on a number of occasions expounded on the meaning of the clause, and the pronouncements are clear. Often the issue has arisen in the context of public education, as it has here. In *Everson v. Board of Education*, 330 U.S. 1, 15–16 (1947), Justice Black stated:

> The "establishment of religion" clause of the First Amendment means at least this: Neither a state nor the Federal Government can set up a church. Neither can pass laws which aid one religion, aid all religions, or prefer one religion over another. Neither can force nor influence a person to go to or to remain away from church against his will or force him to profess a belief or disbelief in any religion. No person can be punished for entertaining or professing religious beliefs or disbeliefs, for church-attendance or non-attendance. No tax, large or small, can be levied to support any religious activities or institutions, whatever they may be called, or what ever form they may adopt to teach or practice religion. Neither a state nor the Federal Government can, openly or secretly, participate in the affairs of any religious organizations or groups and *vice versa*. In the words of Jefferson, the clause...was intended to erect "a wall of separation between church and State."

The Establishment Clause thus enshrines two central values: voluntarism and pluralism. And it is in the area of the public schools that these values must be guarded most vigilantly.

> Designed to serve as perhaps the most powerful agency for promoting cohesion among a heterogeneous democratic people, the public school must keep scrupulously free from entanglement in the strife of sects. The preservation of the community from divisive conflicts, of Government from irreconcilable pressures by religious groups, or religion from censorship and coercion however subtly exercised, requires strict confinement of the State to instruction other than religious, leaving to the individual's church and home, indoctrination in the faith of his choice. [*McCollum v. Board of Education*, 333 U.S. 203, 216–217 (1948) (Opinion of Frankfurter, J., joined by Jackson, Burton, and Rutledge, J. J.)]

The specific formulation of the establishment prohibition has been refined over the years, but its meaning has not varied from the principles articulated by Justice Black in *Everson*. In *Abbington School District v. Schempp*, 374 U.S. 203, 222 (1963), Justice Clark stated that "to withstand the strictures of the Establishment Clause there must be a secular legislative purposed and a primary effect that neither advances nor inhibits religion." The court found it quite clear that the First Amendment does not permit a state to require the daily reading of the Bible in public schools, for "[s]urely the place of the Bible as an instrument of religion cannot be gainsaid." *Id.* at 224. Similarly, in *Engel v. Vitale*, 370 U.S. 421 (1962), the Court held that the First Amendment prohibited the New York Board of Regents from requiring the daily recitation of a certain prayer in the schools. With characteristic succinctness, Justice Black wrote: "Under [the First] Amendment's prohibition against governmental establishment of religion, as reinforced by the provisions of the Fourteenth Amendment, government in this country, be it state or federal, is without power to prescribe by law any particular form of prayer which is to be used as an official prayer in carrying on any program of governmentally sponsored religious activity." *Id.* at 430. Black also identified the objective at which the

Establishment Clause was aimed: "its first and most immediate purpose rested on the belief that a union of government and religion tends to destroy government and to degrade religion." *Id.* at 431.

Most recently, the Supreme Court has held that the clause prohibits a state from requiring the posting of the Ten Commandments in public school classrooms for the same reasons that officially imposed daily Bible reading is prohibited. *Stone v. Graham*, 449 U.S. 39 (1980). The opinion in *Stone* relies on the most recent formulation of the Establishment Clause test, that of *Lemon v. Kurtzman*, 403 U.S. 602, 612–613 (1971):

> First, the statute must have a secular legislative purpose; second, its principal or primary effect must be one that neither advances nor inhibits religion. . .; finally, the statute must not foster "an excessive government entanglement with religion." [*Stone v. Graham*, 449 U.S. at 40.]

It is under this three-part test that the evidence in this case must be judged. Failure on any of these grounds is fatal to the enactment.

II

The religious movement known as Fundamentalism began in nineteenth century America as part of evangelical Protestantism's response to social changes, new religious thought and Darwinism. Fundamentalists viewed these developments as attacks on the Bible and as responsible for a decline in traditional values.

The various manifestations of Fundamentalism have had a number of common characteristics, but a central premise has always been a literal interpretation of the Bible and a belief in the inerrancy of the Scriptures. Following World War I, there was again a perceived decline in traditional morality, and Fundamentalism focused on evolution as responsible for the decline. One aspect of their efforts, particularly in the south, was the promotion of statutes prohibiting the teaching of evolution in public schools. In Arkansas, this resulted in the adoption of Initiated Act 1 of 1929.

Between the 1920's and early 1960's, anti-evolutionary sentiment had a subtle but pervasive influence on the teaching of biology in public schools. Generally, textbooks avoided the topic of evolution and did not mention the name of Darwin. Following the launch of the Sputnik satellite by the Soviet Union in 1957, the National Science Foundation funded several programs designed to modernize the teaching of science in the nation's schools. The Biological Sciences Curriculum Study (BSCS), a nonprofit organization, was among those receiving grants for curriculum study and revision. Working with scientists and teachers, BSCS developed a series of biology texts which, although emphasizing different aspects of biology, incorporated the theory of evolution as a major theme. The success of the BSCS effort is shown by the fact that fifty percent of American school children currently use BSCS books directly and the curriculum is incorporated indirectly in virtually all biology texts. (Testimony of Mayer; Nelkin, Px 1).

In the early 1960's, there was again a resurgence of concern among Fundamentalists about the loss of traditional values and a fear of growing secularism in society. The Fundamentalist movement became more active and has steadily grown in numbers and political influence. There is an emphasis among current Fundamentalists on the literal interpretation of the Bible and the Book of Genesis as the sole source of knowledge about origins.

The term "scientific creationism" first gained currency around 1965 following publication of *The Genesis Flood* in 1961 by Whitcomb and Morris. There is undoubtedly some connection between the appearance of the BSCS texts emphasizing evolutionary thought and efforts of Fundamentalist to attach the theory. (Mayer).

In the 1960's and early 1970's, several Fundamentalist organizations were formed to promote the idea that the Book of Genesis was supported by scientific data. The terms "creation science" and "scientific creationism" have been adopted by these Fundamentalists as descriptive of their study of creation and the origins of man. Perhaps the leading creationist organization is the Institute for Creation Research (ICR), which is affiliated with the Christian Heritage College and supported by the Scott Memorial Baptist Church in San Diego, California. The ICR, through the Creation-Life Publishing Company, is the leading publisher of creation science material. Other creation science organizations include the Creation Science Research Center (CSRC) of San Diego and the Bible Science Association of Minneapolis, Minnesota. In 1963, the Creation Research Society (CRS) was formed from a schism in the American Scientific Affiliation (ASA). It is an organization of literal Fundamentalists who have the equivalent of a master's degree in some recognized area of science. A purpose of the organization is "to reach all people with the vital message of the scientific and historical truth about creation." Nelkin, *The Science Textbook Controversies and the Politics of Equal Time*, 66. Similarly, the CSRC was formed in 1970 from a split in the CRS. Its aim has been "to reach the 63 million children of the United States with the scientific teaching of Biblical creationism." *Id.* at 69.

Among creationist writers who are recognized as authorities in the field by other creationists are Henry M. Morris, Duane Gish, G. E. Parker, Harold S. Slusher, Richard B. Bliss, John W. Moore, Martin E. Clark, W. L. Wysong, Robert E. Kofahl, and Kelly L. Segraves. Morris is Director of ICR, Gish is Associate Director and Segraves is associated with CSRC.

Creationists view evolution as a source of society's ills, and the writings of Morris and Clark are typical expressions of that view.

Evolution is thus not only anti-Biblical and anti-Christian, but it is utterly unscientific and impossible as well. But it has served effectively as the pseudo-scientific basis of atheism, agnosticism, socialism, fascism, and numerous other false and dangerous philosophies over the past century. [Morris and Clark, *The Bible Has the Answer*, (Px 31 and Pretrial Px 89)].

Creationists have adopted the view of Fundamentalists generally that there are only two positions with respect to the origins of the earth and life: belief in the inerrancy of the Genesis story of creation and of a worldwide flood as fact, or a belief in what they call evolution.

Henry Morris has stated, "It is impossible to devise a legitimate means of harmonizing the Bible with evolution." Morris, "evolution and the Bible," *ICR Impact Series* Number 5 (undated, unpaged), quoted in Mayer, Px 8, at 3. This dualistic approach to the subject of origins permeates the creationist literature.

The creationist organizations consider the introduction of creation science into the public schools part of their ministry. The ICR has published at least two pamphlets containing suggested methods for convincing school boards, administrators, and teachers that creationism should be taught in public schools. The ICR has urged its proponents to encourage school officials to voluntarily add creationism to the curriculum.

Citizens For Fairness In Education is an organization based in Anderson, South Carolina, formed by Paul Ellwanger, a respiratory therapist who is trained in neither law nor science. Mr. Ellwanger is of the opinion that evolution is the forerunner of many social ills, including Nazism, racism and abortion (Ellwanger Depo. at 32-34). About 1977, Ellwanger collected several proposed legislative acts with the idea of preparing a model state act requiring the teaching of creationism as science in opposition to evolution. One of the proposals he collected was prepared by Wendell Bird, who is now a staff attorney for ICR. From these various proposals, Ellwanger prepared a "model act" which calls for "balanced treatment" of "scientific creationism" and "evolution" in public schools. He circulated the proposed act to various people and organizations around the country.

Mr. Ellwanger's views on the nature of creation science are entitled to some weight since he personally drafted the model act which became Act 590. His evidentiary deposition with exhibits and unnumbered attachments (produced in response to a subpoena *duces tecum* speaks to both the intent of the Act and the scientific merits of creation science. Mr. Ellwanger does not believe creation science is a science. In a letter to Pastor Robert E. Hays he states, "While neither evolution nor creation can qualify as a scientific theory, and since it is virtually impossible at this point to educate the whole world that evolution is not a true scientific theory, we have freely used these terms—the evolution theory and the theory of scientific creationism—in the bill's text." (Unnumbered attachment to Ellwanger Depo., at 2.) He further states in a letter to Mr. Tom Bethell, "As we examine evolution (remember, we're not making any scientific claims for creation, but we are challenging evolution's claim to be scientific. . ." (Unnumbered attachment to Ellwanger Depo. at 1.)

Ellwanger's correspondence on the subject shows an awareness that Act 590 is a religious crusade, coupled with a desire to conceal this fact. In a letter to State Senator Bill Keith of Louisiana, he says, "I view this whole battle as one between God and anti-God forces, though I know there are a large number of evolutionists who believe in God." And further, ". . .it behooves Satan to do all he can to thwart our efforts and confuse the issue at every turn." Yet Ellwanger suggest to Senator Keith, "If you have a clear choice between having grassroots leaders of this statewide bill promotion effort to be ministerial or non-ministerial, be sure to opt for the non-ministerial. It does the bill effort no good to have ministers out there in the public forum and the adversary will surely pick at this point. . .Ministerial persons can accomplish a tremendous amount of work from behind the scenes, encouraging their congregations to take the organizational and P. R. initiatives. And they can lead their churches in storming Heaven with prayers for help against so tenacious an adversary." (Unnumbered attachment to Ellwanger Depo. at 1.)

Ellwanger shows a remarkable degree of political candor, if not finesse, in a letter to State Senator Joseph Carlucci of Florida:

> It would be very wise, if not actually essential, that all of us who are engaged in this legislative effort be careful not to present our position and our work in a religious framework. For example, in written communications that might somehow be shared with those other persons whom we may be trying to convince, it would be well to exclude our own personal testimony and/or witness for Christ, but rather, if we are so moved, to give that testimony on a separate attached note. (Unnumbered attachment to Ellwanger Depo. at 1.)

The same tenor is reflected in a letter by Ellwanger to Mary Ann Miller, a member of FLAG (Family, Life, America under God) who lobbied the Arkansas Legislature in favor of Act 590:

> . . .we'd like to suggest that you and your co-workers be very cautious about mixing creation-science with creation-religion. . .Please urge your co-workers not to allow themselves to get sucked into the "religion" trap of mixing the two together, for such mixing does incalculable harm to the legislative thrust. It could even bring public opinion to bear adversely upon the

higher courts that will eventually have to pass judgment on the constitutionality of this new law. (Ex. 1 to Miller Depo.)

Perhaps most interesting, however, is Mr. Ellwanger's testimony in his deposition as to his strategy for having the model act implemented:

Q. You're trying to play on other people's religious motives.

A. I'm trying to play on their emotions, love, hate, their likes, dislikes, because I don't know any other way to involve, to get humans to become involved in human endeavors. I see emotions as being a healthy and legitimate means off getting people's feelings into action, and. . .I believe that the predominance of population in America that represents the greatest potential for taking some kind of action in this area is a Christian community. I see the Jewish community as far less potential in taking action. . .but I've seen a lot of interest among Christians and I feel, why not exploit that to get the bill going if that's what it takes. (Ellwanger Depo. at 146-147).

Mr. Ellwanger's ultimate purpose is revealed in the closing of his letter to Mr. Tom Bethell:

"Perhaps all this is old hat to you, Tom, and if so, I'd appreciate your telling me so and perhaps where you've heard it before—the idea of killing evolution instead of playing these debating games that we've been playing for nigh over a decade already." (Unnumbered attachment to Ellwanger Depo. at 3.)

It was out of this milieu that Act 590 emerged. The Reverend W. A. Blount, a Biblical literalist who is a pastor of a church in the Little Rock area and was, in February, 1981, chairman of the Greater Little Rock Evangelical Fellowship, was among those who received a copy of the model act from Ellwanger.

At Reverend Blount's request, the Evangelical Fellowship unanimously adopted a resolution to seek an introduction of Ellwanger's act in the Arkansas Legislature. A committee composed of two ministers, Curtis Thomas and W. A. Young, was appointed to implement the resolution. Thomas obtained from Ellwanger a revised copy of the model act which he transmitted to Carl Hunt, a business associate of Senator James L. Holsted, with the request that Hunt prevail upon Holsted to introduce the act.

Holsted, a self-described "born again" Christian Fundamentalist, introduced the act in the Arkansas Senate. He did not consult the State Department of Education, scientists, science educators or the Arkansas Attorney General. The Act was not referred to any Senate committee for hearing and was passed after only a few minutes' discussion on the Senate floor. In the House of Representatives, the bill was referred to the Education Committee which conducted a perfunctory fifteen-minute hearing. No scientist testified at the hearing, nor was any representative form the State Department of Education called to testify.

Ellwanger's model act was enacted into law in Arkansas as Act 590 with amendment or modification other than minor typographical changes. The legislative "finding of fact" in Ellwanger's act and Act 590 are identical, although no meaningful fact-finding was employed by the General Assembly.

Ellwanger's efforts in preparation of the model act and campaign for its adoption in the states were motivated by his opposition to the theory of evolution and his desire to see the Biblical version of creation taught in the public schools. There is no evidence that the pastors,

Blount, Thomas, Young, or The Greater Little Rock Evangelical Fellowship were motivated by anything other than their religious convictions when proposing its adoption or during their lobbying efforts in its behalf. Senator Holsted's sponsorship and lobbying efforts in behalf of the Act were motivated solely by his religious beliefs and desire to see the Biblical version of creation taught in the public schools.

The State of Arkansas, like a number of states whose citizens have relatively homogeneous religious beliefs, has a long history of official opposition to evolution which is motivated by adherence to Fundamentalist beliefs in the inerrancy of the Book of Genesis. This history is documented in Justice Fortas' opinion in *Epperson v. Arkansas*, 393 U.S. 97 (1968), which struck down Initiated Act 1 of 1929, Ark. Stat. Ann. &&80-1627–1628, prohibiting the teaching of the theory of evolution. To this same tradition may be attributed Initiated Act 1 of 1930, Ark. Stat. Ann. &80-1606 (Repl. 1980), requiring "the reverent daily reading of a portion of the English Bible" in every public school classroom in the State.

It is true, as defendants argue, that courts should look to legislative statements of a statutes purpose in Establishment Clause cases and accord such pronouncements great deference. See, e.g., *Committee for Public Education & Religious Liberty v. Nyquist*, 413 U.S. 756, 773 (1973) and *McGowan v. Maryland*, 366 U.S. 420, 445 (1961). Defendants also correctly state the principle that remarks by the sponsor or author of a bill are not considered controlling in analyzing legislative intent. See, e.g., *United States v. Emmons*, 410 U.S. 396 (1973) and *Chrysler Corp v. Brown*, 441 U.S. 281 (1979).

Courts are not bound, however, by legislative statements of purpose or legislative disclaimers. *Stone v. Graham*, 449 U.S. 39 (1980); *Abbington School Dist. v. Schempp*, 374 U.S. 203 (1963). In determining the legislative purpose of a statute, courts may consider evidence of the historical context of the Act, *Epperson v. Arkansas*, 393 U.S. 97 (1968), the specific sequence of events leading up to passage of the Act, departures from normal procedural sequences, substantive departures from the normal, *Village of Arlington Heights v. Metropolitan Housing Corp.*, 429 U.S. 252 (1977), and contemporaneous statements of the legislative sponsor, *Fed. Energy Admin. v. Algonquin SNG Inc.* 426 U.S. 548, 564 (1976).

The unusual circumstances surrounding the passage of Act 590, as well as the substantive law of the First Amendment warrant an inquiry into the stated legislative purposes. The author of the Act has publicly proclaimed the sectarian purpose of the proposal. The Arkansas residents who sought legislative sponsorship of the bill did so for a purely sectarian purpose. These circumstances alone may not be particularly persuasive, but when considered with the publicly announced motives of the legislative sponsor made contemporaneously with the legislative process; the lack of any legislative investigation, debate or consultation with any educators or scientists; the unprecedented intrusion in school curriculum; and official history of the State of Arkansas on the subject, it is obvious that the statement of purpose has little, if any, support in fact. The State failed to produce any evidence which would warrant an inference or conclusion that at any point in the process anyone considered the legitimate educational value of the Act. It was simply and purely an effort to introduce the Biblical version of creation into the public school curricula. The only inference which can be drawn from these circumstances is that the Act was passed with the specific purpose by the General Assembly of advancing religion. The Act therefore fails the first prong of the three-pronged test, that of secular legislative purpose, as articulated in *Lemon v. Kurtzman, supra*, and *Stone v. Graham, supra*.

III

If the defendants are correct and the Court is limited to an examination of the language of the Act, the evidence is overwhelming that both the purpose and effect of Act 590 is the advancement of religion in the public schools.

Section 4 of the Act provides:

Definitions, as used in this Act:

(a) "Creation-science" means the scientific evidences for creation and inferences from those scientific evidences. Creation-science includes the scientific evidences and related inferences that indicate: (1) Sudden creation of the universe, energy, and life from nothing; (2) The insufficiency of mutation and natural selection in bringing about development of all living kinds from a single organism; (3) Changes only within fixed limits of originally created kinds of plants and animals; (4) Separate ancestry for man and apes; (5) Explanation of the earth's geology by catastrophism, including the occurrence of a worldwide flood; and (6) A relatively recent inception of the earth and living kinds.

(b) "Evolution-science" means the scientific evidences for evolution and inferences from those scientific evidences. Evolution-science includes the scientific evidences and related inferences that indicate: (1) Emergence by naturalistic processes of the universe from disordered matter and emergence of life from nonlife; (2) The sufficiency of mutation and natural selection in bringing about development of present living kinds from simple earlier kinds; (3) Emergence by mutation and natural selection of present living kinds from simple earlier kinds; (4) Emergence of man from a common ancestor with apes; (5) Explanation of the earth's geology and the evolutionary sequence by uniformitarianism; and (6) An inception several billion years ago of the earth and somewhat later of life.

(c) "Public schools" means public secondary and elementary schools.

The evidence establishes that the definition of "creation science" contained in 4(a) has as its unmentioned reference the first 11 chapters of the Book of Genesis. Among the many creation epics in human history, the account of sudden creation from nothing, or *creatio ex nihilo*, and subsequent destruction of the world by flood is unique to Genesis. The concepts of 4(a) are the literal Fundamentalists' view of Genesis. Section 4(a) is unquestionably a statement of religion, with the exception of 4(a) (2) which is a negative thrust aimed at what the creationists understand to be the theory of evolution.

Both the concepts and wording of Section 4(a) convey an inescapable religiosity. Section 4(a) (1) describes "sudden creation of the universe, energy and life from nothing." Every theologian who testified, including defense witnesses, expressed the opinion that the statement referred to a supernatural creation which was performed by God.

Defendants argue that: (1) the fact that 4(a) conveys idea similar to the literal interpretation of Genesis does not make it conclusively a statement of religion; (2) that reference to a creation from nothing is not necessarily a religious concept since the Act only suggests a creator who has power, intelligence and a sense of design and not necessarily the attributes of love, compassion, and justice; and (3) that simply teaching about the concept of a creator is not a religious exercise unless the student is required to make a commitment to the concept of a creator.

The evidence fully answers these arguments. The idea of 4(a) (1) are not merely similar to the literal interpretation of Genesis; they are identical and parallel to no other story of creation.

The argument that creation from nothing in 4(a) (1) does not involve a supernatural deity has no evidentiary or rational support. To the contrary, "creation out of nothing" is a concept unique to Western religions. In traditional Western religious thought, the conception of a creator of the world is a conception of God. Indeed, creation of the world "out of nothing" is the ultimate religious statement because God is the only actor. As Dr. Langdon Gilkey noted, the Act refers to one who has the power to bring all the universe into existence from nothing. The only "one" who has this power is God.

The leading creationist writers, Morris and Gish, acknowledge that the idea of creation described in 4(a) (1) is the concept of creation by God and make no pretense to the contrary. The idea of sudden creation from nothing, or *creatio ex nihilo*, is an inherently religious concept. (Vawter, Gilkey, Geisler, Ayala, Blount, Hicks.)

The argument advanced by defendants' witness, Dr. Norman Geisler, that teaching the existence of God is not religious unless the teaching seeks a commitment, is contrary to common understanding and contradicts settled case law. *Stone v. Graham*, 449 U.S. 39 (1980), *Abbington School District v. Schempp*, 374 U.S. 203, 222 (1963).

The facts that creation science is inspired by the Book of Genesis and that Section 4(a) is consistent with a literal interpretation of Genesis leave no doubt that a major effect of the Act is the advancement of particular religious beliefs. The legal impact of this conclusion will be discussed further at the conclusion of the Court's evaluation of the scientific merit of creation science.

IV(A)

The approach to teaching "creation science" and "evolution-science" found in Act 590 is identical to the two-model approach espoused by the Institute for Creation Research and is taken almost verbatim from ICR writings. It is an extension of Fundamentalists' view that one must either accept the literal interpretation of Genesis or else believe in the godless system of evolution.

The two model approach of the creationists is simply a contrived dualism which has not scientific factual basis or legitimate educational purpose. It assumes only two explanations for the origins of life and existence of man, plants, and animals: it was either the work of a creator or it was not. Application of these two models, according to creationists, and the defendants, dictates that all scientific evidence which fails to support the theory of evolution is necessarily scientific evidence in support of creationism and is, therefore, creation science "evidence" in support of Section 4(a).

IV(B)

The emphasis on origins as an aspect of the theory of evolution is peculiar to the creationist literature. Although the subject of origins of life is within the province of biology, the scientific community does not consider origins of life a part of evolutionary theory. The theory of evolution assumes the existence of life and is directed to an explanation of *how* life evolved. Evolution does not presuppose the absence of a creator or God and the plain inference conveyed by Section 4 is erroneous.

As a statement of the theory of evolution, Section 4(b) is simply a hodgepodge of limited assertions, many of which are factually inaccurate.

For example, although 4(b) (2) asserts, as a tenet of evolutionary theory, "sufficiency of mutation and natural selection in bringing about development of present living kinds from simple earlier kinds," Drs. Ayala and Gould both stated that biologists know that these two processes do not account for all significant evolutionary change. They testified to such phenomena as recombination, the founder effect, genetic drift, and the theory of punctuated equilibrium, which are believed to play important evolutionary roles. Section 4(b) omits any reference to these. Moreover, 4(b) utilizes the term "kinds" which all scientists have said is not a word of science and has no fixed meaning. Additionally, the Act presents both evolution and creation science as "package deals." Thus, evidence critical to some aspect of what the creationists define as evolution is taken as support for a theory which includes a worldwide flood and a relatively young earth.

IV(C)

In addition to the fallacious pedagogy of the two model approach, Section 4(a) lacks legitimate educational value because "creation-science" as defined in that section is simply not science. Several witnesses suggested definitions of science. A descriptive definition was said to be that science is what is "accepted by the scientific community" and is "what scientists do." The obvious implication of this description is that, in a free society, knowledge does not require the imprimatur of legislation in order to become science.

More precisely, the essential characteristics of science are:

(1) It is guided by natural law;

(2) It has to be explanatory by reference to nature law;

(3) It is testable against the empirical world;

(4) Its conclusions are tentative, i.e., are not necessarily the final word; and

(5) It is falsifiable. (Ruse and other science witnesses).

Creation science as described in Section 4(a) fails to meet these essential characteristics. First, the section revolves around 4(a) (1) which asserts a sudden creation "from nothing." Such a concept is not science because it depends upon a supernatural intervention which is not guided by natural law. It is not explanatory by reference to natural law, is not testable and is not falsifiable.

If the unifying idea of supernatural creation by God is removed from Section 4, the remaining parts of the section explain nothing and are meaningless assertions.

Section 4(a) (2), relating to the "insufficiency of mutation and natural selection in bringing about development of all living kinds from a single organism," is an incomplete negative generalization directed at the theory of evolution.

Section 4(a) (3) which describes "changes only within fixed limits of originally created kinds of plants and animals" fails to conform to the essential characteristics of science for several reasons. First, there is no scientific definition of "kinds" and none of the witnesses was able to point to any scientific authority which recognized the term or knew how many "kinds" existed. One defense witness suggested there may be 100 to 10,000 different "kinds." Another believes there were "about 10,000, give or take a few thousand." Second, the assertion appears to be an effort to establish outer limits of changes within species. There is no scientific explanation for these limits which is guided by natural law and the limitations, whatever they are, cannot be explained by natural law.

The statement in 4(a) (4) of "separate ancestry of man and apes" is a bald assertion. It explains nothing and refers to no scientific fact or theory.

Section 4(a) (5) refers to "explanation of the earth's geology by catastrophism, including the occurrence of a worldwide flood." This assertion completely fails as science. The Act is referring to the Noachian flood described in the Book of Genesis. The creationist writers concede that *any* kind of Genesis Flood depends upon supernatural intervention. A worldwide flood as an explanation of the world's geology is not the product of natural law, nor can its occurrence be explained by natural law.

Section 4(a) (6) equally fails to meet the standards of science. "Relatively recent inception" has no scientific meaning. It can only be given in reference to creationist writings which place the age at between 6,000 and 20,000 years because of the genealogy of the Old Testament. See, e.g., Px 78, Gish (6,000 to 10,000); Px 87, Segraves (6,000 to 20,000). Such a reasoning process is not the product of natural law; not explainable by natural law; nor is it tentative.

Creation science as defined in Section 4(a), not only fails to follow the canons of dealing with scientific theory, it also fails to fit the more general descriptions of "what scientists think" and "what scientists do." The scientific community consists of individuals and groups, nationally and internationally, who work independently in such varied fields as biology, paleontology, geology, and astronomy. Their work is published and subject to review and testing by their peers. The journals for publication are both numerous and varied. There is, however, not one recognized scientific journal which has published an article espousing the creation science theory described in Section 4(a). Some of the State's witnesses suggested that the scientific community was "close-minded" on the subject of creationism and that explained the lack of acceptance of the creation science arguments. Yet no witness produced a scientific article for which publication has been refused. Perhaps some members of the scientific community are resistant to new ideas. It is, however, inconceivable that such a loose knit group of independent thinkers in all the varied fields of science could, or would, so effectively censor new scientific thought.

The creationists have difficulty maintaining among their ranks consistency in the claim that creationism is science. The author of Act 590, Ellwanger, said that neither evolution or creationism was science. He thinks that both are religious. Duane Gish recently responded to an article in *Discover* critical of creationism by stating:

> Stephen Jay Gould states that creationists claim creation is a scientific theory. This is a
> false accusation. Creationists have repeatedly stated that neither creation nor evolution is a
> scientific theory (and each is equally religious). (Gish, letter to editor of *Discover*, July, 1981,
> App. 30 to Plaintiffs' Pretrial Brief).

The methodology employed by creationists is another factor which is indicative that their work is not science. A scientific theory must be tentative and always subject to revision or abandonment in light of facts that are inconsistent with, or falsify, the theory. A theory that is by its own terms dogmatic, absolutist, and never subject to revision is not a scientific theory.

The creationists' methods do not take data, weigh it against the opposing scientific data, and thereafter reach the conclusions stated in Section 4(a). Instead, they take the literal wording of the Book of Genesis and attempt to find scientific support for it. The method is best explained in the language of Morris in his book (Px 31) *Studies in The Bible and Science* at page 114:

> ...it is...quite impossible to determine anything about Creation through a study of present
> processes, because present processes are not creative in character. If man wished to know
> anything about Creation (the time of Creation, the duration of Creation, the order of Creation,
> the methods of Creation, or anything else) his sole source of true information is that of
> divine revelation. God was there when it happened. We were not there...Therefore, we are

completely limited to what God has seen fit to tell us, and this information is in His written Word. This is our textbook on the science of Creation!

The Creation Research Society employs the same unscientific approach to the issue of creationism. Its applicants for membership must subscribe to the belief that the Book of Genesis is "historically and scientifically true in all of the original autographs." The Court would never criticize or discredit any person's testimony based on his or her religious beliefs. While anybody is free to approach a scientific inquiry in any fashion they choose, they cannot properly describe the methodology as scientific, if they start with the conclusion and refuse to change it regardless of the evidence developed during the course of the investigation.

IV(D)

In efforts to establish "evidence" in support of creation science, the defendants relied upon the same false premise as the two-model approach contained in Section 4, i.e., all evidence which criticized evolutionary theory was proof in support of creation science. For example, the defendants established that the mathematical probability of a chance chemical combination resulting in life from non-life is as remote that such an occurrence is almost beyond imagination. Those mathematical facts, the defendants argue, are scientific evidences that life was the product of a creator. While the statistical figures may be impressive evidence against the theory of chance chemical combinations as an explanation of origins, it requires a leap of faith to interpret those figures so as to support a complex doctrine which includes a sudden creation from nothing, a worldwide flood, separate ancestry of man and apes, and a young earth.

The defendants' argument would be more persuasive if, in fact, there were only two theories or idea about the origins of life and the world. That there are a number of theories was acknowledged by the State's witnesses, Dr. Wickramasinghe and Dr. Geisler. Dr. Wickramasinghe testified at length in support of a theory that life on earth was "seeded" by comets which delivered genetic material and perhaps organisms to the earth's surface from interstellar dust far outside the solar system. The "seeding" theory further hypothesizes that the earth remains under the continuing influence of genetic material from space which continues to affect life. While Wickramasinghe's theory about the origins of life on earth has not received general acceptance within the scientific community, he has, at least, used scientific methodology to produce a theory of origins which meets the essential characteristics of science.

The Court is at a loss to understand why Dr. Wickramasing was called in behalf of the defendants. Perhaps it was because he was generally critical of the theory of evolution and the scientific community, a tactic consistent with the strategy of the defense. Unfortunately for the defense, he demonstrated that the simplistic approach of the two-model analysis of the origins of life is false. Furthermore, he corroborated the plaintiffs' witnesses by concluding that "no rational scientist" would believe the earth's geology could be explained by reference to a worldwide flood or that the earth was less than one million years old.

The proof in support of creation science consisted almost entirely of efforts to discredit the theory of evolution through a rehash of data and theories which have been before the scientific community for decades. The arguments asserted by the creationists are not based upon new scientific evidence or laboratory data which has been ignored by the scientific community.

Robert Gentry's discovery of radioactive polonium haloes in granite and coalified woods is, perhaps, the most recent scientific work which the creationists use as argument for a "relatively recent inception" of the earth and a "worldwide flood." The existence of polonium haloes in granite and coalified wood is thought to be inconsistent with radiometric dating methods based upon constant radioactive decay rates. Mr. Gentry's findings were published almost ten years ago and have been the subject of some discussion in the scientific community. The discoveries have not, however, led to the formulation of any scientific hypothesis or theory which would explain a relatively recent inception of the earth or a worldwide flood. Gentry's discovery has been treated as a minor mystery which will eventually be explained. It may deserve further investigation, but the National Science Foundation has not deemed it to be of sufficient import to support further funding.

The testimony of Marianne Wilson was persuasive evidence that creation science is not science. Ms. Wilson is in charge of the science curriculum for Pulaski County Special School District, the largest school district in the State of Arkansas. Prior to the passage of Act 590, Larry Fisher, a science teacher in the District, using materials from the ICR convinced the School Board that it should voluntarily adopt creation science as part of its science curriculum. The District Superintendent assigned Ms. Wilson the job of producing a creation science curriculum guide. Ms. Wilson's testimony about the project was particularly convincing because she obviously approached the assignment with an open mind and no preconceived notions about the subject. She had not heard of creation science until about a year ago and did not know its meaning before she began her research.

Ms. Wilson worked with a committee of science teachers appointed from the District. They reviewed practically all of the creationist literature. Ms. Wilson and the committee members reached the unanimous conclusion that creationism is not science; it is religion. They so reported to the Board. The Board ignored the recommendation and insisted that a curriculum guide be prepared.

In researching the subject, Ms. Wilson sought the assistance of Mr. Fisher who initiated the Board action and asked professors in the science departments of the University of Arkansas at Little Rock and the University of Central Arkansas for reference material and assistance, and attended a workshop conducted at Central Baptist College by Dr. Richard Bliss of the ICR staff. Act 590 became law during the course of her work so she used Section 4(a) as a format for her curriculum guide.

Ms. Wilson found all available creationists' materials unacceptable because they were permeated with religious references and reliance upon religious beliefs.

It is easy to understand why Ms. Wilson and other educators find the creationists' textbook material and teaching guides unacceptable. The materials misstate the theory of evolution in the same fashion as Section 4(b) of the Act, with emphasis on the alternative mutually exclusive nature of creationism and evolution. Students are constantly encouraged to compare and make a choice between the two models, and the material is not presented in an accurate manner.

A typical example is *Origins* (Px 76) by Richard B. Bliss, Directory of Curriculum Development of the ICR. The presentation begins with a chart describing "preconceived idea about origins" which suggests that some people believe that evolution is atheistic. Concepts of evolution, such as "adaptative radiation" are erroneously presented. At page 11, Figure 1.6 of the text, a chart purports to illustrate this "very important" part of the evolution model. The chart conveys the idea that such diverse mammals as a whale, bear, bat, and monkey all evolved from a shrew through the process of adaptive radiation. Such a suggestion is, of course, a totally erroneous and misleading application of the theory. Even more objectionable, especially when viewed in light of the emphasis on asking the student to elect one of the models, is the chart presentation at page 17, Figure 1.6. That chart purports to illustrate the evolutionists' belief that man evolved from bacteria to fish to reptile to mammals and, thereafter, into man. The illustration indicates, however, that the mammal which evolved was *a rat*.

Biology, A Search For Order in Complexity is a high school biology text typical of creationists' materials. The following quotations are illustrative:

Flowers and roots do not have a mind to have purpose of their own: therefore, this planning must have been done for them by the Creator (at page 12).

The exquisite beauty of color and shape in flowers exceeds the skill of poet, artist, and king. Jesus said (from Matthew's gospel), "Consider the lilies in the field, how they grow; they toil not, neither do they spin. . ." (Px 129 at page 363).

The "public school edition" texts written by creationists simply omit Biblical references but the content and message remain the same. For example, *Evolution—The Fossils Say No!* contains the following:

> Creation. By creation we mean the bringing into being by a supernatural Creator of the basic kinds of plants and animals by the process of sudden, or fiat, creation.
>
> We do not know how the Creator created, what processes He used, *for he used processes which are not now operating anywhere in the natural universe*. This is why we refer to creation as Special Creation. We cannot discover by scientific investigation anything about the creative processes used by the Creator (page 40).

Gish's book also portrays the large majority of evolutionists as "materialistic atheists or agnostics."

Scientific Creationism (Public School Edition) by Morris, is another text reviewed by Ms. Wilson's committee and rejected as unacceptable. The following quotes illustrate the purpose and theme of the text:

Forward

Parents and youth leaders today, and even many scientists and educators, have become concerned about the prevalence and influence of evolutionary philosophy in modern curriculum. Not only is the system inimical to orthodox Christianity and Judaism, but also, as many are convinced, to a healthy society and true science as well (at page iii).

The rationalist of course finds the concept of special creation insufferably naive, even "incredible". Such a judgment, however, is warranted only if one categorically dismisses the existence of an omnipotent God (at page 17).

Without using creationist literature, Ms. Wilson was unable to locate one genuinely scientific article or work which supported Section 4(a). In order to comply with the mandate of the Board she used such materials as an article from *Readers Digest* about "atomic clocks" which inferentially suggested that the earth was less than $4^1/_2$ billion years old. She was unable to locate any substantive teaching material for some parts of Section 4 such as the worldwide flood. The curriculum guide which she prepared cannot be taught and has no education value as science. The defendants did not produce any text or writing in response to this evidence which they claimed was usable in the public school classroom.

The conclusion that creation science has no scientific merit or educational value as science has legal significance in light of the Court's previous conclusion that creation science has, as one major effect, the advancement of religion. The second part of the three-pronged test for establishment reaches only those statutes as having their *primary* effect the advancement of religion. Secondary effects which advance religion are not constitutionally fatal. Since creation science is not science, the conclusion is inescapable that the *only* real effect of Act 590 is the advancement of religion. The Act therefore fails both the first and second portions of the test in *Lemon v. Kurtzman*, 403 U.S. 602 (1971).

IV(E)

Act 590 mandates "balanced treatment" for creation science and evolution science. The Act prohibits instruction in any religious doctrine or references to religious writings. The Act is self-contradictory and compliance is impossible unless the public schools elect to forego significant potions of subjects such as biology, world history, geology, zoology, botany, psychology, anthropology, sociology, philosophy, physics, and chemistry. Presently, the concepts of evolutionary theory as described in 4(b) permeate the public textbooks. There is no way teachers can teach the Genesis account of creation in a secular manner.

The State Department of Education, through its textbook selection committee, school boards and school administrators will be required to constantly monitor materials to avoid using religious references. The school boards, administrators, and teachers face an impossible task. How is the teacher to respond to questions about a creation suddenly and out of nothing? How will a teacher explain the occurrence of a worldwide flood? How will a teacher explain the concept of a relatively recent age of the earth? The answer is obvious because the only source of this information is ultimately contained in the Book of Genesis.

References to the pervasive nature of religious concepts in creation science texts amply demonstrate why State entanglement with religion is inevitable under Act 590. Involvement of the State in screening texts for impermissible religious references will require State officials to make delicate religious judgments. The need to monitor classroom discussion in order to uphold the Act's prohibition against religious instruction will necessarily involve administrators in questions concerning religion. These continuing involvements of State officials in questions and issues of religion create an excessive and prohibited entanglement with religion. *Brandon v. Board of Education*, 487 F.Supp 1219, 1230 (N.D.N.Y.), *aff'd.*, 635 F.2d 971 (2nd Cir. 1980).

V

These conclusions are dispositive of the case and there is no need to reach legal conclusions with respect to the remaining issues. The plaintiffs raised two other issues questioning the constitutionality of the Act and, insofar as the factual findings relevant to these issues are not covered in the preceding discussion, the Court will address these issues. Additionally, the defendants raise two other issues which warrant discussion.

V(A)

First, plaintiff teachers argue the Act is unconstitutionally vague to the extent that they cannot comply with its mandate of "balanced" treatment without jeopardizing their employment. The argument centers around the lack of a precise definition in the Act for the word "balanced." Several witnesses expressed opinions that the word has such meanings as equal time, equal weight, or equal legitimacy. Although the Act could have been more explicit, "balanced" is a word subject to ordinary understanding. The proof is not convincing that a teacher using a reasonably acceptable understanding of the word and making a good faith effort to comply with the Act will be in jeopardy of termination. Other portions of the Act are arguably vague, such as the "relatively recent" inception of the earth and life. The evidence establishes, however, that relatively recent means from 6,000 to 20,000 years, as commonly understood in creation science literature. The meaning of this phrase, like Section 4(a) generally, is, for purposes of the Establishment Clause, all too clear.

V(B)

The plaintiffs' other argument revolves around the alleged infringement by the defendants upon the academic freedom of teachers and students. It is contended this unprecedented intrusion in the curriculum by the State prohibits teachers from teaching what they believe should be taught or requires them to teach that which they do not believe is proper. The evidence reflects that traditionally the State Department of Education, local school boards and administration officials exercise little, if any, influence upon the subject matter taught by classroom teachers. Teachers have been given freedom to teach and emphasize those portions of subjects the individual teacher considered important. The limits to this discretion have generally been derived from the approval of textbooks by the State Department and preparation of curriculum guides by the school districts.

Several witnesses testified that academic freedom for the teacher means, in substance, that the individual teacher should be permitted unlimited discretion subject only to the bounds of professional ethics. The Court is not prepared to adopt such a broad view of academic freedom in the public schools.

In any event, if Act 590 is implemented, many teachers will be required to teach materials in support of creation science which they do not consider academically sound. Many teachers will simply forego teaching subjects which might trigger the "balanced treatment" aspects of Act 590 even though they think the subjects are important to a proper presentation of a course.

Implementation of Act 580 will have serious and untoward consequences for students, particularly those planning to attend college. Evolution is the cornerstone of modern biology, and many courses in public schools contain subject matter relating to such varied topics as the age of the earth, geology and relationships among living things. Any student who is deprived of instruction as to the prevailing scientific thought on these topics will be denied a significant part of science education. Such a deprivation through the high school level would undoubtedly have an impact upon the quality of education in the State's colleges and universities, especially including the pre-professional and professional programs in the health sciences.

V(C)

The defendants argue in their brief that evolution is, in effect, a religion, and that by teaching a religion which is contrary to some students' religious views, the State is infringing upon the student's free exercise rights under the First Amendment. Mr. Ellwanger's legislative findings, which were adopted as a finding of fact by the Arkansas Legislature in Act 590, provides:

> Evolution-science is contrary to the religious convictions or moral values or philosophical beliefs of many students and parents, including individuals of many different religious faiths and with diverse moral and philosophical beliefs, Act 590, &7(d).

The defendants argue that the teaching of evolution alone presents both a free exercise problem and an establishment problem which can only be redressed by giving balanced treatment to creation science, which is admittedly consistent with some religious beliefs. This argument appears to have its genesis in a student note written by Mr. Wendell Bird, "Freedom of Religion and Science Instruction in Public Schools," 87 Yale L.J. 515 (1978). The argument has no legal merit.

If creation science is, in fact, science and not religion, as the defendants claim, it is difficult to see how the teaching of such a science could "neutralize" the religious nature of evolution.

Assuming for the purposes of argument, however, that evolution is a religion or religious tenet, the remedy is to stop the teaching of evolution, not establish another religion in opposition to it. Yet it is clearly established in the case law, and perhaps also in common sense, that evolution is not a religion and that teaching evolution does not violate the Establishment Clause, *Epperson v. Arkansas, supra, Willoughby v. Stever*, No. 15574-75 (D.D.C. May 18, 1973); *aff'd.* 504 F.2d 271 (D.C. Cir. 1974), *cert. denied*, 420 U.S. 924 (1975); *Wright v. Houston Indep. School Dist.*, 366 F. Supp. 1208 (S.D. Tex 1978), *aff.d.* 486 F.2d 137 (5th Cir. 1973), *cert. denied* 417 U.S. 969 (1974).

V(D)

The defendants presented Dr. Larry Parker, a specialist in devising curricula for public schools. He testified that the public school's curriculum should reflect the subjects the public wants in schools. The witness said that polls indicated a significant majority of the American public thought creation science should be taught if evolution was taught. The point of this testimony was never placed in a legal context. No doubt a sizeable majority of Americans believe in the concept of a Creator or, at least, are not opposed to the concept and see nothing wrong with teaching school children the idea.

The application and content of First Amendment principles are not determined by public opinion polls or by a majority vote. Whether the proponents of Act 590 constitute the majority or the minority is quite irrelevant under a constitutional system of government. No group, no matter how large or small, may use the organs of government, of which the public schools are the most conspicuous and influential, to foist its religious beliefs on others.

The Court closes this opinion with a thought expressed eloquently by the great Justice Frankfurter:

> We renew our conviction that "we have stake the very existence of our country on the faith that complete separation between the state and religion is best for the state and best for religion." *Everson v. Board of Education*, 330 U.S. at 59. If nowhere else, in the relation between Church and State, "good fences make good neighbors." [*McCollum v. Board of Education*, 333 U.S. 203, 232 (1948)].

An injunction will be entered permanently prohibiting enforcement of Act 590.

It is ordered this January 5, 1982.

SOURCE: Overton, William R. *McLean v. Arkansas Board of Education*, 1982.

LARRY LAUDAN, "SCIENCE AT THE BAR—CAUSES FOR CONCERN" (1982)[2]

In reviewing the ruling against creationism in Arkansas, Larry Laudan determined to uphold the field of philosophy of science for philosophy's sake. While he acknowledged that the outcome of the trial served the purpose of science, and even the purpose of truth as he

[2]Larry Laudan, *Science Technology & Human Values* (7), pp. 16–19, copyright 1982 by Larry Laudan. Reprinted by permission of Sage Publications.

saw it, the reasoning might well prove problematic in the long run. He worried that savvy creationists could read between the lines of what scientists and philosophers had constructed as science and proceed to define creation science more carefully to meet that construction. If Laudan appreciated the constitutional basis for defining the current view of creation science as religion, he saw the effort that had gone into defeating it as shortsighted. He would prefer to defeat creation science on the merits of the evidence, keeping the boundaries of science unrestricted. While he wrote as one convinced of the outcome of such an ongoing inquiry, the process would serve all parties better than the conclusions reached in this particular court case. Many scientists ultimately agreed with this view, although those who saw the inquiry process played out—at many levels in the classroom and on school boards by non-specialists—preferred to have the law on the side of science.

In the wake of the decision in the Arkansas Creationism trial (*McLean v. Arkansas*), the friends of science are apt to be relishing the outcome. The creationists quite clearly made a botch of their case and there can be little doubt that the Arkansas decision may, at least for a time, blunt legislative pressure to enact similar laws in other states. Once the dust has settled, however, the trial in general and Judge William R. Overton's ruling in particular may come back to haunt us; for, although the verdict itself is probably to be commended, it was reached for all the wrong reasons and by a chain of argument which is hopelessly suspect. Indeed, the ruling rests on a host of misrepresentations of what science is and how it works.

The heart of Judge Overton's Opinion is a formulation of "the essential characteristics of science." These characteristics serve as touchstones for contrasting evolutionary theory with Creationism; they lead Judge Overton ultimately to the claim, specious in its own right, that since Creationism is not "science," it must be religion. The Opinion offers five essential properties that demarcate scientific knowledge from other things: "(1) It is guided by natural law; (2) it has to be explanatory by reference to natural law; (3) it is testable against the empirical world; (4) its conclusions are tentative, i.e., are not necessarily the final word; and (5) it is falsifiable."

These fall naturally into two families: properties (1) and (2) have to do with lawlikeness and explanatory ability; the other three properties have to do with the fallibility and testability of scientific claims. I shall deal with the second set of issues first, because it is there that the most egregious errors of fact and judgment are to be found.

At various key points in the Opinion, Creationism is charged with being untestable, dogmatic (and thus non-tentative), and unfalsifiable. All three charges are of dubious merit. For instance, to make the interlinked claims that Creationism is neither falsifiable nor testable is to assert that Creationism makes no empirical assertions whatever. That is surely false. Creationists make a wide range of testable assertions about empirical matters of fact. Thus, as Judge Overton himself grants (apparently without seeing its implications), the creationists say that the earth is of very recent origin (say 6,000 to 20,000 years old); they argue that most of the geological features of the earth's surface are diluvial in character (i.e., products of the postulated worldwide Noachian deluge); they are committed to a large number of factual historical claims with which the Old Testament is replete; they assert the limited variability of species. They are committed to the view that, since animals and man were created at the same time, the human fossil record must be paleontologically co-extensive with the record of lower animals. It is fair to say that no one has shown how to reconcile such claims with the available evidence—evidence which speaks persuasively to a long earth history, among other things.

In brief, these claims are testable, they have been tested, and they have failed those tests. Unfortunately, the logic of the Opinion's analysis precludes saying any of the above. By arguing that the tenets of Creationism are neither testable nor falsifiable, Judge Overton (like those scientists who similarly charge Creationism with being untestable) deprives science of its strongest argument against Creationism. Indeed, if any doctrine in the history of science has ever been falsified, it is the set of claims associated with "creation-science." Asserting that Creationism makes no empirical claims plays directly, if inadvertently, into the hands of the creationists by immunizing their ideology from empirical confrontation. The correct way to combat Creationism is to confute the empirical claims it does make, not to pretend that it makes no such claims at all.

It is true, of course, that some tenets of Creationism are not testable in isolation (e.g., the claim that man emerged by a direct supernatural act of creation). But that scarcely makes Creationism "unscientific." It is now widely acknowledged that many scientific claims are not testable in isolation, but only when embedded in a larger system of statements, some of whose consequences can be submitted to test.

Judge Overton's third worry about Creationism centers on the issue of revisability. Over and over again, he finds Creationism and its advocates "unscientific" because they have "refuse[d] to change it regardless of the evidence developed during the course of the[ir] investigation." In point of fact, the charge is mistaken. If the claims of modern-day creationists are compared with those of their nineteenth-century counterparts, significant shifts in orientation and assertion are evident. One of the most visible opponents of Creationism, Stephen Gould, concedes that creationists have modified their views about the amount of variability allowed at the level of species change. Creationists do, in short, change their minds from time to time. Doubtless they would create these shifts to their efforts to adjust their views to newly emerging evidence, in what they imagine to be a scientifically respectable way.

Perhaps what Judge Overton had in mind was the fact that some of Creationism's core assumptions (e.g., that there was a Noachian flood, that man did not evolve from lower animals, or that God created the world) seem closed off from any serious modification. But historical and sociological researches on science strongly suggest that the scientists of any epoch likewise regard some of their beliefs as so fundamental as not to be open to repudiation or negotiation. Would Newton, for instance, have been tentative about the claim that there were forces in the world? Are quantum mechanicians willing to contemplate giving up the uncertainty relation? Are physicists willing to specify circumstances under which they would give up energy conservation? Numerous historians and philosophers of science (e.g., Kuhn, Mitroff, Feyerabend, Lakatos) have documented the existence of a certain degree of dogmatism about core commitments in scientific research and have argued that such dogmatism plays a constructive role in promoting the aims of science. I am not denying that there may be subtle but important differences between the dogmatism of scientists and that exhibited by many creationists; but one does not even begin to get at those differences by pretending that science is characterized by an uncompromising open-mindedness.

Even worse, the *ad hominem* charge of dogmatism against Creationism egregiously confuses doctrines with the proponents of those doctrines. Since no law mandates that creationists should be invited into the classroom, it is quite irrelevant whether they themselves are close-minded. The Arkansas statute proposed that Creationism be taught, not that creationists should teach it. What counts is the epistemic status of Creationism, not the cognitive idiosyncrasies of the creationists. Because many of the theses of Creationism are testable, the mind set of creationists has no bearing in law or in the fact on the merits of Creationism.

What about the other pair of essential characteristics which the *McLean* Opinion cites, namely, that science is a matter of natural law and explainable by natural law? I find the formulation in the Opinion to be rather fuzzy; but the general idea appears to be that it is inappropriate and unscientific to postulate the existence of any process or fact which cannot be explained in terms of some known scientific laws—for instance, the creationists' assertion that there are outer limits to the change of species "cannot be explained by natural law." Earlier in the Opinion, Judge Overton also writes "there is no scientific explanation for these limits which is guided by natural law," and thus concludes that such limits are unscientific. Still later, remarking on the hypothesis of the Noachian flood, he says, "A worldwide flood as an explanation of the world's geology is not the product of natural law, nor can its occurrence be explained by natural law." Quite how Judge Overton knows that a worldwide flood "cannot" be explained by the laws of science is left opaque; and even if we did not know how to reduce a universal flood to the familiar laws of physics, this requirement is an altogether inappropriate standard for ascertaining whether a claim is scientific. For centuries scientists have recognized a difference between establishing the existence of a phenomenon and explaining that phenomenon in a lawlike way. Our ultimate goal, no doubt, is to do both. But to suggest, as the *McLean* Opinion does repeatedly, that an existence claim (e.g., there was a worldwide flood) is unscientific until we have found the laws on which the alleged phenomenon depends is simply outrageous. Galileo and Newton took themselves to have established the existence of gravitational phenomena, long before anyone was able to give a causal or explanatory account of gravitation. Darwin took himself to have established the existence of natural selection almost a half-century before geneticists were able to lay out the laws of heredity on which natural selection depended. If we took the *McLean* Opinion criterion seriously, we should have to say that Newton and Darwin were unscientific; and, to take an example from our own time, it would follow that plate tectonics is unscientific because we have not yet identified the laws of physics and chemistry which account for the dynamics of crustal motion.

The real objection to such creationist claims as that of the (relative) invariability of species is not that such invariability has not been explained by scientific laws, but rather that the evidence for invariability is less robust than the evidence for its contrary, variability. But to say as much requires renunciation of the Opinion's order charge—to wit, that Creationism is not testable.

I could continue with this tale of woeful fallacies in the Arkansas ruling, but that is hardly necessary. What is worrisome is that the Opinion's line of reasoning—which neatly coincides with the predominant tactic among scientists who have entered the public fray on this issue—leaves many loopholes for the creationists to exploit. As numerous authors have shown, the requirements of testability, revisability, and falsifiability are exceedingly weak requirements. Leaving aside the fact that (as I pointed out above) it can be argued that Creationism already satisfies these requirements, it would be easy for a creationist to say the following: "I will abandon my views if we find a living specimen of a species intermediate between man and apes." It is, of course, extremely unlikely that such an individual will be discovered. But, in that statement the creationist would satisfy, in one fell swoop, all the formal requirements of testability, falsifiability, and revisability. If we set very weak standards for scientific status— and, let there be no mistake, I believe that all of the Opinion's last three criteria fall in this category—then it will be quite simple for Creationism to qualify as "scientific."

Rather than taking on the creationists obliquely and in wholesome fashion by suggesting that what they are doing is "unscientific" *tout court* (which is doubly silly because few authors can even agree on what makes an activity scientific), we should confront their claims directly and in piecemeal fashion by asking what evidence and arguments can be marshaled for and against each of them. The core issue is not whether Creationism satisfies some undemanding and highly controversial definitions of what is scientific; the real question is whether the existing evidence provides stronger arguments for evolutionary theory than for Creationism. Once that question is settled, we will know what belongs in the classroom and what does not. Debating the scientific status of Creationism (especially when "science" is construed in such an unfortunate manner) is a red herring that diverts attention away from the issues that should concern us.

Some defenders of the scientific orthodoxy will probably say that my reservations are just nit-picking ones, and that—at least to a first order of approximation—Judge Overton has correctly identified what is fishy about Creationism. The apologists for science, such as the editor of *The Skeptical Inquirer*, have already objected to those who criticize this whitewash of science "on arcane, semantic grounds...[drawn] from the most remote reaches of the academic philosophy of science." But let us be clear about what is at stake. In setting out in the *McLean* Opinion to characterize the "essential" nature of science, Judge Overton was explicitly venturing into philosophical terrain. His *obiter dicta* are about as remote from well-founded opinion in the philosophy of science as Creationism is from respectable geology. It simply will not do for the defenders of science to invoke philosophy of science when it suits them (e.g., their much-loved principle of falsifiability comes directly from the philosopher Karl Popper) and to dismiss it as "arcane" and "remote" when it does not. However noble the motivation, bad philosophy makes for bad law.

The victory in the Arkansas case was hollow, for it was achieved only at the expense of perpetuating and canonizing a false stereotype of what science is and how it works. If it goes unchallenged by the scientific community, it will raise grave doubts about that community's intellectual integrity. No one familiar with the issues can really believe that anything important was settled through anachronistic efforts to revive a variety of discredited criteria for distinguishing between the scientific and the non-scientific. Fifty years ago, Clarence Darrow asked, *à propos* the Scopes trial, "Isn't it difficult to realize that a trial of this kind is possible in the twentieth century in the United States of America?" We can raise that question anew, with the added irony that, this time, the pro-science forces are defending a philosophy of science which is, in its way, every bit as outmoded as the "science" of the creationists.

SOURCE: Laudan, Larry. "Science at the Bar—Causes for Concern," (1982) in *But Is It Science?* Ed. Michael Ruse. Amherst, NY: Prometheus Books, 1996.

MICHAEL RUSE, "PRO JUDICE" (1982)[3]

Michael Ruse (1940–) served as an expert witness in the controversial Arkansas case over the teaching of evolution. He took a notably different stance than his colleague, Larry Laudan.

---

[3]Michael Ruse, *Science Technology & Human Values* (7), pp. 29–23, copyright 1982 by Michael Ruse. Reprinted by permission of Sage Publications Inc.

Ultimately, Ruse agreed more fully with the final ruling, probably because his position proved more influential to the deciding judge and ultimately more damning of the case for creation science. While some scientists and philosophers of science would grant certain merits to the arguments devised by creationists, especially as they portrayed their beliefs as science, Ruse would make no such space in his view of science. On many levels, he used the philosophy of science to draw clear and unequivocal demarcations around science that excluded creation science. He recognized that an expert witness served his or her purpose only when the broader insights of the field of expertise exposed the case in sharp relief. In a courtroom, shades of gray undermined the expert's authority, even if those shades might more accurately illuminate certain details of an argument. As a witness, then, and in commentary afterwards, Ruse attempted to erase any doubt that creation science derived inextricably from fundamentalist Christianity.

As always, my friend Larry Laudan writes in an entertaining and provocative manner, but, in his complaint against Judge William Overton's ruling in *McLean v. Arkansas*, Laudan is hopelessly wide of the mark. Laudan's outrage centers on the criteria for the demarcation of science which Judge Overton adopted, and the judge's conclusion that, evaluated by these criteria, creation-science fails as science. I shall respond directly to this concern—after making three preliminary remarks.

First, although Judge Overton does not need defense from me or anyone else, as one who participated in the Arkansas trial, I must go on record as saying that I was enormously impressed by his handling of the case. His written judgment is a first-class piece of reasoning. With cause, many have criticized the State of Arkansas for passing the "Creation-Science Act," but we should not ignore that, to the state's credit, Judge Overton was born, raised, and educated in Arkansas.

Second, Judge Overton, like everyone else, was fully aware that proof that something is not science is not the same as proof that it is religion. The issue of what constitutes science arose because the creationists claim that their ideas qualify as genuine science rather than as fundamentalist religion. The attorneys developing the American Civil Liberties Union (ACLU) case believed it important to show that creation-science is not genuine science. Of course, this demonstration does raise the question of what creation-science really is. The plaintiffs claimed that creation-science always was (and still is) religion. The plaintiffs' lawyers went beyond the negative argument (against science) to make the positive case (for religion). They provided considerable evidence for the religious nature of creation-science, including such things as the creationists' explicit reliance on the Bible in their various writings. Such arguments seem about as strong as one could wish, and they were duly noted by Judge Overton and used in support of his ruling. It seems a little unfair, in the context, therefore, to accuse him of "specious" argumentation. He did not adopt the naive dichotomy of "science or religion but nothing else."

Third, whatever the merits of the plaintiffs' case, the kinds of conclusions and strategies apparently favored by Laudan are simply not strong enough for legal purposes. His strategy would require arguing that creation-science is weak science and therefore ought not to be taught:

The core issue is not whether Creationism satisfies some undemanding and highly controversial definitions of what is scientific; the real question is whether the existing evidence provides stronger arguments for evolutionary theory than for Creationism. Once that question is settled, we will know what belongs in the classroom and what does not.

Unfortunately, the U.S. Constitution does not bar the teaching of weak science. What it bars (through the Establishment Clause of the First Amendment) is the teaching of religion. The plaintiffs' tactic was to show that creation-science is less than weak or bad science. It is not science at all.

Turning now to the main issue, I see three questions that must be addressed. Using the five criteria listed by Judge Overton, can one distinguish science from non-science? Assuming a positive answer to the first question, does creation-science fail as genuine science when it is judged by these criteria? And, assuming a positive answer to the second, does the Opinion in *McLean* make this case?

The first question has certainly tied philosophers of science in knots in recent years. Simple criteria that supposedly give a clear answer to every case—for example, Karl Popper's single stipulation of falsifiability—will not do. Nevertheless, although there may be many gray areas, white does seem to be white and black does seem to be black. Less metaphorically, something like psychoanalytic theory may or may not be science, but there do appear to be clear-cut cases of real science and of real non-science. For instance, an explanation of the fact that my son has blue eyes, given that both parents have blue eyes, done in terms of dominant and recessive genes and with an appeal to Mendel's first law, is scientific. The Catholic doctrine of transubstantiation (i.e., that in the Mass the bread and wine turn into the body and blood of Christ) is not scientific.

Furthermore, the five cited criteria of demarcation do a good job of distinguishing the Mendelian example from the Catholic example. Law and explanation through law come into the first example. They do not enter the second. We can test the first example, rejecting it if necessary. In this case, it is tentative, in that something empirical might change our minds. The case of transubstantiation is different. God may have His own laws, but neither scientist nor priest can tell us about those which turn bread and wine into flesh and blood. There is no explanation through law. No empirical evidence is pertinent to the miracle. Nor would the believer be swayed by any empirical facts. Microscopic examination of the Host is considered irrelevant. In this sense, the doctrine is certainly not tentative.

One pair of examples certainly do not make for a definitive case, but at least they do suggest that Judge Overton's criteria are not quite as irrelevant as Laudan's critique implies. What about the types of objections (to the criteria) that Laudan does or could make? As far as the use of law is concerned, he might complain that scientists themselves have certainly not always been that particular about reference to law. For instance, consider the following claim by Charles Lyell in his *Principles of Geology* (1830/3): "We are not, however, contending that a real departure from the antecedent course of physical events cannot be traced in the introduction of man." All scholars agree that in this statement Lyell was going beyond law. The coming of man required special divine intervention. Yet, surely the *Principles* as a whole qualify as a contribution to science.

Two replies are open: either one agrees that the case of Lyell shows that science has sometimes mingled law with non-law; or one argues that Lyell (and others) mingled science and non-science (specifically, religion at this point). My inclination is to argue the latter. Insofar as Lyell acted as scientist, he appealed only to law. A century and a half ago, people were not as conscientious as today about separating science and religion. However, even if one argues the former alternative-that some science has allowed place for non-lawbound events—this hardly makes Laudan's case. Science, like most human cultural phenomena, has evolved. What was allowable in the early nineteenth century is not necessarily allowable in the late twentieth century. Specifically, science today does not break with law. And this is what counts for us. We want criteria of science for today, not for yesterday. (Before I am accused of making my case by fiat, let me challenge Laudan to find one point within the modem geological theory of plate tectonics where appeal is made to miracles, that is, to breaks with law. Of course, saying that science appeals to law is not asserting that we know all of the laws. But, who said that we did? Not Judge Overton in his Opinion.)

What about the criterion of tentativeness, which involves a willingness to test and reject if necessary? Laudan objects that real science is hardly all that tentative: "[H]istorical and sociological researches on science strongly suggest that the scientists of any epoch likewise regard some of their beliefs as so fundamental as not to be open to repudiation or negotiation."

It cannot be denied that scientists do sometimes—frequently—hang on to their views, even if not everything meshes precisely with the real world. Nevertheless, such tenacity can be exaggerated. Scientists, even Newtonians, have been known to change their minds. Although I would not want to say that the empirical evidence is all-decisive, it plays a major role in such mind changes. As an example, consider a major revolution of our own time, namely, that which occurred in geology. When I was an undergraduate in 1960, students were taught that continents do not move. Ten years later, they were told that they do move. Where is the dogmatism here? Furthermore, it was the new empirical evidence—*e.g.*, about the nature of the sea-bed—which persuaded geologists. In short, although science may not be as open-minded as Karl Popper thinks it is, it is not as close-minded as, say, Thomas Kuhn thinks it is.

Let me move on to the second and third questions, the status of creation-science and Judge Overton's treatment of the problem. The slightest acquaintance with the creation-science literature and Creationism movement shows that creation-science fails abysmally as science. Consider the following passage, written by one of the leading creationists, Duane T. Gish, in *Evolution: The Fossils Say No!*

> CREATION. By creation we mean the bringing into being by a supernatural Creator of the basic kinds of plants and animals by the process of sudden, or fiat, creation.
>
> We do not know how the Creator created, what processes He used, *for He used processes which are not operating anywhere in the natural universe.* This is why we refer to creation as Special Creation. We cannot discover by scientific investigations anything about the creative processes used by the Creator.

The following similar passage was written by Henry M. Morris, who is considered to be the founder of the creation-science movement:

> ...it is...quite impossible to determine anything about Creation through a study of present processes, because present processes are not created in character. If man wishes to know anything about Creation (the time of Creation, the duration of Creation, the order of Creation, the methods of Creation, or anything else) his sole source of true information is that of divine revelation. God was there when it happened. We were not there...therefore, we are completely limited to what God has seen fit to tell us, and this information is in His written Word. This is our textbook on the science of Creation!

By their own words, therefore, creation-scientists admit that they appeal to phenomena not covered or explicable by any laws that humans can grasp as laws. It is not simply that the pertinent laws are not yet known. Creative processes stand outside law as humans know it (or could know it) on Earth—at least—there is no way that scientists can know Mendel's laws through observation and experiment. Even if God did use His own laws, they are necessarily veiled from us forever in this life, because Genesis says nothing of them.

Furthermore, there is nothing tentative or empirically checkable about the central claims of creation-science. Creationists admit as much when they join the Creation Research Society (the leading organization of the movement). As a condition of membership applicants must sign a document specifying that they now believe and will continue to believe:

> (1) The Bible is the written Word of God, and because we believe it to be inspired throughout, all of its assertions are historically and scientifically true in all of the original autographs. To the student of nature, this means that the account of origins in Genesis is a factual presentation of simple historical truths. (2) All basic types of living things, including man, were made by direct creative acts of God during Creation Week as described in Genesis. Whatever biological changes have occurred since Creation have accomplished only changes within the original created kinds. (3) The great Flood described in Genesis, commonly referred to as the Noachian Deluge, was an historical event, worldwide in its extent and effect. (4) Finally, we are an organization of Christian men of science, who accept Jesus Christ as our Lord and Savior. The account of the special creation of Adam and Eve as one man and one woman, and their subsequent fall into sin, is the basis for our belief in the necessity of a Savior for all mankind. Therefore, salvation can come only thru accepting Jesus Christ as our Savior.

It is difficult to imagine evolutionists signing a comparable statement, that they will never deviate from the literal text of Charles Darwin's *On the Origin of Species*. The non-scientific nature of creation-science is evident for all to see, as is also its religious nature. Moreover, the quotes I have used above were all used by Judge Overton, in the *McLean* Opinion, to make exactly the points I have just made. Creation-science is not genuine science, and Judge Overton showed this.

Finally, what about Laudan's claim that some parts of creation-science (e.g., claims about the Flood) are falsifiable and that other parts (e.g., about the originally created "kinds") are revisable? Such parts are not falsifiable or revisable in a way indicative of genuine science. Creation-science is not like physics, which exists as part of humanity's common cultural heritage and domain. It exists solely in the imaginations and writing of a relatively small group of people. Their publications (and stated intentions) show that, for example, there is no way they will relinquish belief in the Flood, whatever the evidence. In this sense, their doctrines are truly unfalsifiable.

Furthermore, any revisions are not genuine revisions, but exploitations of the gross ambiguities in the creationists' own position. In the matter of origins, for example, some elasticity could be perceived in the creationist position, given the conflicting claims about the possibility of (degenerative) change within the originally created "kinds." Unfortunately, any open-mindedness soon proves illusory for creationists have no real idea about what God is supposed to have created in the beginning, except that man was a separate species. They rely solely on the Book of Genesis:

> And God said, Let the waters bring forth abundantly the moving creature that hath life, and the fowl that may fly above the earth in the open firmament of heaven.
> And God created great whales, and every living creature that moveth, which the waters brought forth abundantly, after their kind, and every winged fowl after his kind: and God saw that it was good.

And God blessed them, saying Be fruitful, and multiply, and fill the waters in the seas, and let fowl multiply in the earth.

And the evening and the morning were the fifth day.

And God said, Let the earth bring forth the living creature after his kind, cattle, and creeping thing, and beast of the earth after his kind: and it was so.

And God made the beast of the earth after his kind, and cattle after their kind, and everything that creepeth upon the earth after his kind: and God saw that it was good.

But the *definition* of "kind," what it really is, leaves creationists as mystified as it does evolutionists. For example, creationist Duane Gish makes this statement on the subject:

[W]e have defined a basic kind as including all of those variants which have been derived from a single stock…We cannot always be sure, however, what constitutes a separate kind. The division into kinds is easier the more the divergence observed. It is obvious, for example, that among invertebrates the protozoa, sponges, jellyfish, worms, snails, trilobites, lobsters, and bees are all different kinds. Among the vertebrates, the fishes, amphibians, reptiles, birds, and mammals are obviously different basic kinds.

Among the reptiles, the turtles, crocodiles, dinosaurs, pterosaurs (flying reptiles), and ichthyosaurs (aquatic reptiles) would be placed in different kinds. Each one of these major groups of reptiles could be further subdivided into the basic kinds within each.

Within the mammalian class, duck-billed platypus, bats, hedgehogs, rats, rabbits, dogs, cats, lemurs, monkeys, apes, and men are easily assignable to different basic kinds. Among the apes, the gibbons, orangutans, chimpanzees, and gorillas would each be included in a different basic kind.

Apparently, a "kind" can be anything from humans (one species) to trilobites (literally thousands of species). The term is flabby to the point of inconsistency. Because humans are mammals, if one claims (as creationists do) that evolution can occur within but not across kinds, then humans could have evolved from common mammalian stock—but because humans themselves are kinds such evolution is impossible.

In brief, there is no true resemblance between the creationists' treatment of their concept of "kind" and the openness expected of scientists. Nothing can be said in favor of creation-science or its inventors. Overton's judgment emerges unscathed by Laudan's complaints.

SOURCE: Michael Ruse, *Science Technology & Human Values* (7), pp. 29–23, copyright 1982 by Michael Ruse. Reprinted by permission of Sage Publications Inc.

POPE JOHN PAUL II, "TRUTH CANNOT CONTRADICT TRUTH" (1996)[4]

The Roman Catholic Church represented a somewhat unique religious perspective in the late twentieth century with respect to the debate between evolution and creation. While Galileo's imprisonment at the hands of the Church marked a low point in the relationship between science and religion, Catholic theology generally allowed scientists complete liberty

---

[4]Pope John Paul II, "Truth Cannot Contradict Truth," October 30, 1996. Reproduced by permission. © Libreria Editrice Vaticana 2006.

in exploring and explaining the natural world in the twentieth century. While Protestant denominations became increasingly skeptical of evolution and served as the core of opposition, Catholics remained somewhat indifferent. The exceptions varied widely in local areas. In one community, a particular leader might join with other Fundamentalists against evolution; in another, Catholics might join more liberal sects in supporting the teaching of evolution in public schools. This situation lasted until 1996, when the Pope issued a statement that clearly supported evolution as a scientific theory that satisfied modern science and did not conflict with modern Catholic teachings. From that point, the unique existence of the human soul remained a question beyond the realm of evolutionary science, but the Pope suggested all biological explanations could extend to humans through evolution. American Catholics remained somewhat divided between Fundamentalist and liberal camps.

With great pleasure I address cordial greeting to you, Mr. President, and to all of you who constitute the Pontifical Academy of Sciences, on the occasion of your plenary assembly. I offer my best wishes in particular to the new academicians, who have come to take part in your work for the first time. I would also like to remember the academicians who died during the past year, whom I commend to the Lord of life.

1. In celebrating the 60th anniversary of the academy's refoundation, I would like to recall the intentions of my predecessor Pius XI, who wished to surround himself with a select group of scholars, relying on them to inform the Holy See in complete freedom about developments in scientific research, and thereby to assist him in his reflections.

   He asked those whom he called the Church's "senatus scientificus" to serve the truth. I again extend this same invitation to you today, certain that we will be able to profit from the fruitfulness of a trustful dialogue between the Church and science (cf. Address to the Academy of Sciences, No. 1, Oct. 28, 1986; *L'Osservatore Romano*, Eng. ed., Nov. 24, 1986, p. 22).

2. I am pleased with the first theme you have chosen, that of the origins of life and evolution, an essential subject which deeply interests the Church, since revelation, for its part, contains teaching concerning the nature and origins of man. How do the conclusions reached by the various scientific disciplines coincide with those contained in the message of revelation? And if, at first sight, there are apparent contradictions, in what direction do we look for their solution? We know, in fact, that truth cannot contradict truth (cf. Leo XIII, encyclical *Providentissimus Deus*). Moreover, to shed greater light on historical truth, your research on the Church's relations with science between the 16th and 18th centuries is of great importance. During this plenary session, you are undertaking a "reflection on science at the dawn of the third millennium," starting with the identification of the principal problems created by the sciences and which affect humanity's future. With this step you point the way to solutions which will be beneficial to the whole human community. In the domain of inanimate and animate nature, the evolution of science and its applications give rise to new questions. The better the Church's knowledge is of their essential aspects, the more she will understand their impact. Consequently, in accordance with her specific mission she will be able to offer criteria for discerning the moral conduct required of all human beings in view of their integral salvation.

3. Before offering you several reflections that more specifically concern the subject of the origin of life and its evolution, I would like to remind you that the magisterium of the Church has already made pronouncements on these matters within the framework of her own competence. I will cite here two interventions.

   In his encyclical *Humani Generis* (1950), my predecessor Pius XII had already stated that there was no opposition between evolution and the doctrine of the faith about man and his vocation, on condition that one did not lose sight of several indisputable points.

For my part, when I received those taking part in your academy's plenary assembly on October 31, 1992, I had the opportunity with regard to Galileo to draw attention to the need of a rigorous hermeneutic for the correct interpretation of the inspired word. It is necessary to determine the proper sense of Scripture, while avoiding any unwarranted interpretations that make it say what it does not intend to say. In order to delineate the field of their own study, the exegete and the theologian must keep informed about the results achieved by the natural sciences (cf. AAS 85 1/81993 3/8, pp. 764–772; address to the Pontifical Biblical Commission, April 23, 1993, announcing the document on the *Interpretation of the Bible in the Church*: AAS 86 1/81994 3/8, pp. 232–243).

4. Taking into account the state of scientific research at the time as well as of the requirements of theology, the encyclical *Humani Generis* considered the doctrine of "evolutionism" a serious hypothesis, worthy of investigation and in-depth study equal to that of the opposing hypothesis. Pius XII added two methodological conditions: that this opinion should not be adopted as though it were a certain, proven doctrine and as though one could totally prescind from revelation with regard to the questions it raises. He also spelled out the condition on which this opinion would be compatible with the Christian faith, a point to which I will return. Today, almost half a century after the publication of the encyclical, new knowledge has led to the recognition of the theory of evolution as more than a hypothesis. [*Aujourdhui, près dun demi-siècle après la parution de l'encyclique, de nouvelles connaissances conduisent à reconnaitre dans la théorie de l'évolution plus qu'une hypothèse.*] It is indeed remarkable that this theory has been progressively accepted by researchers, following a series of discoveries in various fields of knowledge. The convergence, neither sought nor fabricated, of the results of work that was conducted independently is in itself a significant argument in favor of this theory.

What is the significance of such a theory? To address this question is to enter the field of epistemology. A theory is a metascientific elaboration, distinct from the results of observation but consistent with them. By means of it a series of independent data and facts can be related and interpreted in a unified explanation. A theory's validity depends on whether or not it can be verified; it is constantly tested against the facts; wherever it can no longer explain the latter, it shows its limitations and unsuitability. It must then be rethought.

Furthermore, while the formulation of a theory like that of evolution complies with the need for consistency with the observed data, it borrows certain notions from natural philosophy.

And, to tell the truth, rather than the theory of evolution, we should speak of several theories of evolution. On the one hand, this plurality has to do with the different explanations advanced for the mechanism of evolution, and on the other, with the various philosophies on which it is based. Hence the existence of materialist, reductionist and spiritualist interpretations. What is to be decided here is the true role of philosophy and, beyond it, of theology.

5. The Church's magisterium is directly concerned with the question of evolution, for it involves the conception of man: Revelation teaches us that he was created in the image and likeness of God (cf. Gn 1:27–29). The conciliar constitution *Gaudium et Spes* has magnificently explained this doctrine, which is pivotal to Christian thought. It recalled that man is "the only creature on earth that God has wanted for its own sake" (No. 24). In other terms, the human individual cannot be subordinated as a pure means or a pure instrument, either to the species or to society; he has value *per se*. He is a person. With his intellect and his will, he is capable of forming a relationship of communion, solidarity and self-giving with his peers. St. Thomas observes that man's likeness to God resides especially in his speculative intellect, for his relationship with the object of his knowledge resembles God's relationship with what he has created (Summa Theologica I-II:3:5, ad 1). But even more, man is called to enter into a relationship of knowledge and love with God himself, a relationship which will find its complete fulfillment beyond time, in eternity. All the depth and grandeur of this vocation are revealed to us in the mystery of the

risen Christ (cf. *Gaudium et Spes*, 22). It is by virtue of his spiritual soul that the whole person possesses such a dignity even in his body. Pius XII stressed this essential point: If the human body take its origin from pre-existent living matter, the spiritual soul is immediately created by God ("animas enim a Deo immediate creari catholica fides nos retinere iubei"; "Humani Generis," 36). Consequently, theories of evolution which, in accordance with the philosophies inspiring them, consider the spirit as emerging from the forces of living matter or as a mere *epiphenomenon* of this matter, are incompatible with the truth about man. Nor are they able to ground the dignity of the person.

6. With man, then, we find ourselves in the presence of an ontological difference, an ontological leap, one could say. However, does not the posing of such ontological discontinuity run counter to that physical continuity which seems to be the main thread of research into evolution in the field of physics and chemistry? Consideration of the method used in the various branches of knowledge makes it possible to reconcile two points of view which would seem irreconcilable. The sciences of observation describe and measure the multiple manifestations of life with increasing precision and correlate them with the time line. The moment of transition to the spiritual cannot be the object of this kind of observation, which nevertheless can discover at the experimental level a series of very valuable signs indicating what is specific to the human being. But the experience of metaphysical knowledge, of self-awareness and self-reflection, of moral conscience, freedom, or again of aesthetic and religious experience, falls within the competence of philosophical analysis and reflection, while theology brings out its ultimate meaning according to the Creator's plans.

7. In conclusion, I would like to call to mind a Gospel truth which can shed a higher light on the horizon of your research into the origins and unfolding of living matter. The Bible in fact bears an extraordinary message of life. It gives us a wise vision of life inasmuch as it describes the loftiest forms of existence. This vision guided me in the encyclical which I dedicated to respect for human life, and which I called precisely "Evangelium Vitae."

It is significant that in St. John's Gospel *life* refers to the divine light which Christ communicates to us. We are called to enter into eternal life, that is to say, into the eternity of divine beatitude. To warn us against the serious temptations threatening us, our Lord quotes the great saying of Deuteronomy: "Man shall not live by bread alone, but by every word that proceeds from the mouth of God" (Dt 8:3; cf. Mt 4:4). Even more, "life" is one of the most beautiful titles which the Bible attributes to God. He is the living God.

I cordially invoke an abundance of divine blessings upon you and upon all who are close to you.

SOURCE: Pope John Paul II. "Truth Cannot Contradict Truth," (1996) in Catholic Library (online). Retrieved February 24, 2007, from http://www.newadvent.org/library/docs_jp02tc.htm.

MICHAEL RUSE, AMERICAN ASSOCIATION FOR THE ADVANCEMENT OF SCIENCE SYMPOSIUM, "THE NEW ANTIEVOLUTIONISM" (1993)

When Eugenie Scott introduced Ruse before he gave the below talk, she noted that he almost needed no introduction. His testimony at the Arkansas trial and many publications arguing against creation science had made him famous in this debate. His personal style and sometimes irreverent comments, along with his deep commitment to liberal ideals and philosophical pursuits drew supportive colleagues and opponents alike into the dialogue. He wrote book after book on the different perspectives and issues raised by creationists, scientists, and philosophers of science. He provided thorough critiques of creationist views, thoughtful analysis of the science, and often scathing reviews of philosophical contributions. Ruse consistently sought new insights for debate, and when creationists disappointed him by merely putting a new spin on an old argument, he could be ruthless in his dismissal.

Eugenie Scott: Our next speaker is Dr. Michael Ruse, from the Department of Philosophy at the University of Guelph in Ontario. I thought I saw him a little earlier today. Michael, hello. Michael is actually doing a couple of sessions today, he's been a very busy fellow. And we're very pleased that he was able to make ours as well.

    Michael Ruse is a philosopher of science, particularly of the evolutionary sciences. He's almost a person who needs no introduction in this context. He's the author of several books on Darwinism and evolutionary theory, including an analysis of scientific creationism entitled *But Is It Science?* No. I don't think I've spoiled the plot. I mean, I would recommend that you read this book, it's really quite good. But that is his conclusion. He'll be speaking today about "Nonliteralist Antievolution." Michael? Would you like some more light?

[The speaker's podium is dark.]

Ruse: It's the first time I've actually sort of given a lecture literally in the dark, as opposed to just metaphorically. Actually, the title of my book *But Is It Science? the Evolution-Creation Controversy*, is intended very much to raise the question about both evolution and creationism, and, in a way, that's the theme of what I want to say today. I've noticed that we're moving right along, so I'm not going to say very much at all, but I am going to throw out one or two ideas, which, in the words of Father Huddleston, who of course got them from somewhere else, "I trust they're not to your comfort."

[The microphone is moved closer to Ruse.]

    God, not only am I in the dark, I've got this bloody great thing sticking in my face too! Even if you can see me, I can't see you anymore. Talk about non-intelligent design going on here. I was intending to come along, when I was asked to participate in this colloquium, I was intending to come along and talk about the book by the California lawyer Phillip Johnson, the title of the book I'm glad to say has thankfully escaped me just at the moment. *Darwin on Trial*, okay. What happened was I was asked to review Phillip Johnson's book a couple of years ago, and it was an exercise in what not to do, from my point of view, what not to do if you're a book reviewer. Namely, if you write such a critical review of a book, the editor who has commissioned the review might look at your review and say, obviously that book is so lousy I don't think it's worth talking about in our journal. And that's what happened to my review of Phillip Johnson. It became a non-review, not I think in any sense because it was being censored, but simply because the editor, the book review editor, said, well frankly, I've got a lot more interesting books that we could talk about, so we'll just drop it.

    In fact, when I read Phillip Johnson's book, I mean, at one level, it's a very impressively put together piece of work. Phillip Johnson is certainly I think a very good lawyer, he's got a good legal mind, and he does a good slick job of packaging. I think that when you look, when you dig down underneath, you do start to see many of the same sorts of themes and the ideas coming across which have been expressed—perhaps more crudely, let's put it—by some of the friends who have been mentioned earlier, people like Duane Gish and Henry Morris. Like everybody who reads a book who's written anything themselves, I looked up my own name in the index first, and then went to the passages which refer to me, and thank God, I am—it's not just Stephen Jay Gould who's being referred to these days—but there were a couple of comments about me—regretfully in footnotes. And I was able to satisfy myself quite readily that in fact Phillip Johnson was playing much the same trick that everybody else was. I was quoted as putting forward some fairly hard-line social Darwinian views, in East Germany, of all places, a country which as you know no longer exists. And, in fact, fortunately the comments I had in fact made in what was East Germany in those days were taken down and in fact are printed. And I went and I checked, and, I must say, not to my—to my great relief, anyway—I was saying the exact opposite of what Phillip Johnson was saying. I mean, I'm much given to contradiction, but, thank God, this was one of those—thank God, well, thank Darwin, anyhow, as we've just heard—this wasn't one of those occasions.

So, I was intending, as I say, to come along and talk about Phillip Johnson. What happened between then and now, on the way, was that a few months ago I was invited to participate by some evangelicals in what was a sort of weekend session that they'd got, and Phillip Johnson and I were put face to face. And as I always find when I meet creationists or non-evolutionists or critics or whatever, I find it a lot easier to hate them in print than I do in person. And in fact I found—I must confess—I found Phillip Johnson to be a very congenial person, with a fund of very funny stories about Supreme Court justices, some of which may even be true, unlike his scientific claims. We did debate, and in fact I thought that we had, as others said afterwards, both evolutionists and non-evolutionists, I thought that we had what was really quite, and I want to be quite fair about this, I thought we had a really quite constructive interchange. Because basically we didn't talk so much about creationism. We certainly didn't so much talk about his particular arguments in his book, or arguments that I've put forward in *Darwinism Defended*, or these sorts of things.

But we did talk much more about the whole question of metaphysics, the whole question of philosophical bases. And what Johnson was arguing was that, at a certain level, the kind of position of a person like myself, an evolutionist, is metaphysically based at some level, just as much as the kind of position of let us say somebody, some creationist, someone like Gish or somebody like that. And to a certain extent, I must confess, in the ten years since I performed, or I appeared, in the creationism trial in Arkansas, I must say that I've been coming to this kind of position myself. And, in fact, when I first thought of putting together my collection *But Is It Science?*, I think Eugenie was right, I was inclined to say, well, yes, creationism is not science and evolution is, and that's the end of it, and you know just trying to prove that. Now I'm starting to feel—I'm no more of a creationist now than I ever was, and I'm no less of an evolutionist now that I ever was—but I'm inclined to think that we should move our debate now onto another level, or move on. And instead of just sort of, just—I mean I realize that when one is dealing with people, say, at the school level, or these sorts of things, certain sorts of arguments are appropriate. But those of us who are academics, or for other reasons pulling back and trying to think about these things, I think that we should recognize, both historically and perhaps philosophically, certainly that the science side has certain metaphysical assumptions built into doing science, which—it may not be a good thing to admit in a court of law—but I think that in honesty that we should recognize, and that we should be thinking about some of these sorts of things.

Certainly, I think that philosophers like myself have been much more sensitized to these things, over the last ten years, by trends and winds and whatever the right metaphor is, in the philosophy of science. That we've become aware, thanks to Marxists and to feminists, criticisms—the criticisms of historians and sociologists and others—that science is a much more idealistic, in the *a priori* sense, enterprise, than one would have got from reading the logical positivists, or even the great philosophers. The people like Popper and Hempel and Nagel, of the 1950s and 1960s, which was when my generation entered the field and started to grow up.

Certainly, historically, that if you look at, say, evolutionary theory, and of course this was brought out I think rather nicely by the talk just before me, it's certainly been the case that evolution has functioned, if not as a religion as such, certainly with elements akin to a secular religion. Those of us who teach philosophy of religion always say there's no way of defining religion by a neat, necessary, and sufficient condition. The best that you can do is list a number of characteristics, some of which all religions have, and none of which any religion, whatever or however you sort of put it. And certainly, there's no doubt about it, that in the past, and I think also in the present, for many evolutionists, evolution has functioned as something with elements which are, let us say, akin to being a secular religion.

I think, for instance, of the most famous family in the history of evolution, namely, the Huxleys. I think of Thomas Henry Huxley, the grandfather, and of Julian Huxley, the grandson. Certainly, if you read Thomas Henry Huxley, when he's in full flight, there's no question but that for Huxley at some very important level, evolution and science generally, but certainly evolution in particular, is functioning a bit as a kind of secular religion. Interestingly, Huxley—and I've gone through his own lectures, I've gone through two complete sets of lecture notes that Huxley gave to his students—Huxley never talked about evolution when he was actually teaching. He kept evolution for affairs like this, and when he was talking at a much more popular sort of level. Certainly, though, as I say, for Thomas Henry Huxley, I don't think there's any question but that evolution functioned, at a level, as a kind of secular religion.

And there's no question whatsoever that for Julian Huxley, when you read *Evolution, the Modern Synthesis*, that Julian Huxley saw evolution as a kind of progressive thing upwards. I think Julian Huxley was certainly an atheist, but he was at the same time a kind of neo-vitalist, and he bound this up with his science. If you look both at his printed stuff, and if you go down to Rice University which has got all his private papers, again and again in the letters, it comes through very strongly that for Julian Huxley evolution was functioning as a kind of secular religion.

I think that this—and I'm not saying this now particularly in a critical sense, I'm just saying this in a matter-of-fact sense—I think that today also, for more than one eminent evolutionist, evolution in a way functions as a kind of secular religion. And let me just mention my friend Edward O. Wilson. Certainly, I think that if you look at some of the stuff which caused some much controversy in the 1970s, what is interesting is not so much the fact that Wilson was talking about trying to include humans in the evolutionary scenario. Everybody was doing that. It was not so much even the fact that he was using what is now called sexist language, like "Man," because I went to look at Richard Lewontin's book, which he published the year before Wilson, and in the index it says "Homo sapiens, see 'Man'"—so, I mean, we were all committing that sort of mistake, as it is now judged. But certainly, if you look for instance in *On Human Nature*, Wilson is quite categorical about wanting to see evolution as the new myth, and all sorts of language like this. That for him, at some level, it's functioning as a kind of metaphysical system.

So, as I say, historically I think, however we're going to deal with creationism, or new creationism, or these sorts of things, whether you think that this is—that what I've just been saying means that we'd better put our house in order, or whatever—I think at least we must recognize the historical facts. I think also, and I am going to speak very, very briefly, because time is so short, is I think that we should also look at evolution and science, in particular, biology, generally philosophically I think a lot more critically—and I don't say negatively, please understand that—I think a lot more critically than we were doing ten years ago. Sensitized, I say, by the work of the social constructivists and others, historians, sociologists, and these sorts of people.

And it seems to me very clear that at some very basic level, evolution as a scientific theory makes a commitment to a kind of naturalism, namely, that at some level one is going to exclude miracles and these sorts of things, come what may. Now, you might say, does this mean it's just a religious assumption, does this mean it's irrational to do something like this. I would argue very strongly that it's not. At a certain pragmatic level, the proof of the pudding is in the eating. And that if certain things do work, you keep going with this, and that you don't change in midstream, and so on and so forth. I think that one can in fact defend a scientific and naturalistic approach, even if one recognizes that this does include a metaphysical assumption to the regularity of nature, or something of this nature.

So as I say, I think that one can defend it as reasonable, but I don't think it helps matter by denying that one is making it. And I think that once one has made such an assumption, one has perfect powers to turn to, say, creation science, which claims to be naturalistic also, and point out that it's wrong. I think one has every right to show that evolutionary theory in various forms certainly seems to be the most reasonable position, once one has taken a naturalistic position. So I'm not coming here and saying, give up evolution, or anything like that.

But I am coming here and saying, I think that philosophically that one should be sensitive to what I think history shows, namely, that evolution, just as much as religion—or at least, leave "just as much," let me leave that phrase—evolution, akin to religion, involves making certain a priori or metaphysical assumptions, which at some level cannot be proven empirically. I guess we all knew that, but I think that we're all much more sensitive to these facts now. And I think that the way to deal with creationism, but the way to deal with evolution also, is not to deny these facts, but to recognize them, and to see where we can go, as we move on from there.

Well, I've been very short, but that was my message, and I think it's an important one.

Scott: Any questions?

[*There is a momentary silence.*]

Ruse: State of shock! Yes, Ed Manier. [*Manier is on the faculty of Notre Dame University, in the Department of History and Philosophy of Science.*]

Manier: Well, congratulations. I mean, you took less time than Bill Clinton. I think—maybe not quite. But you made a remark about Stephen Gould. I earlier made a remark about Stephen Gould. I think there is perhaps some sense in which you and Stephen disagree, either scientifically or metaphysically. I wonder if you could comment on that.

Ruse: That we agree or disagree?

Manier: That you disagree. I'm always more interested in disagreement.

Ruse: Certainly I think that Steve Gould and I, we certainly disagree about the nature of evolution, there's no question about that. At some level, I'm a hard-line Darwinian. That means, you know, I'm somewhere to the right of Archdeacon Paley when it comes to design. I mean, when I look, even at you, Ed, when I look even at you, I'm already speculating why you've got a bald head, and, you know, why this makes you sexually attractive, and so on. So, I mean, yeah—whereas I think that Gould falls very much into the other, much more Germanic Naturphilosophie tradition, which stresses form over function. I don't think there's any question about that. And at a certain level, I'd be inclined to say that these are, if you like, metaphysical assumptions, paradigms, or something like that, *a priori* constraints that we're putting on the ways that we're looking at the world and all those sorts of things. Certainly, at that level, we do differ.

Where else do we differ? Gould says that he thinks that science is simply, you know, disinterested reflection of reality, then again we differ also. But of course the thing is that Gould, although he denies being a Marxist or anything like this, certainly if you look at Gould's work, for instance, when he's praising stuff, even apart from when he's criticizing stuff, I think that Gould—as much as anybody, more than most—has long been sensitive to the fact that science involves a kind of metaphysical assumption. I use the word "metaphysical" because I don't look on the word "metaphysical" as a dirty word. Like I don't look upon "teleology" as a dirty word. He may, you know, he may very ardently say don't call me a metaphysician, but I suspect that we agree, whatever we call the terms. I mean, the trouble is, metaphysics, you know, people think of metaphysics and Scottish idealists and Hegelians and all those sorts of things. So he may not want to use my language. But I suspect that about the nature of science—I suspect, but ask him—I suspect that we don't differ there. But we do differ about how we want to cash it out in the actual evolutionary realm.

Manier: Well, if I could just pursue that, for just a minute, he may very well be more of a Naturphilosoph than you. And perhaps, although I suspect that you deny this in almost every context, more of a Romantic than you. But I'm wondering–

Ruse: How can you say that about me? After the things you said last night over drinks, but go on.

Manier: You made reference to my baldness, and I'm sensitive about that.

Ruse: I was trying to give it an adaptive function. It's okay, I don't think it's a mistake. I mean, you know, I think God designed it that way. Go on.

Manier: But you say that about everything.

Ruse: That's right. I'm somewhere to the right of Archdeacon Paley on this, I really am.

Manier: Well, pardon me if I'm not flattered. What I'm curious about is the extent to which your talk suggests a strategy to the National Society of Science Teachers to have something like a pluralistic approach to these issues. That is, it's one thing to be snide about them–

Ruse: Yes, I think that's a point well taken. The trouble is, you know, Ed, I mean, everybody, I mean, the trouble is, we're balancing, we're trying to juggle so many balls in the air.

On the one hand, we're trying to do some philosophy. Another ball is trying to be science educators, both at the university level, but more particularly, at the schools level. At another level, we've got the actual political facts, of how do you fight school boards, and that sort of thing. At another level you've got the legal questions of, you know, your laws are different from my laws, for instance. Up in Canada we don't have a Constitution in that sort of way. Or at least, we've got a Constitution which has a weasel clause, you know, "in a democratic and fair society" which means that it can all be altered, if they want to, and it often is.

So, I mean you've got all of these sorts of issues, and I'm very sensitive to the fact that if a philosopher tries out, say, ideas and thinks those sorts of things, people might well say, well I hope to God you don't say this outside in public, because we're going to run into problems with the third or fourth ball, and I'm very sensitive to that.

And, to a certain extent, I think I personally have for many years used, to a certain extent, self-censorship, you know just basically not talking too much on these sorts of lines. But at the same time, I'm not sure that the way forward is by simply not thinking about things philosophically or responding to ideas, or saying, well gosh, I find what the social constructivists are saying very interesting, but, by God, I'd better not believe or accept any of this— because it's going to get us into trouble at the school board level. I mean, that's a tension. But I think that somehow, it seems to me, well, maybe two wrongs don't make a right, or do make a right. But I just don't want to do that.

As I hope I said right at the end, I don't come here preaching creationism or preaching, you know, some message of negativism: folks give up, modern philosophy of science is now showing that science is just as much a religion as creation science, so frankly folks there's nothing that you could do, and if I could go back ten years to Arkansas I'd just reverse everything. I think that you can do it. I mean I think you can't do it in just a gung-ho, straightforward, neo-Popperian way: here we've got science on the one side, here we've got religion on the other side, evolution falls on the science side, creationism falls on the other side, and, you know, never the twain shall meet. I think you've got to go at different ways, things like, as I mentioned, pragmatism, for instance. Taking some sort of coherence theory of truth, or something like that. I still think that one can certainly exclude creation science on those grounds. Now, whether or not—how that fits in with your laws—one has to ask the lawyers, those sorts of things. I certainly think that's something that you can do.

[Applause]

Scott: Wait a minute, just–

Ruse: Before you start applauding, she's going to cut off all of my buttons, and drum me out of the society.

Scott: Not a bit, but he's not done yet. I'm going to take my chairman's prerogative, to ask a question, if I may. I wonder whether it might be useful to distinguish between the naturalism or materialism that is necessary to perform science as we do it in the twentieth century, as opposed to the Baconian approach, etc., and distinguish that from philosophical attitudes that we as individuals may or may not have regarding materialism or naturalism. And perhaps some of this confusion that we find at the practical level, at the school board level, and in dealing with people with Johnson, is that Johnson, for example, does confuse these two things. He assumes that if you are a scientist then you therefore are a philosophical materialist, in addition to being a practical materialist, in the operation of your work.

Ruse: Oh yes, I think that point is well taken. I think to sort of redress some of the rather flip comments I made, I think that's absolutely true. Let me end certainly by saying that although I got on quite well with Johnson at the personal level, I still think that his book is a slippery piece of work. And you're absolutely right that he, like any lawyer, is out to win. That's the name of the game in law. And certainly he can get points by shifting back and forth on meanings of naturalism, or if he can get a report on what Ed Manier and I were doing, and then sort of take it out of context, I've no reason to think that he wouldn't do that sort of thing. Don't misunderstand me. I'm not saying, I'm not denying the power or the importance of the sort of thing he's doing, or the importance of combating that sort of thing.

What I am saying, nevertheless, and I will sit down now, is I don't think that we're going—well, I don't know whether we're going to serve—I mean, the easy thing is we're not going to serve our purpose by—let me just simply say that I as a philosopher of science am worried about what I think were fairly crude neo-positivistic attitudes that I had about science, even as much as ten years ago, when I was fighting in Arkansas. This doesn't mean to say that I don't want to stand up for evolution, I certainly do. But I do think that philosophy of science, history of science, moves on, and I think it's incumbent upon us who take this particular creationism – evolution debate seriously, to be sensitive to these facts, and not simply put our heads in the sand, and say, well, if we take this sort of stuff seriously, we're in deep trouble. Perhaps we are. But I don't think that the solution is just by simply ignoring them.

Scott: Now you can applaud, he's done.

SOURCE: Ruse, Michael. American Association for the Advancement of Science Symposium, "The New Antievolutionism" (1993) in Access Research Network (online). Retrieved February 24, 2007, from http://www.arn.org/docs/orpages/or151/mr93tran.htm. Used by permission.

TERMS

Falsifiability—A philosophical criterion for science, where a hypothesis can be tested and proven false. While many people expect scientists to spend most of their time proving ideas to be true, in fact, this is extremely difficult to achieve in any science. Most of the, scientists gather data and conduct experiments in the hope of supporting their hypotheses, but often refuting them. Any hypothesis that cannot be challenged by data in the natural world is generally seen as unscientific due to its unfalsifiability, while one that is repeatedly supported, despite consistent efforts to challenge it, may be adopted as a theory that provides explanatory power to scientists' understanding of the universe. Proposed as a criterion for science by the philosopher Karl Popper, the term became a crucial and sometimes convenient boundary marker for science in this debate.

8

INTELLIGENT DESIGN AND THE SCHOOL BOARD DEBATES

When in 1982 the *McLean v. Arkansas* decision clearly identified scientific creation as religion and therefore inappropriate to teach in public school science classes, opponents of the teaching of evolution in public schools found themselves unable to continue to press successfully for the inclusion of creation science in the high school science curriculum. For over a decade the issue of teaching evolution in public school science classrooms, and more importantly teaching some form of creation in those classes, appeared to be a settled issue. This began to change in the mid-1990s as a cadre of antievolutionists emerged with what to many appeared to be a valid argument against evolution and a secular alternative to evolution.

In 1996 Michael Behe's *Darwin's Black Box* brought the argument from design to bear once again on the issue, arguing that a careful study of nature revealed evidence of what was increasingly called intelligent design. His book helped spawn a new wave of attacks on the teaching of evolution in public schools with demands to include the teaching of intelligent design as a scientific, rather than a religious, viewpoint. Behe made no claims about precisely who or what intentionally designed the universe and the laws that governed it, but the vast majority of his advocates assumed that it was the Christian God. Other, less popular authors, claimed that the designer could just as easily have been technologically advanced aliens, which begged the question, who designed the aliens?

Just as Behe published *Darwin's Black Box*, the Seattle-based Discovery Institute grew increasingly involved in attacking the teaching of evolution in public schools. A conservative think tank, the Discovery Institute distributed grants and advanced generally right-wing causes. In advancing the cause of intelligent design, officials at the institute believed they were encouraging the demise of a theory that they claimed was both badly crippled and inherently antireligious.

Just as *Epperson v. Arkansas* and *McLean v. Arkansas* settled similar issues in previous decades, belligerents in the contest over evolution and creationism again found themselves before a judge debating their claims. The 2005 *Kitzmiller v. Dover* case put the claim that intelligent design was religion masquerading as science to the test. The majority opinion from the cause is an encyclopedic analysis of the influences and claims within the intelligent design movement as well as a critical evaluation of its scientific merits. In the end, the court ruled that intelligent design was essentially religion, just as creation science had been ruled nearly a quarter century earlier, and therefore could not be required as part of the science curriculum in public school classrooms.

MICHAEL BEHE, *DARWIN'S BLACK BOX: A BIOCHEMICAL CHALLENGE TO EVOLUTION* (1996)[1]

Michael Behe (1952–) is an American biochemist, professor at Lehigh University, and one of the leading advocates of intelligent design. After reading Michael Denton's attacks on evolution (*Evolution: A Theory in Crisis*) he came to believe that Darwin's theory of evolution by natural selection and its adaptation into the modern evolutionary synthesis in the mid-twentieth century was unsupportable at the biochemical level. His principle contribution to the intelligent design movement was his 1996 book *Darwin's Black Box*, which laid out his argument against evolution by natural selection and introduced the concept of irreducible complexity. Behe argued that the complex and apparently purposeful structure of even the simplest living things suggested the existence of a purposeful designer, rather than the naturalistic mechanism of evolution by natural selection. By refusing to name the designer, he maintained a claim to being secular and appropriately scientific, but the mainstream scientific community generally dismissed his arguments as merely a form of argument from personal incredulity. However, his work added a secular veneer to the creationist movement and is widely appreciated by antievolutionists.

Irreducible Complexity and the Nature of Mutation

Darwin knew that his theory of gradual evolution by natural selection carried a heavy burden:

> If it could be demonstrated that any complex organ existed which could not possibly have been formed by numerous, successive, slight modifications, my theory would absolutely break down.

It is safe to say that most of the scientific skepticism about Darwinism in the past century has centered on this requirement. From Mivart's concern over the incipient stages of new structures to Margulis's dismissal of gradual evolution, critics of Darwin have suspected that his criterion of failure had been met. But how can we be confident? What type of biological system could not be formed by "numerous, successive, slight modifications"?

Well, for starters, a system that is irreducibly complex. By *irreducibly complex* I mean a single system composed of several well-matched, interacting parts that contribute to the basic function, wherein the removal of any one of the parts causes the system to effectively cease functioning. An irreducibly complex system cannot be produced directly (that is, by continuously improving the initial function, which continues to work by the same mechanism) by slight, successive modifications of a precursor system, because any precursor to an irreducibly complex system that is missing a part is by definition nonfunctional. An irreducibly complex

---

[1]Reprinted with permission of The Free Press, a Division of Simon & Schuster, Inc., from *Darwin's Black Box: The Biochemical Challenge to Evolution* by Michael J. Behe. Copyright © 1996 by Michael J. Behe.

biological system, if there is such a thing, would be a powerful challenge to Darwinian evolution. Since natural selection can only choose systems that are already working, then if a biological system cannot be produced gradually it would have to arise as an integrated unit, in one fell swoop, for natural selection to have anything to act on.

Even if a system is irreducibly complex (and thus cannot have been produced directly), however; one can not definitively rule out the possibility of an indirect, circuitous route. As the complexity of an interacting system increases, though, the likelihood of such an indirect route drops precipitously. And as the number of unexplained, irreducibly complex biological systems increases, our confidence that Darwin's criterion of failure has been met skyrockets toward the maximum that science allows.

In the abstract, it might be tempting to imagine that irreducible complexity simply requires multiple simultaneous mutations—that evolution might be far chancier than we thought, but still possible. Such an appeal to brute luck can never be refuted. Yet it is an empty argument. One may as well say that the world lucidly popped into existence yesterday with all the features it now has. Luck is metaphysical speculation; scientific explanations invoke causes. It is almost universally conceded that such sudden events would be irreconcilable with the gradualism Darwin envisioned. Richard Dawkins explains the problem well:

> Evolution is very possibly not, in actual fact, always gradual. But it must be gradual when it is being used to explain the coming into existence of complicated, apparently designed objects, like eyes. For if it is not gradual in these cases, it ceases to have any explanatory power at all. Without gradualness in these cases, we are back to miracle, which is simply a synonym for the total absence of explanation.
>
> The reason why this is so rests in the nature of mutation.

In biochemistry, a mutation is a change in DNA. To be inherited, the change must occur in the DNA of a reproductive cell. The simplest mutation occurs when a single nucleotide (nucleotides are the "building blocks" of DNA) in a creature's DNA is switched to a different nucleotide. Alternatively, a single nucleotide can be added or left out when the DNA is copied during cell division. Sometimes, though, a whole region of DNA—thousands or millions of nucleotides—is accidentally deleted or duplicated. That counts as a single mutation, too, because it happens at one time, as a single event. Generally a single mutation can, at best, make only a small change in a creature—even if the change impresses us as a big one. For example, there is a well-known mutation called *antennapedia* that scientists can produce in a laboratory fruit fly: the poor mutant creature has legs growing out of its head instead of antennas. Although that strikes us as a big change, it really isn't. The legs on the head are typical fruit-fly legs, only in a different location.

An analogy may be useful here: Consider a step-by-step list of instructions. A mutation is a change in *one* of the lines of instructions. So instead of saying, "Take a 1/4-inch nut," a mutation might say, "Take a 3/8-inch nut." Or instead of "Place the round peg in the round hole," we might get "Place the round peg in the square hole." Or instead of "Attach the seat to the top of the engine," we might get "Attach the seat to the handlebars" (but we could only get this if the nuts and bolts could be attached to the handlebars). What a mutation cannot do is change all the instructions in one step—say, to build a fax machine instead of a radio.

Thus, to go back to the bombardier beetle and the human eye, the question is whether the numerous anatomical changes can be accounted for by many small mutations. The frustrating answer is that we can't tell. Both the bombardier beetle's defensive apparatus and the vertebrate eye contain so many molecular components (on the order of tens of thousands of different types of molecules) that listing them—and speculating on the mutations that might have produced them—is currently impossible. Too many of the nuts and bolts (and screws, motor parts, handlebars, and so on) are unaccounted for. For us to debate whether Darwinian evolution could produce such large structures is like nineteenth century scientists debating whether cells could arise spontaneously. Such debates are fruitless because not all the components are known.

We should not, however, lose our perspective over this; other ages have been unable to answer many questions that interested them. Furthermore, because we can't yet evaluate the question of eye evolution or beetle evolution does not mean we can't evaluate Darwinism's claims for any biological structure. When we descend from the level of a whole animal (such as a beetle) or whole organ (such as an eye) to the molecular level, then in many cases we can make a judgment on evolution because all of the parts of many discrete molecular systems are known. In the next five chapters we will meet a number of such systems—and render our judgment.

Now, let's return to the notion of irreducible complexity. At this point in our discussion *irreducible complexity* is just a term whose power resides mostly in its definition. We must ask how we can recognize an irreducibly complex system. Given the nature of mutation, when can we be sure that a biological system is irreducibly complex?

The first step in determining irreducible complexity is to specify both the function of the system and all system components. An irreducibly complex object will be composed of several parts, all of which contribute to the function. To avoid the problems encountered with extremely complex objects (such as eyes, beetles, or other multicellular biological systems) I will begin with a simple mechanical example: the humble mousetrap.

The function of a mousetrap is to immobilize a mouse so that it can't perform such unfriendly acts as chewing through sacks of flour or electrical cords, or leaving little reminders of its presence in unswept corners. The mousetraps that my family uses consist of a number of parts (Figure 8.1): (1) a flat wooden platform to act as a base; (2) a metal hammer, which does the actual job of crushing the little mouse; (3) a spring with extended ends to press against the platform and the hammer when the trap is charged; (4) a sensitive catch that releases when slight pressure is applied, and (5) a metal bar that connects to the catch and holds the hammer back when the trap is charged. (There are also assorted staples to hold the system together.)

The second step in determining if a system is irreducibly complex is to ask if all the components are required for the function. In this example, the answer is clearly yes. Suppose that while reading one evening, you hear the patter of little feet in the pantry, and you go to the utility drawer to get a mousetrap. Unfortunately, due to faulty manufacture, the trap is missing one of the parts listed above. Which part could be missing and still allow you to catch a mouse? If the wooden base were gone, there would be no platform for attaching the other components. If the hammer were gone, the mouse could dance all night on the platform without becoming pinned to the wooden base. If there were no spring, the hammer and platform would jangle loosely and again the rodent would be unimpeded. If there were no catch or metal

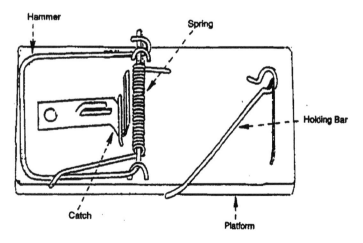

A Household Mousetrap.

holding bar, then the spring would snap the hammer shut as soon as you let go of it; in order to use a trap like that you would have to chase the mouse around while holding the trap open.

To feel the full force of the conclusion that a system is irreducibly complex and therefore has no functional precursors, we need to distinguish between a *physical* precursor and a *conceptual* precursor. The trap described above is not the only system that can immobilize a mouse. On other occasions my family has used a glue trap. In theory, at least, one can use a box propped open with a stick that could be tripped. Or one can simply shoot the mouse with a BB gun. These are not physical precursors to the standard mousetrap, however; since they cannot be transformed, step by Darwinian step, into a trap with a base, hammer, spring, catch, and holding bar.

To clarify the point, consider this sequence: skateboard, toy wagon, bicycle, motorcycle, automobile, airplane, jet plane, space shuttle. It seems like a natural progression, both because it is a list of objects that all can be used for transportation and also because they are lined up in order of complexity. They can be conceptually connected and blended together into a single continuum. But is, say, a bicycle a physical (and potentially Darwinian) precursor of a motorcycle? No. It is only a *conceptual* precursor. No motorcycle in history, not even the first, was made simply by modifying a bicycle in a stepwise fashion. It might easily be the case that a teenager on a Saturday afternoon could take an old bicycle, an old lawnmower engine, and some spare parts and (with a couple of hours of effort) build himself a functioning motorcycle. But this only shows that humans can design irreducibly complex systems, which we knew already. To be a precursor in Darwin's sense we must show that a motorcycle can be built from "numerous, successive, slight modifications" to a bicycle.

So let us attempt to evolve a bicycle into a motorcycle by the gradual accumulation of mutations. Suppose that a factory produced bicycles, but that occasionally there was a mistake in manufacture. Let us further suppose that if the mistake led to an improvement in the bicycle, then the friends and neighbors of the lucky buyer would demand similar bikes, and the factory would retool to make the mutation a permanent feature. So, like biological mutations, successful mechanical mutations would reproduce and spread. If we are to keep our analogy relevant to biology, however; each change can only be a slight modification, duplication, or rearrangement of a preexisting component, and the change must improve the function of the bicycle. So if the factory mistakenly increased the size of a nut or decreased the diameter of a rear tire, bolt, or added an extra wheel onto the front axle or left off the rear tire, or put a pedal on the handlebars or added extra spokes, and if any these slight changes improved the bike ride, then the improvement would immediately be noticed by the buying public and the mutated bikes would, in true Darwinian fashion, dominate the market.

Given these conditions, can we evolve a bicycle into a motorcycle? We can move in the right direction by making the seat more comfortable in small steps, the wheels bigger, and even (assuming our customers prefer the "biker" look) imitating the overall shape in various ways. But a motorcycle depends on a source of fuel, and a bicycle has nothing that can be slightly modified to become a gasoline tank. And what part of the bicycle could be duplicated to begin building a motor? Even if a lucky accident brought a lawnmower engine from a neighboring factory into the bicycle factory, the motor would have to be mounted on the bike and be connected in the right way to the drive chain. How could this be done step-by-step from bicycle parts? A factory that made bicycles simply could not produce a motorcycle by natural selection acting on variation—by "numerous, successive, slight modifications"—and in fact there is no example in history of a complex change in a product occurring in this manner.

A bicycle thus may be a conceptual precursor to a motorcycle, but it is not a physical one. Darwinian evolution requires physical precursors.

SOURCE: Behe, Michael J. *Darwin's Black Box: A Biochemical Challenge to Evolution.* New York: Simon and Schuster, 1996.

MICHAEL RUSE, "THEY'RE HERE!" (1989)

Michael Ruse (1940–) is a philosopher of biology who has done considerable work on the history of the evolution/creation debate and has been a participant in it. As a witness in the 1981 *McLean v. Arkansas Board of Education* trial in which equal time laws were declared unconstitutional because they violated the separation between church and state, Ruse offered a Popperian definition of science that relied on the concept of falsifiability to argue that creation science was not science. He has been an active proponent of the claim that Darwinism and Christianity are in fact reconcilable, but opposes most creationists attempts to fashion a theistic form of evolution that coincides with an active, ever-present God.

Several years ago, I was a witness in Arkansas testifying against a bill passed into law mandating the teaching of biblical literalism, alongside evolution, in state schools. The ACLU brought suit on the grounds that this violated the separation of Church and State. The law was thrown out, as was right and proper, but I still remember what one of the ACLU lawyers said: "Don't think the Creationists will go away. They won't! They'll just regroup and be smarter and sneakier next time."

I am afraid the lawyer was right, as this book under review—whose title has the gall to echo a work of one of the other ACLU witnesses, Stephen Jay Gould—shows only too well. The early creationist publications were crude affairs, not the least in their physical appearance. The leader of the pack, Duane T. Gish's *Evolution, the Fossils Say No!*, may well have sold over 150,000 copies as its cover proudly proclaimed. Yet, it was a cheap-looking thing on second-rate paper with flawed print and with contents that were little better.

Now, we have *Of Pandas and People: The Central Question of Biological Origins*—clean, crisp, beautifully-illustrated, with attractive photographs. There is even the "mandatory" Gary Larson cartoon (a good one, too!). But, it is as bogus as ever, maybe even more so than were the original productions of Gish and friends. They, at least, let you know their thesis clearly—6,000 years of Earth history, six days of Creation, and an enormous flood. In *Of Pandas and People*, the real (same old) message is carefully concealed until you are well and truly hooked. You are, as with a judicious lecture, given an apparently disinterested discussion of rival views—evolution and design (more on this latter term in a moment). You are, as in a real textbook, taken through the various branches of biological science. (This, at least, is an improvement, since previously one tended to begin and end with the fossil record.) You are, as in a scholarly text, given plausible quotes from eminent scientists in the field—Ernst Mayr and the like. You are, however, in these types of books, plunked down on the side of evangelical religion.

Any view of theory of origins must be held in spite of unsolved problems; proponents of both views acknowledge this. Such uncertainties are part of the healthy dynamic that drives science. However, without exaggeration, there is impressive and consistent evidence, from each area we have studied, for the view that living things are the product of intelligent design.

Beneath the gloss of the text, one soon starts to turn up familiar pseudo-arguments.

Many design proponents and some evolutionists believe *Archaeopteryx* was a true bird, capable of powered flight. The fact that it possessed reptilian features not found in most other birds does not require a relationship between birds and reptiles, anymore than the duck-billed platypus, a mammal, must be related to a duck.

Well, if you believe that, you will believe anything! You might believe, for instance, that, as we read earlier in the book, Ernst Mayr thinks so poorly of the *Origin of Species* that his views merit being included in a section on "The failure of natural selection." Or you might believe that Stephen Gould and David Raup think that the fossil record gaps spell "Big Trouble" for evolutionists. Or you might believe that the molecules tell nothing of homology and evolution, as the book suggests.

This book is worthless and dishonest—but slick and appealing, so be on your guard. And, let me (for what seems the millionth time in my life) protest at the Creationists appropriating exclusively unto themselves the mantle of religion. The world of life may or may not be designed. But the argument is not that the choice is between an exclusive disjunction of evolution and design. I believe that if God chooses to do things through unbroken law, then that is God's business, not ours. What is our business is the proper use of our God-given powers of sense and reason, to follow fearlessly where the quest for truth leads. Where it does not lead is to the pages of the book *Of Pandas and People*.

SOURCE: Ruse, Michael. "They're Here!" *Bookwatch Reviews* 2:11 (1989). Used by permission.

PHILLIP E. JOHNSON, *DEFEATING DARWINISM BY OPENING MINDS* (1997)[2]

Phillip Johnson is a retired University of California Berkeley law professor and author of several books that attack Darwinian evolutionary theory. His law credentials are impressive and include experience as a law clerk for Earl Warren, Chief Justice of the United States Supreme Court and over thirty years as a Berkeley law professor. Beginning in the early 1990s, he turned his attention to evolutionary theory, although he has no formal background in the biological sciences. Johnson's description of the "wedge strategy" in his 1997 *Defeating Darwinism by Opening Minds* formed the core of the new attack on the teaching of evolution in public schools. Combined with Behe's *Darwin's Black Box*, it initiated the latest wave of antievolution in the United States, which gave birth to the intelligent design movement. Johnson and other advocates of intelligent design believe that it could serve as a wedge to crack open first the subject of evolutionary theory and eventually the ideology of scientific materialism, which they believed caused "destructive consequences that we can see all around us."

Building the Wedge

We call our strategy "the wedge." A log is a seeming solid object, but a wedge can eventually split it by penetrating a crack and gradually widening the split. In this case the ideology of

[2]Taken from *Defeating Darwinism by Opening Minds* by Philip Johnson. Copyright © 1997 by Phillip E. Johnson. Used with permission of InterVarsity Press, PO Box 1400, Downers Grove, IL 60515. www.ivpress.com.

scientific materialism is the apparently solid log. The widening crack is the important but seldom-recognized difference between the fact revealed by scientific investigation and the materialist philosophy that dominate the scientific culture. What happens when the facts cast doubt on the philosophy? Will scientists and philosophers allow materialism to be questioned, or will they rely on Microphone Man to suppress the facts and protect the philosophy?

My own books (including this one) represent the sharp edge of the wedge. I had two goals in writing those books and in pursuing the program of public speaking that followed their publication. First, I wanted to make it possible to question naturalistic assumptions in the secular academic community. Second, I wanted to redefine what is at issue in the creation-evolution controversy so that Christians, and other believers in God, could find common ground in the most fundamental issue—the reality of God as our true Creator.

Protestants will disagree on various issues among themselves, Catholics will disagree with Protestants, and observant Jews will disagree with Christians. What all these should agree on is that God—not some purposeless material process—is our true Creator. Given that we inhabit a culture whose intellectual leaders deny this fundamental fact, we should unite our energies to affirm the reality of God. After we have had that positive experience of unity and affirmation, we may be able to talk about the remaining points of disagreement with renewed goodwill. This is the program I call *theistic realism*.

Michael Behe's book *Darwin's Black Box*, which I described briefly in chapter five, represents the first broadening of the initial crack in the scientific materialism "log." I first became acquainted with Behe when he wrote a letter to the editor of the journal *Science*, which had published a dismissive news article about me. I was naturally pleased to receive support from a reputable biochemist, and even more pleased that the letter was very well written. Subsequently, friends who were interested in promoting my ideas arranged an academic conference at Southern Methodist University, to which they invited ten scientists and philosophers, including Behe, to discuss the relationship between evolutionary science and philosophical naturalism.

Behe presented a paper on proteins at the conference and formed the idea of writing a book to demonstrate that biologists who study cells and molecular systems constantly see examples of irreducibly complex systems that cannot have formed by Darwinian evolution. A major New York trade publisher (Free Press) brought out Behe's book, indicating that our small movement was breaking of out the Christian ghetto and into the cultural mainstream.

The wedge is continuing to broaden. With the assistance of some generous donors and the staff of Christian Leadership Ministries, we put on a major conference on "Mere Creation" at Biola University in November 1996. Approximately two hundred persons attended, including scientists, philosophers, and potential academic and financial supporters. Most were Christians, but the only requirement for attenders [sic] was a willingness and ability to contribute to the theme of the conference, which was that "the first step for a twenty-first-century science of origins is to separate materialist philosophy from empirical science." Sixteen persons gave papers, and of course there was extensive discussion about the next steps on our intellectual agenda.

What is next on the agenda? Scientifically, there is the question of how far the reconsideration of Darwinism is going to take us. We know that the Darwinian mechanism doesn't work and that complex biological systems never were put together by the accumulation of random mutations through natural selection. This is not a mere gap in a theory that is sound in other respects. It isn't just that the Darwinists have failed to provide a complete explanation; they've failed even to understand what needs to be explained. Their theory assumes that *variation* is all they need to explain and that the accumulation of small variations over immense amounts of time can produce complex organisms from simple beginnings. That is why they think that finch-beak variation illustrates the process that created birds in the first place.

Once the problems of informational content and irreducible complexity are out on the table in plain view, well-informed people are going to be amazed that scientists took so long to see that random mutation is not an information creator and that the Darwinian mechanism is therefore irrelevant to the real problem of biological creation. A few scientific materialists are aware of this and hope to rescue the situation by discovering new information-creating laws of physics and chemistry. Good luck to them, but the prospects are about as promising as the prospects of finding new laws of ink and paper that can create Shakespeare's plays.

Granted that the materialist mechanism has to be discarded, what does this imply for what scientists call the "fact of evolution," the concept that all organisms share a common ancestor? Universal common ancestry is as much a product of materialist philosophy as is the mutation/selection mechanism. Consider the proposition that a single ancestral bacterium gave birth to distant descendants as divers as trees, insects, and birds. If materialism is true, then universal common ancestry virtually has to be true also. The only materialist alternative is that life arose from nonliving chemicals many separate times, and this seems not only improbable but inconsistent with the observable fact that all living organisms share a common biochemistry. Life seems to have arisen from a single source, and if materialism is true, that source must have been a material ancestor.

Put aside the materialism, however and the common ancestry thesis is as dubious as the Darwinian mechanism. There is no known process by which a bacterial species can evolve the immense complexity of plants and animals—in fact there is only a beginning of an understanding of what that complexity involves. There is no fossil history of single-celled organisms changing step by step into complex plants and animals. On the contrary, the major groups of animals all appear suddenly in the rocks of the Cambrian era—and no new groups appear thereafter. (High-school textbooks either fail to mention this fundamental fact of the fossil record or refer to it so obliquely that students don't see the implications.) The fossil problems are only the beginning, however, because evidence from embryology and genetics is adding to the difficulties.

This is not the place to develop the scientific ideas further; my purpose here is just to give a hint of the excitement that animated the scientists and philosophers who attended the Mere Creation Conference. The British scientific materialist J. B. Haldane wrote years ago, "My own suspicion is that the universe is not only queerer than we suppose, but queerer than we *can* suppose." For some obscure reason, Darwinists like to quote that statement, although Darwinism asserts that the realm of life is not queerer than we can suppose but at bottom very simple and commonsensical. All it takes to make a world of living things, according to the Theory, is variation, natural selection, changing environments, and long periods of time. But that is nineteenth-century science, and it won't survive the opening of *Darwin's Black Box*. When biology finally has its quantum revolution, our view of life and its origin will change profoundly.

Even more exiting than the scientific part of the theistic realism agenda, at least to me, is the new understanding of rationality that it promises. (This is the subject of my book *Reason in the Balance*.) Materialism tells us, incredibly, that the universe can be rational only if it is the product of impersonal laws, and not if it is the creation of a supreme mind. Materialists tend to think the only alternative to materialism is some form of primitive superstition, where science would be impossible because all events would be produced by the whimsy of capricious gods. This is nonsense, of course. Intelligent design does not mean unintelligent chaos. Computers and space rockets are designed, but they work according to lawlike principles.

The real objection scientific materialist have to design is that the Designer would be something outside of science and hence not subject to human control. The attraction of a materialistic universe is that it feeds the imperialism of science by seeming to promise that everything can in principle be understood (and controlled) by science. There is an immense price to be paid for this illusion that we can have a "theory of everything," however. There can be no science of value, or of beauty, or of goodness. The whole realm of value is left to the subjective imagination, with destructive consequences that we can see all around us. Eventually materialist philosophy undermines the reliability of the mind itself—and hence even the basis for science. The true foundation of rationality is not found in particles and impersonal laws but in the mind of the Creator who formed us in his image.

Probably many readers of this book feel that pursuing the intellectual program of theistic realism, or even completely understanding it, is beyond them. No matter if it is; nearly everyone know some young person who has the necessary gifts. I find as time goes by that my greatest satisfaction comes not from the work I can do myself but from the accomplishments of younger people to whom I have given encouragement and for whom I have opened doors. If you know a gifted young person, help him or her to see the vision. Those who are called to it won't need any further encouragement. Once they have seen their calling, you had better step out of the way because you won't be able to stop them even if you try.

SOURCE: Johnson, Phillip E. *Defeating Darwinism by Opening Minds*. Downers Grove, IL: InterVarsity Press, 1997.

THE DISCOVERY INSTITUTE, CENTER FOR THE RENEWAL OF SCIENCE & CULTURE, "THE WEDGE DOCUMENT" (1998)

Founded in 1990 as a non-profit educational foundation and think tank, the Discovery Institute is a Seattle-based organization based on the Christian apologetics of C. S. Lewis. Established as a branch of the conservative Hudson Institute, it is opposed to what it views as widespread and corrosive materialism in American politics and culture. In the mid-1990s the institute created under it the Center for the Renewal of Science and Culture, which has served as the primary advocate of the intelligent design movement in the United States. It aggressively promotes acceptance, especially among American policymakers, of intelligent design by arguing that students should be taught to the shortcoming of modern evolutionary theory. Called "teach the controversy," this approach seeks to portray evolution as a theory on the verge of collapse, a claim widely and assertively denied by the vast majority of biologists. The "Wedge Document" was created by and circulated within the Discovery Institute in 1998 and has since found its way onto the Internet. It outlined a plan to "drive a wedge" into "scientific materialism" and demonstrates the melding of religious, philosophical, and political ideologies at work in the Discovery Institutes' attacks on evolution.

The editors had hoped to include the "Wedge Document" in this volume, but were unable to obtain permission from the Discovery Institute to do so. For anyone interested in the evolution/creation debate today, the "Wedge Document" is vital to understanding the political and cultural influences that motivate some of the most ardent opponents to the teaching of evolution as well as some of the most aggressive advocates of intelligent design. Lacking the ability to publish it here, the editors can only encourage interested readers to locate a copy of the document on the internet, where it is widely available.

JOHN E. JONES III, *KITZMILLER v. DOVER* (2005)

The most recent in a long line of court cases that have resulted in a court-ordered prohibition against the requirement that science teachers instruct their students in creation science, the majority opinion in *Kitzmiller v. Dover* affirmed the claim that intelligent design is in fact creation science.

Introduction:

On October 18, 2004, the Defendant Dover Area School Board of Directors passed by a 6–3 vote the following resolution:

Students will be made aware of gaps/problems in Darwin's theory and of other theories of evolution including, but not limited to, intelligent design. Note: Origins of Life is not taught.

On November 19, 2004, the Defendant Dover Area School District announced by press release that, commencing in January 2005, teachers would be required to read the following statement to students in the ninth grade biology class at Dover High School:

The Pennsylvania Academic Standards require students to learn about Darwin's Theory of Evolution and eventually to take a standardized test of which evolution is a part.

Because Darwin's Theory is a theory, it continues to be tested as new evidence is discovered. The Theory is not a fact. Gaps in the Theory exist for which there is no evidence. A theory is defined as a well-tested explanation that unifies a broad range of observations.

Intelligent Design is an explanation of the origin of life that differs from Darwin's view. The reference book, *Of Pandas and People*, is available for students who might be interested in gaining an understanding of what Intelligent Design actually involves.

With respect to any theory, students are encouraged to keep an open mind. The school leaves the discussion of the Origins of Life to individual students and their families. As a Standards-driven district, class instruction focuses upon preparing students to achieve proficiency on Standards-based assessments.

Background and Procedural History

On December 14, 2004, Plaintiffs filed the instant suit challenging the constitutional validity of the October 18, 2004 resolution and November 19, 2004 press release (collectively, "the ID Policy"). It is contended that the ID Policy constitutes an establishment of religion prohibited by the First Amendment to the United States Constitution, which is made applicable to the states by the Fourteenth Amendment, as well as the Constitution of the Commonwealth of Pennsylvania. Plaintiffs seek declaratory and injunctive relief, nominal damages, costs, and attorneys' fees. . . .

For the reasons that follow, we hold that the ID Policy is unconstitutional pursuant to the Establishment Clause of the First Amendment of the United States Constitution and Art. I, § 3 of the Pennsylvania Constitution.

The Parties to the Action

We will now introduce the individual Plaintiffs and provide information regarding their acquaintance with the biology curriculum controversy. Tammy Kitzmiller, resident of Dover, Pennsylvania, is a parent of a child in the ninth grade and a child in the eleventh grade at Dover High School. She did not attend any Board meetings until November 2004 and first learned of the biology curriculum controversy from reading the local newspapers. Bryan and Christy Rehm, residents of Dover, Pennsylvania, are parents of a child in the eighth grade, a child in the second grade, a child in kindergarten in the Dover Area School District, and a child of pre-school age. They intend for their children to attend Dover High School. Bryan Rehm learned of the biology curriculum controversy by virtue of being a member of the science faculty at Dover Area High School. Before and after his resignation, he regularly attended Board meetings. His wife, fellow Plaintiff Christy Rehm learned of the biology curriculum controversy by virtue of discussions she had with her husband and also regularly attended Board meetings in 2004. Deborah F. Fenimore and Joel A. Leib, residents of Dover, Pennsylvania, are the parents of a child in the twelfth grade at Dover High School and a child in the seventh grade in the Dover Area School District. They intend for their seventh-grade child to attend Dover High School. Leib first learned of a change in the biology curriculum by reading local newspapers. Steven Stough, resident of Dover, Pennsylvania, is a parent of a child in the eighth grade in the Dover Area School District and intends for his child to attend Dover High School. Stough did not attend any Board meetings until December 2004 and prior to that, he had learned of the biology curriculum change by reading the local newspapers. Beth A. Eveland, resident of York, Pennsylvania, is a parent of a child in the first grade in the Dover Area School District and a child of pre-school age who intends for her children to attend Dover High School. Eveland attended her first Board meeting on June 14, 2004. Prior to that, she had learned of the issues relating to the purchase of the biology books from reading the *York Daily Record* newspaper. Cynthia Sneath, resident of Dover, Pennsylvania, is a parent of a child in the first grade in the Dover Area School District and a child of pre-school age who intends for her children to attend Dover High School. Sneath attended her first Board meeting on October 18, 2004 and prior to that, she had learned of the biology curriculum controversy from reading the local newspapers. Julie Smith, resident of York, Pennsylvania, is a parent of a child in the tenth grade at Dover High School. Smith did not attend a Board meeting in 2004; she learned of and followed the biology curriculum controversy by reading the local newspapers. Aralene (hereinafter "Barrie") Callahan and Frederick B. Callahan, residents of Dover, Pennsylvania, are parents of a child in the tenth grade at Dover High School. Barrie Callahan learned of the biology curriculum controversy by virtue of her status of a former Board member and from attending Board meetings. Fred

Callahan learned of the biology curriculum controversy based upon discussions with his wife Barrie and from attending Board meetings.

The Defendants include the Dover Area School District (hereinafter "DASD") and Dover Area School District Board of Directors (hereinafter "the Board") (collectively "Defendants"). Defendant DASD is a municipal corporation governed by a board of directors, which is the Board. The DASD is comprised of Dover Township, Washington Township, and Dover Borough, all of which are located in York County, Pennsylvania. There are approximately 3,700 students in the DASD, with approximately 1,000 attending Dover High School.

The trial commenced September 26, 2005 and continued through November 4, 2005. This Memorandum Opinion constitutes the Court's findings of fact and conclusions of law which are based upon the Court's review of the evidence presented at trial, the testimony of the witnesses at trial, the parties' proposed findings of fact and conclusions of law with supporting briefs, other documents and evidence in the record, and applicable law. Further orders and judgments will be in conformity with this opinion. . ...

An Objective Observer Would Know that ID and Teaching about "Gaps" and "Problems" in Evolutionary Theory are Creationist. Religious Strategies that Evolved from Earlier Forms of Creationism

The history of the intelligent design movement (hereinafter "IDM") and the development of the strategy to weaken education of evolution by focusing students on alleged gaps in the theory of evolution is the historical and cultural background against which the Dover School Board acted in adopting the challenged ID Policy. As a reasonable observer, whether adult or child, would be aware of this social context in which the ID Policy arose, and such context will help to reveal the meaning of Defendants' actions, it is necessary to trace the history of the IDM.

It is essential to our analysis that we now provide a more expansive account of the extensive and complicated federal jurisprudential legal landscape concerning opposition to teaching evolution, and its historical origins. As noted, such opposition grew out of a religious tradition, Christian Fundamentalism that began as part of evangelical Protestantism's response to, among other things, Charles Darwin's exposition of the theory of evolution as a scientific explanation for the diversity of species. Subsequently, as the United States Supreme Court explained in *Epperson*, in an "upsurge of fundamentalist religious fervor of the twenties," 393 U.S. at 98 (citations omitted), state legislatures were pushed by religiously motivated groups to adopt laws prohibiting public schools from teaching evolution. Between the 1920s and early 1960s, anti-evolutionary sentiment based upon a religious social movement resulted in formal legal sanctions to remove evolution from the classroom. . ...

As we previously noted, the legal landscape radically changed in 1968 when the Supreme Court struck down Arkansas's statutory prohibition against teaching evolution in *Epperson*. Although the Arkansas statute at issue did not include direct references to the Book of Genesis or to the fundamentalist view that religion should be protected from science, the Supreme Court concluded that "the motivation of the [Arkansas] law was the same. . .: to suppress the teaching of a theory which, it was thought, 'denied' the divine creation of man. . . .."

Post-*Epperson*, evolution's religious opponents implemented "balanced treatment" statutes requiring public school teachers who taught evolution to devote equal time to teaching the biblical view of creation; however, such statutes did not pass constitutional muster under the Establishment Clause. . . . In *Daniel*, the Sixth Circuit Court of Appeals held that by assigning a "preferential position for the Biblical version of creation" over "any account of the development of man based on scientific research and reasoning," the challenged statute officially promoted religion, in violation of the Establishment Clause.

Next, and as stated, religious opponents of evolution began cloaking religious beliefs in scientific sounding language and then mandating that schools teach the resulting "creation science" or "scientific creationism" as an alternative to evolution. However, this tactic was likewise unsuccessful under the First Amendment. "Fundamentalist organizations were formed to promote the idea that the Book of Genesis was supported by scientific data. The terms 'creation science' and 'scientific creationism' have been adopted by these Fundamentalists as descriptive of their study of creation and the origins of man." In 1982, the district court in *McLean* reviewed Arkansas's balanced-treatment law and evaluated creation science in light of *Scopes*, *Epperson*, and the long history of Fundamentalism's attack on the scientific theory of evolution, as well as the statute's legislative history and historical context. The court found that creation science organizations were fundamentalist religious entities that "consider[ed] the introduction of creation science into the public schools part of their ministry." The court in *McLean* stated that creation science rested on a "contrived dualism" that recognized only two possible explanations for life, the scientific theory of evolution and biblical creationism, treated the two as mutually exclusive such that "one must either accept the literal interpretation of Genesis or else believe in the godless system of evolution," and accordingly viewed any critiques of evolution as evidence that necessarily supported biblical creationism. The court concluded that creation science "is simply not science" because it depends upon "supernatural intervention," which cannot be explained by natural causes, or be proven through empirical investigation, and is therefore neither testable nor falsifiable. Accordingly, the United States District Court for the Eastern District of Arkansas deemed creation science as merely biblical creationism in a new guise and held that Arkansas' balanced-treatment statute could have no valid secular purpose or effect, served only to advance religion, and violated the First Amendment. . . .

Five years after *McLean* was decided, in 1987, the Supreme Court struck down Louisiana's balanced-treatment law in *Edwards* for similar reasons. After a thorough analysis of the history of fundamentalist attacks against evolution, as well as the applicable legislative history including statements made by the statute's sponsor, and taking the character of organizations advocating for creation science into consideration, the Supreme Court held that the state violated the Establishment Clause by "restructur[ing] the science curriculum to conform with a particular religious viewpoint."

Among other reasons, the Supreme Court in *Edwards* concluded that the challenged statute did not serve the legislature's professed purposes of encouraging academic freedom and making the science curriculum more comprehensive by "teaching all of the evidence" regarding origins of life because: the state law already allowed schools to teach any scientific theory, which responded to the alleged purpose of academic freedom; and if the legislature really had intended to make science education more comprehensive, "it would have encouraged the teaching of all scientific theories about the origins of humankind" rather than permitting schools to forego teaching evolution, but mandating that schools that teach evolution must also teach creation science, an inherently religious view. The Supreme Court further held that the belief that a supernatural creator was responsible for the creation of human kind is a religious viewpoint and that the Act at issue "advances a religious doctrine by requiring either the banishment of the theory of evolution from public school classrooms or the presentation of a religious viewpoint that rejects evolution in its entirety." Therefore, as noted, the import of *Edwards* is that the Supreme Court made national the prohibition against teaching creation science in the public school system.

The concept of intelligent design (hereinafter "ID"), in its current form, came into existence after the *Edwards* case was decided in 1987. For the reasons that follow, we conclude that the religious nature of ID would be readily apparent to an objective observer, adult, or child.

We initially note that John Haught, a theologian who testified as an expert witness for Plaintiffs and who has written extensively on the subject of evolution and religion, succinctly explained to the Court that the argument for ID is not a new scientific argument, but is rather an old religious argument for the existence of God. He traced this argument back to at least Thomas Aquinas in the 13th century, who framed the argument as a syllogism: Wherever complex design exists, there must have been a designer; nature is complex; therefore nature must have had an intelligent designer. Dr. Haught testified that Aquinas was explicit that this intelligent designer "everyone understands to be God." The syllogism described by Dr. Haught is essentially the same argument for ID as presented by defense expert witnesses Professors Behe and Minnich who employ the phrase "purposeful arrangement of parts."

Dr. Haught testified that this argument for the existence of God was advanced early in the 19th century by Reverend Paley and defense expert witnesses Behe and Minnich admitted that their argument for ID based on the "purposeful arrangement of parts" is the same one that Paley made for design. The only apparent difference between the argument made by Paley and the argument for ID, as expressed by defense expert witnesses Behe and Minnich, is that ID's "official position" does not acknowledge that the designer is God. However, as Dr. Haught testified, anyone familiar with Western religious thought would immediately make the association that the tactically unnamed designer is God, as the description of the designer in *Of Pandas and People* (hereinafter "*Pandas*") is a "master intellect," strongly suggesting a supernatural deity as opposed to any intelligent actor known to exist in the natural world. Moreover, it is notable that both Professors Behe and Minnich admitted their personal view is that the designer is God and Professor Minnich testified that he understands many leading advocates of ID to believe the designer to be God.

Although proponents of the IDM occasionally suggest that the designer could be a space alien or a time-traveling cell biologist, no serious alternative to God as the designer has been proposed by members of the IDM, including Defendants' expert witnesses.

In fact, an explicit concession that the intelligent designer works outside the laws of nature and science and a direct reference to religion is *Pandas'* rhetorical statement, "what kind of intelligent agent was it [the designer]" and answer: "On its own science cannot answer this question. It must leave it to religion and philosophy. . ."

A significant aspect of the IDM is that despite Defendants' protestations to the contrary, it describes ID as a religious argument. In that vein, the writings of leading ID proponents reveal that the designer postulated by their argument is the God of Christianity. Dr. Barbara Forrest, one of Plaintiffs' expert witnesses, is the author of the book *Creationism's Trojan Horse*. She has thoroughly and exhaustively chronicled the history of ID in her book and other writings for her testimony in this case. Her testimony, and the exhibits which were admitted with it, provide a wealth of statements by ID leaders that reveal ID's religious, philosophical, and cultural content. The following is a representative grouping of such statements made by prominent ID proponents.

The evidence that Defendants are asking this Court to ignore is exactly the sort that the court in *McLean* considered and found dispositive concerning the question of whether creation science was a scientific view that could be taught in public schools, or a religious one that could not. The *McLean* court considered writings and statements by creation science advocates like Henry Morris and Duane Gish, as well as the activities and mission statements of creationist Phillip Johnson, considered to be the father of the IDM, developer of ID's "Wedge Strategy," which will be discussed below, and author of the 1991 book entitled *Darwin on Trial*, has written that "theistic realism" or "mere creation" are defining concepts of the IDM. This means "that God is objectively real as Creator and recorded in the biological evidence...." In addition, Phillip Johnson states that the "Darwinian theory of evolution contradicts not just the Book of Genesis, but every word in the Bible from beginning to end. It contradicts the idea that we are here because a creator brought about our existence for a purpose." ID proponents Johnson, William Dembski, and Charles Thaxton, one of the editors of *Pandas*, situate ID in the Book of John in the New Testament of the Bible, which begins, "In the Beginning was the Word, and the Word was God." Dembski has written that ID is a "ground clearing operation" to allow Christianity to receive serious consideration, and "Christ is never an addendum to a scientific theory but always a completion." Moreover, in turning to Defendants' lead expert, Professor Behe, his testimony at trial indicated that ID is only a scientific, as opposed to a religious, project for him; however, considerable evidence was introduced to refute this claim. Consider, to illustrate, that Professor Behe remarkably and unmistakably claims that the *plausibility of the argument for ID depends upon the extent to which one believes in the existence of God* (emphasis added). As no evidence in the record indicates that any other scientific proposition's validity rests on belief in God, nor is the Court aware of any such scientific propositions, Professor Behe's assertion constitutes substantial evidence that in his view, as is commensurate with other prominent ID leaders, ID is a religious and not a scientific proposition.

Dramatic evidence of ID's religious nature and aspirations is found in what is referred to as the "Wedge Document." The Wedge Document, developed by the Discovery Institute's Center for Renewal of Science and Culture (hereinafter "CRSC"), represents from an institutional standpoint, the IDM's goals and objectives, much as writings from the Institute for Creation Research did for the earlier creation-science movement, as discussed in *McLean*. Wedge Document states in its "Five Year Strategic Plan Summary" that the IDM's goal is to replace science as currently practiced with "theistic and Christian science." As posited in the Wedge Document, the IDM's "Governing Goals" are to "defeat scientific materialism and its destructive moral, cultural, and political legacies" and "to replace materialistic explanations with the theistic understanding that nature and human beings are created by God." The CSRC expressly announces, in the Wedge Document, a program of Christian apologetics to promote ID. A careful review of the Wedge Document's goals and language throughout the document reveals cultural and religious goals, as opposed to scientific ones. ID aspires to change the ground rules of science to make room for religion, specifically, beliefs consonant with a particular version of Christianity.

In addition to the IDM itself describing ID as a religious argument, ID's religious nature is evident because it involves a supernatural designer. The courts in *Edwards*, and *McLean* expressly found that this characteristic removed creationism from the realm of science and made it a religious proposition. Prominent ID proponents have made abundantly clear that the designer is supernatural.

Defendants' expert witness ID proponents confirmed that the existence of a supernatural designer is a hallmark of ID. First, Professor Behe has written that by ID he means "not designed by the laws of nature," and that it is "implausible that the designer is a natural entity." Second, Professor Minnich testified that for ID to be considered science, the ground rules of science have to be broadened so that supernatural forces can be considered. Third, Professor Steven William Fuller testified that it is ID's project to change the ground rules of science to include the supernatural. Turning from defense expert witnesses to leading ID proponents, Johnson has concluded that science must be redefined to include the supernatural if religious challenges to evolution are to get a hearing. Additionally, Dembski agrees that science is ruled by methodological naturalism and argues that this rule must be overturned if ID is to prosper.

Further support for the proposition that ID requires supernatural creation is found in the book *Pandas*, to which students in Dover's ninth grade biology class are directed. *Pandas*, indicates that there are two kinds of causes, natural and intelligent, which demonstrate that intelligent causes are beyond nature. Professor Haught, who as noted was the only theologian to testify in this case, explained that in Western intellectual tradition, non-natural causes occupy a space reserved for ultimate religious explanations. Robert Pennock, Plaintiffs' expert in the philosophy of science, concurred with Professor Haught and concluded that because its basic proposition is that the features of the natural world are produced by a transcendent, immaterial, non-natural being, ID is a religious proposition regardless of whether that religious proposition is given a recognized religious label. It is notable that not one defense expert was able to explain how the supernatural action suggested by ID could be anything other than an inherently religious proposition. Accordingly, we find that ID's religious nature would be further evident to our objective observer because it directly involves a supernatural designer.

A "hypothetical reasonable observer," adult or child, who is "aware of the history and context of the community and forum" is also presumed to know that ID is a form of creationism. *Child Evangelism*. . . . The evidence at trial demonstrates that ID is nothing less than the progeny of creationism. What is likely the strongest evidence supporting the finding of ID's creationist nature is the history and historical pedigree of the book to which students in Dover's ninth grade biology class are referred, *Pandas*. *Pandas* is published by an organization called FTE, as noted, whose articles of incorporation and filings with the Internal Revenue Service describe it as a religious, Christian organization. . . . *Pandas* was written by Dean Kenyon and Percival Davis, both acknowledged creationists, and Nancy Pearcey, a Young Earth Creationist, contributed to the work.

As Plaintiffs meticulously and effectively presented to the Court, *Pandas* went through many drafts, several of which were completed prior to and some after the Supreme Court's decision in *Edwards*, which held that the Constitution forbids teaching creationism as science. By comparing the pre and post *Edwards* drafts of *Pandas*, three astonishing points emerge: (1) the definition for creation science in early drafts is identical to the definition of ID; (2) cognates of the word creation (creationism and creationist), which appeared approximately 150 times were deliberately and systematically replaced with the phrase ID; and (3) the changes occurred shortly after the Supreme Court held that creation science is religious and cannot be taught in public school science classes in *Edwards*. This word substitution is telling, significant, and reveals that a purposeful change of *words* was effected without any corresponding change in *content*, which directly refutes FTE's argument that by merely disregarding the words "creation" and "creationism," FTE expressly rejected creationism in

Pandas. In early pre *Edwards* drafts of *Pandas*, the term "creation" was defined as "various forms of life that began abruptly through an intelligent agency with their distinctive features intact—fish with fins and scales, birds with feathers, beaks, and wings, etc.," the very same way in which ID is defined in the subsequent published versions. This definition was described by many witnesses for both parties, notably including defense experts Minnich and Fuller, as "special creation" of kinds of animals, an inherently religious and creationist concept. Professor Behe's assertion that this passage was merely a *description* of appearances in the fossil record is illogical and defies the weight of the evidence that the passage is a conclusion about how life began based upon an *interpretation* of the fossil record, which is reinforced by the content of drafts of *Pandas.*

The weight of the evidence clearly demonstrates, as noted, that the systemic change from "creation" to "intelligent design" occurred sometime in 1987, *after* the Supreme Court's important *Edwards* decision. This compelling evidence strongly supports Plaintiffs' assertion that ID is creationism re-labeled. Importantly, the objective observer, whether adult or child, would conclude from the fact that *Pandas* posits a master intellect that the intelligent designer is God.

Further evidence in support of the conclusion that a reasonable observer, adult or child, who is "aware of the history and context of the community and forum" is presumed to know that ID is a form of creationism concerns the fact that ID uses the same, or exceedingly similar arguments as were posited in support of creationism. One significant difference is that the words "God," "creationism," and "Genesis" have been systematically purged from ID explanations, and replaced by an unnamed "designer." Dr. Forrest testified and sponsored exhibits showing six arguments common to creationists. ... Demonstrative charts introduced through Dr. Forrest show parallel arguments relating to the rejection of naturalism, evolution's threat to culture and society, "abrupt appearance" implying divine creation, the exploitation of the same alleged gaps in the fossil record, the alleged inability of science to explain complex biological information like DNA, as well as the theme that proponents of each version of creationism merely aim to teach a scientific alternative to evolution to show its "strengths and weaknesses," and to alert students to a supposed "controversy" in the scientific community. In addition, creationists made the same argument that the complexity of the bacterial flagellum supported creationism as Professors Behe and Minnich now make for ID. The IDM openly welcomes adherents to creationism into its "Big Tent," urging them to postpone biblical disputes like the age of the earth. Moreover and as previously stated, there is hardly better evidence of ID's relationship with creationism than an explicit statement by defense expert Fuller that ID is a form of creationism. . .

Although contrary to Fuller, defense experts Professors Behe and Minnich testified that ID is not creationism, their testimony was primarily by way of bare assertion and it failed to directly rebut the creationist history of *Pandas* or other evidence presented by Plaintiffs showing the commonality between creationism and ID. The sole argument Defendants made to distinguish creationism from ID was their assertion that the term "creationism" applies only to arguments based on the Book of Genesis, a young earth, and a catastrophic Noaich flood; however, substantial evidence established that this is only one form of creationism, including the chart that was distributed to the Board Curriculum Committee, as will be described below. . .

Having thus provided the social and historical context in which the ID Policy arose of which a reasonable observer, either adult or child would be aware, we will now focus on what the objective student alone would know. We will accordingly determine whether an objective student would view the disclaimer read to the ninth grade biology class as an official endorsement of religion...

WHETHER ID IS SCIENCE

After a searching review of the record and applicable caselaw, we find that while ID arguments may be true, a proposition on which the Court takes no position, ID is not science. We find that ID fails on three different levels, any one of which is sufficient to preclude a determination that ID is science. They are: (1) ID violates the centuries-old ground rules of science by invoking and permitting supernatural causation; (2) the argument of irreducible complexity, central to ID, employs the same flawed and illogical contrived dualism that doomed creation science in the 1980s; and (3) ID's negative attacks on evolution have been refuted by the scientific community. As we will discuss in more detail below, it is additionally important to note that ID has failed to gain acceptance in the scientific community, it has not generated peer-reviewed publications, nor has it been the subject of testing and research.

Expert testimony reveals that since the scientific revolution of the 16th and 17th centuries, science has been limited to the search for natural causes to explain natural phenomena. This revolution entailed the rejection of the appeal to authority, and by extension, revelation, in favor of empirical evidence. Since that time period, science has been a discipline in which testability, rather than any ecclesiastical authority or philosophical coherence, has been the measure of a scientific idea's worth. In deliberately omitting theological or "ultimate" explanations for the existence or characteristics of the natural world, science does not consider issues of "meaning" and "purpose" in the world. While supernatural explanations may be important and have merit, they are not part of science. This self-imposed convention of science, which limits inquiry to testable, natural explanations about the natural world, is referred to by philosophers as "methodological naturalism" and is sometimes known as the scientific method. Methodological naturalism is a "ground rule" of science today which requires scientists to seek explanations in the world around us based upon what we can observe, test, replicate, and verify.

As the National Academy of Sciences (hereinafter "NAS") was recognized by experts for both parties as the "most prestigious" scientific association in this country, we will accordingly cite to its opinion where appropriate. NAS is in agreement that science is limited to empirical, observable, and ultimately testable data: "Science is a particular way of knowing about the world. In science, explanations are restricted to those that can be inferred from the confirmable data—the results obtained through observations and experiments that can be substantiated by other scientists. Anything that can be observed or measured is amenable to scientific investigation. Explanations that cannot be based upon empirical evidence are not part of science."

This rigorous attachment to "natural" explanations is an essential attribute to science by definition and by convention. We are in agreement with Plaintiffs' lead expert, Dr. Miller, that from a practical perspective, attributing unsolved problems about nature to causes and forces that lie outside the natural world is a "science stopper." As Dr. Miller explained, once you attribute a cause to an untestable supernatural force, a proposition that cannot be disproven, there is no reason to continue seeking natural explanations as we have our answer.

ID is predicated on supernatural causation, as we previously explained and as various expert testimony revealed. ID takes a natural phenomenon and, instead of accepting or seeking a natural explanation, argues that the explanation is supernatural. Further support for the conclusion that ID is predicated on supernatural causation is found in the ID reference book to which ninth grade biology students are directed, *Pandas*. *Pandas*, states, in pertinent part, as follows:

> Darwinists object to the view of intelligent design because *it does not give a natural cause explanation* of how the various forms of life started in the first place. Intelligent design means that various forms of life began abruptly, through an intelligent agency, with their distinctive features already intact—fish with fins and scales, birds with feathers, beaks, and wings, etc. (emphasis added).

Stated another way, ID posits that animals did not evolve naturally through evolutionary means but were created abruptly by a non-natural, or supernatural, designer. Defendants' own expert witnesses acknowledged this point. . . .

It is notable that defense experts' own mission, which mirrors that of the IDM itself, is to change the ground rules of science to allow supernatural causation of the natural world, which the Supreme Court in *Edwards* and the court in *McLean* correctly recognized as an inherently religious concept. First, defense expert Professor Fuller agreed that ID aspires to "change the ground rules" of science and lead defense expert Professor Behe admitted that his broadened definition of science, which encompasses ID, would also embrace astrology. Moreover, defense expert Professor Minnich acknowledged that for ID to be considered science, the ground rules of science have to be broadened to allow consideration of supernatural forces.

Prominent IDM leaders are in agreement with the opinions expressed by defense expert witnesses that the ground rules of science must be changed for ID to take hold and prosper. William Dembski, for instance, an IDM leader, proclaims that science is ruled by methodological naturalism and argues that this rule must be overturned if ID is to prosper. . . .

The Discovery Institute, the think tank promoting ID whose CRSC developed the Wedge Document, acknowledges as "Governing Goals" to "defeat scientific materialism and its destructive moral, cultural and political legacies" and "replace materialistic explanations with the theistic understanding that nature and human beings are created by God." In addition, and as previously noted, the Wedge Document states in its "Five Year Strategic Plan Summary" that the IDM's goal is to replace science as currently practiced with "theistic and Christian science. . . ."

Notably, every major scientific association that has taken a position on the issue of whether ID is science has concluded that ID is not, and cannot be considered as such. Initially, we note that NAS, the "most prestigious" scientific association in this country, views ID as follows:

> Creationism, intelligent design, and other claims of supernatural intervention in the origin of life or of species are not science because they are not testable by the methods of science. These claims subordinate observed data to statements based on authority, revelation, or religious belief. Documentation offered in support of these claims is typically limited to the special publications of their advocates. These publications do not offer hypotheses subject to change in light of new data, new interpretations, or demonstration of error. This contrasts with science, where any hypothesis or theory always remains subject to the possibility of rejection or modification in the light of new knowledge.

Additionally, the American Association for the Advancement of Science (hereinafter "AAAS"), the largest organization of scientists in this country, has taken a similar position on ID, namely, that it "has not proposed a scientific means of testing its claims" and that "the lack of scientific warrant for so-called 'intelligent design theory' makes it improper to include as part of science education...." Not a single expert witness over the course of the six-week trial identified one major scientific association, society, or organization that endorsed ID as science. What is more, defense experts concede that ID is not a theory as that term is defined by the NAS and admit that ID is at best "fringe science" which has achieved no acceptance in the scientific community.

It is therefore readily apparent to the Court that ID fails to meet the essential ground rules that limit science to testable, natural explanations. Science cannot be defined differently for Dover students than it is defined in the scientific community as an affirmative action program, as advocated by Professor Fuller, for a view that has been unable to gain a foothold within the scientific establishment. Although ID's failure to meet the ground rules of science is sufficient for the Court to conclude that it is not science, out of an abundance of caution and in the exercise of completeness, we will analyze additional arguments advanced regarding the concepts of ID and science.

ID is at bottom premised upon a false dichotomy, namely, that to the extent evolutionary theory is discredited, ID is confirmed. This argument is not brought to this Court anew, and in fact, the same argument, termed "contrived dualism" in *McLean*, was employed by creationists in the 1980s to support "creation science." The court in *McLean* noted the "fallacious pedagogy of the two model approach" and that "[i]n efforts to establish 'evidence' in support of creation science, the defendants relied upon the same false premise as the two model approach...all evidence which criticized evolutionary theory was proof in support of creation science." We do not find this false dichotomy any more availing to justify ID today than it was to justify creation science two decades ago.

ID proponents primarily argue for design through negative arguments against evolution, as illustrated by Professor Behe's argument that "irreducibly complex" systems cannot be produced through Darwinian, or any natural, mechanisms. However, we believe that arguments against evolution are not arguments for design. Expert testimony revealed that just because scientists cannot explain today how biological systems evolved does not mean that they cannot, and will not, be able to explain them tomorrow. As Dr. Padian aptly noted, "absence of evidence is not evidence of absence." To that end, expert testimony from Drs. Miller and Padian provided multiple examples where *Pandas* asserted that no natural explanations exist, and in some cases that none could exist, and yet natural explanations have been identified in the intervening years. It also bears mentioning that as Dr. Miller stated, just because scientists cannot explain every evolutionary detail does not undermine its validity as a scientific theory as no theory in science is fully understood.

As referenced, the concept of irreducible complexity is ID's alleged scientific centerpiece. Irreducible complexity is a negative argument against evolution, not proof of design, a point conceded by defense expert Professor Minnich (irreducible complexity "is not a test of intelligent design; it's a test of evolution"). Irreducible complexity additionally fails to make a positive scientific case for ID, as will be elaborated upon below.

We initially note that irreducible complexity as defined by Professor Behe in his book *Darwin's Black Box* and subsequently modified in his 2001 article entitled, "Reply to My Critics," appears as follows:

> By irreducibly complex I mean a single system which is composed of several well-matched,
> interacting parts that contribute to the basic function, wherein the removal of any one of the

parts causes the system to effectively cease functioning. An irreducibly complex system cannot be produced directly by slight, successive modifications of a precursor system, because any precursor to an irreducibly complex system that is missing a part is by definition nonfunctional. . . Since natural selection can only choose systems that are already working, then if a biological system cannot be produced gradually it would have to arise as an integrated unit, in one fell swoop, for natural selection to have anything to act on.

Professor Behe admitted in "Reply to My Critics" that there was a defect in his view of irreducible complexity because, while it purports to be a challenge to natural selection, it does not actually address "the task facing natural selection." Professor Behe specifically explained that "[t]he current definition puts the focus on removing a part from an already functioning system," but "[t]he difficult task facing Darwinian evolution, however, would not be to remove parts from sophisticated pre-existing systems; it would be to bring together components to make a new system in the first place." In that article, Professor Behe wrote that he hoped to "repair this defect in future work;" however, he has failed to do so even four years after elucidating his defect.

In addition to Professor Behe's admitted failure to properly address the very phenomenon that irreducible complexity purports to place at issue, natural selection, Drs. Miller and Padian testified that Professor Behe's concept of irreducible complexity depends on ignoring ways in which evolution is known to occur. Although Professor Behe is adamant in his definition of irreducible complexity when he says a precursor "missing a part is by definition nonfunctional," what he obviously means is that it will not function in the same way the system functions when all the parts are present. For example in the case of the bacterial flagellum, removal of a part may prevent it from acting as a rotary motor. However, Professor Behe excludes, by definition, the possibility that a precursor to the bacterial flagellum functioned not as a rotary motor, but in some other way, for example as a secretory system.

As expert testimony revealed, the qualification on what is meant by "irreducible complexity" renders it meaningless as a criticism of evolution. In fact, the theory of evolution proffers exaptation as a well-recognized, well-documented explanation for how systems with multiple parts could have evolved through natural means. Exaptation means that some precursor of the subject system had a different, selectable function before experiencing the change or addition that resulted in the subject system with its present function For instance, Dr. Padian identified the evolution of the mammalian middle ear bones from what had been jawbones as an example of this process. By defining irreducible complexity in the way that he has, Professor Behe attempts to exclude the phenomenon of exaptation by definitional fiat, ignoring as he does so abundant evidence which refutes his argument.

Notably, the NAS has rejected Professor Behe's claim for irreducible complexity by using the following cogent reasoning:

[S]tructures and processes that are claimed to be 'irreducibly' complex typically are not on closer inspection. For example, it is incorrect to assume that a complex structure or biochemical process can function only if all its components are present and functioning as we see them today. Complex biochemical systems can be built up from simpler systems through natural selection. Thus, the 'history' of a protein can be traced through simpler organisms. . . The evolution of complex molecular systems can occur in several ways. Natural selection can bring together parts of a system for one function at one time and then, at a later time, recombine those parts with other systems of components to produce a system that has a different function. Genes can be duplicated, altered, and then amplified through natural selection. The complex biochemical cascade resulting in blood clotting has been explained in this fashion.

As irreducible complexity is only a negative argument against evolution, it is refutable and accordingly testable, unlike ID, by showing that there are intermediate structures with selectable functions that could have evolved into the allegedly irreducibly complex systems. Importantly, however, the fact that the negative argument of irreducible complexity is testable does not make testable the argument for ID. Professor Behe has applied the concept of irreducible complexity to only a few select systems: (1) the bacterial flagellum; (2) the blood-clotting cascade; and (3) the immune system. Contrary to Professor Behe's assertions with respect to these few biochemical systems among the myriad existing in nature, however, Dr. Miller presented evidence, based upon peer-reviewed studies, that they are not in fact irreducibly complex.

First, with regard to the bacterial flagellum, Dr. Miller pointed to peer-reviewed studies that identified a possible precursor to the bacterial flagellum, a subsystem that was fully functional, namely the Type-III Secretory System. Moreover, defense expert Professor Minnich admitted that there is serious scientific research on the question of whether the bacterial flagellum evolved into the Type-III Secretary System, the Type-III Secretory System into the bacterial flagellum, or whether they both evolved from a common ancestor. None of this research or thinking involves ID. In fact, Professor Minnich testified about his research as follows: "we're looking at the function of these systems and how they could have been derived one from the other. And it's a legitimate scientific inquiry."

Second, with regard to the blood-clotting cascade, Dr. Miller demonstrated that the alleged irreducible complexity of the blood-clotting cascade has been disproven by peer-reviewed studies dating back to 1969, which show that dolphins' and whales' blood clots despite missing a part of the cascade, a study that was confirmed by molecular testing in 1998. Additionally and more recently, scientists published studies showing that in puffer fish, blood clots despite the cascade missing not only one, but three parts. Accordingly, scientists in peer-reviewed publications have refuted Professor Behe's predication about the alleged irreducible complexity of the blood-clotting cascade. Moreover, cross-examination revealed that Professor Behe's redefinition of the blood-clotting system was likely designed to avoid peer-reviewed scientific evidence that falsifies his argument, as it was not a scientifically warranted redefinition.

The immune system is the third system to which Professor Behe has applied the definition of irreducible complexity. Although in *Darwin's Black Box*, Professor Behe wrote that not only were there no natural explanations for the immune system at the time, but that natural explanations were impossible regarding its origin. However, Dr. Miller presented peer-reviewed studies refuting Professor Behe's claim that the immune system was irreducibly complex. Between 1996 and 2002, various studies confirmed each element of the evolutionary hypothesis explaining the origin of the immune system. In fact, on cross-examination, Professor Behe was questioned concerning his 1996 claim that science would never find an evolutionary explanation for the immune system. He was presented with fifty-eight peer-reviewed publications, nine books, and several immunology textbook chapters about the evolution of the immune system; however, he simply insisted that this was still not sufficient evidence of evolution, and that it was not "good enough."

We find that such evidence demonstrates that the ID argument is dependent upon setting a scientifically unreasonable burden of proof for the theory of evolution. As a further example, the test for ID proposed by both Professors Behe and Minnich is to grow the bacterial flagellum in the laboratory; however, no one inside or outside of the 1DM, including those who propose the test, has conducted it. Professor Behe conceded that the proposed test could not approximate real world conditions and even if it could, Professor Minnich admitted that it would merely be a test of evolution, not design.

We therefore find that Professor Behe's claim for irreducible complexity has been refuted in peer-reviewed research papers and has been rejected by the scientific community at large. Additionally, even if irreducible complexity had not been rejected, it still does not support ID as it is merely a test for evolution, not design.

We will now consider the purportedly "positive argument" for design encompassed in the phrase used numerous times by Professors Behe and Minnich throughout their expert testimony, which is the "purposeful arrangement of parts." Professor Behe summarized the argument as follows: We infer design when we see parts that appear to be arranged for a purpose. The strength of the inference is quantitative; the more parts that are arranged, the more intricately they interact, the stronger is our confidence in design. The appearance of design in aspects of biology is overwhelming. Since nothing other than an intelligent cause has been demonstrated to be able to yield such a strong appearance of design, Darwinian claims notwithstanding, the conclusion that the design seen in life is real design is rationally justified. As previously indicated, this argument is merely a restatement of the Reverend William Paley's argument applied at the cell level. Minnich, Behe, and Paley reach the same conclusion, that complex organisms must have been designed using the same reasoning, except that Professors Behe and Minnich refuse to identify the designer, whereas Paley inferred from the presence of design that it was God. Expert testimony revealed that this inductive argument is not scientific and as admitted by Professor Behe, can never be ruled out.

Indeed, the assertion that design of biological systems can be inferred from the "purposeful arrangement of parts" is based upon an analogy to human design. Because we are able to recognize design of artifacts and objects, according to Professor Behe, that same reasoning can be employed to determine biological design. Professor Behe testified that the strength of the analogy depends upon the degree of similarity entailed in the two propositions; however, if this is the test, ID completely fails.

Unlike biological systems, human artifacts do not live and reproduce over time. They are non-replicable, they do not undergo genetic recombination, and they are not driven by natural selection. For human artifacts, we know the designer's identity, human, and the mechanism of design, as we have experience based upon empirical evidence that humans can make such things, as well as many other attributes including the designer's abilities, needs, and desires. With ID, proponents assert that they refuse to propose hypotheses on the designer's identity, do not propose a mechanism, and the designer, he/she/it/they, has never been seen. In that vein, defense expert Professor Minnich agreed that in the case of human artifacts and objects, we know the identity and capacities of the human designer, but we do not know any of those attributes for the designer of biological life. In addition, Professor Behe agreed that for the design of human artifacts, we know the designer and its attributes and we have a baseline for human design that does not exist for design of biological systems. Professor Behe's only response to these seemingly insurmountable points of disanalogy was that the inference still works in science fiction movies.

It is readily apparent to the Court that the only attribute of design that biological systems appear to share with human artifacts is their complex appearance, i.e. if it looks complex or designed, it must have been designed. This inference to design based upon the appearance of a "purposeful arrangement of parts" is a completely subjective proposition, determined in the eye of each beholder and his/her viewpoint concerning the complexity of a system. Although both Professors Behe and Minnich assert that there is a quantitative aspect to the inference, on cross-examination they admitted that there is no quantitative criteria for determining the degree of complexity or number of parts that bespeak design, rather than a natural process. As Plaintiffs aptly submit to the Court, throughout the entire trial only one piece of evidence generated by Defendants addressed the strength of the ID inference: the argument is less plausible to those for whom God's existence is in question, and is much less plausible for those who deny God's existence.

Accordingly, the purported positive argument for ID does not satisfy the ground rules of science which require testable hypotheses based upon natural explanations. ID is reliant upon forces acting outside of the natural world, forces that we cannot see, replicate, control, or test, which have produced changes in this world. While we take no position on whether such forces exist, they are simply not testable by scientific means and therefore cannot qualify as part of the scientific process or as a scientific theory.

It is appropriate at this juncture to address ID's claims against evolution. ID proponents support their assertion that evolutionary theory cannot account for life's complexity by pointing to real gaps in scientific knowledge, which indisputably exist in all scientific theories, but also by misrepresenting well-established scientific propositions.

Before discussing Defendants' claims about evolution, we initially note that an overwhelming number of scientists, as reflected by every scientific association that has spoken on the matter, have rejected the ID proponents' challenge to evolution. Moreover, Plaintiffs' expert in biology, Dr. Miller, a widely recognized biology professor at Brown University who has written university-level and high school biology textbooks used prominently throughout the nation, provided unrebutted testimony that evolution, including common descent and natural selection, is "overwhelmingly accepted" by the scientific community and that every major scientific association agrees. As the court in *Selman* explained, "evolution is more than a *theory* of origin in the context of science. To the contrary, evolution is the dominant *scientific* theory of origin accepted by the majority of scientists." Despite the scientific community's overwhelming support for evolution, Defendants and ID proponents insist that evolution is unsupported by empirical evidence. Plaintiffs' science experts, Drs. Miller and Padian, clearly explained how ID proponents generally and *Pandas* specifically, distort and misrepresent scientific knowledge in making their anti-evolution argument.

In analyzing such distortion, we turn again to *Pandas*, the book to which students are expressly referred in the disclaimer. Defendants hold out *Pandas* as representative of ID and Plaintiffs' experts agree in that regard. A series of arguments against evolutionary theory found in *Pandas* involve paleontology, which studies the life of the past and the fossil record. Plaintiffs' expert Professor Padian was the only testifying expert witness with any expertise in paleontology. His testimony therefore remains unrebutted. Dr. Padian's demonstrative slides, prepared on the basis of peer-reviewing scientific literature, illustrate how *Pandas* systematically distorts and misrepresents established, important evolutionary principles.

We will provide several representative examples of this distortion. First, *Pandas* misrepresents the "dominant form of understanding relationships" between organisms, namely, the tree of life, represented by classification determined via the method of cladistics. Second, *Pandas* misrepresents "homology," the "central concept of comparative biology" that allowed scientists to evaluate comparable parts among organisms for classification purposes for hundreds of years. Third, *Pandas* fails to address the well-established biological concept of exaptation, which involves a structure changing function, such as fish fins evolving fingers and bones to become legs for weight-bearing land animals. Dr. Padian testified that ID proponents fail to address exaptation because they deny that organisms change function, which is a view necessary to support abrupt appearance. Finally, Dr. Padian's unrebutted testimony demonstrates that *Pandas* distorts and misrepresents evidence in the fossil record about pre-Cambrian-era fossils, the evolution of fish to amphibians, the evolution of small carnivorous dinosaurs into birds, the evolution of the mammalian middle ear, and the evolution of whales from land animals.

In addition to Dr. Padian, Dr. Miller also testified that *Pandas* presents discredited science. Dr. Miller testified that *Pandas*' treatment of biochemical similarities between organisms is "inaccurate and downright false" and explained how *Pandas* misrepresents basic molecular biology concepts to advance design theory through a series of demonstrative slides. Consider, for example, that he testified as to how *Pandas* misinforms readers on the standard evolutionary relationships between different types of animals, a distortion which Professor Belie, a "critical reviewer" of *Pandas* who wrote a section within the book, affirmed. In addition, Dr. Miller refuted *Pandas*' claim that evolution cannot account for new genetic information and pointed to more than three dozen peer-reviewed scientific publications showing the origin of new genetic information by evolutionary processes. In summary, Dr. Miller testified that *Pandas* misrepresents molecular biology and genetic principles, as well as the current state of scientific knowledge in those areas in order to teach readers that common descent and natural selection are not scientifically sound.

Accordingly, the one textbook to which the Dover ID Policy directs students contains outdated concepts and badly flawed science, as recognized by even the defense experts in this case.

A final indicator of how ID has failed to demonstrate scientific warrant is the complete absence of peer-reviewed publications supporting the theory. Expert testimony revealed that the peer review process is "exquisitely important" in the scientific process. It is a way for scientists to write up their empirical research and to share the work with fellow experts in the field, opening up the hypotheses to study, testing, and criticism. In fact, defense expert Professor Behe recognizes the importance of the peer review process and has written that science must "publish or perish." Peer review helps to ensure that research papers are scientifically accurately, meet the standards of the scientific method, and are relevant to other scientists in the field. Moreover, peer review involves scientists submitting a manuscript to a scientific journal in the field, journal editors soliciting critical reviews from other experts in the field and deciding whether the scientist has followed proper research procedures, employed up-to-date methods, considered and cited relevant literature and generally, whether the researcher has employed sound science.

The evidence presented in this case demonstrates that ID is not supported by any peer-reviewed research, data, or publications. Both Drs. Padian and Forrest testified that recent literature reviews of scientific and medical-electronic databases disclosed no studies supporting a biological concept of ID. On cross-examination, Professor Behe admitted that: "There are no peer reviewed articles by anyone advocating for intelligent design supported by pertinent experiments or calculations which provide detailed rigorous accounts of how intelligent design of any biological system occurred." Additionally, Professor Behe conceded that there are no peer-reviewed papers supporting his claims that complex molecular systems, like the bacterial flagellum, the blood-clotting cascade, and the immune system, were intelligently designed. In that regard, there are no peer-reviewed articles supporting Professor Behe's argument that certain complex molecular structures are "irreducibly complex." In addition to failing to produce papers in peer-reviewed journals, ID also features no scientific research or testing.

After this searching and careful review of ID as espoused by its proponents, as elaborated upon in submissions to the Court, and as scrutinized over a six-week trial, we find that ID is not science and cannot be adjudged a valid, accepted scientific theory as it has failed to publish in peer-reviewed journals, engage in research and testing, and gain acceptance in the scientific community. ID, as noted, is grounded in theology, not science. Accepting for the sake of argument its proponents', as well as Defendants' argument that to introduce ID to students will encourage critical thinking, it still has utterly no place in a science curriculum. Moreover, ID's backers have sought to avoid the scientific scrutiny which we have now determined that it cannot withstand by advocating that the *controversy*, but not ID itself, should be taught in science class. This tactic is at best disingenuous, and at worst a canard. The goal of the IDM is not to encourage critical thought, but to foment a revolution which would supplant evolutionary theory with ID.

To conclude and reiterate, we express no opinion on the ultimate veracity of ID as a supernatural explanation. However, we commend to the attention of those who are inclined to superficially consider ID to be a true "scientific" alternative to evolution without a true understanding of the concept the foregoing detailed analysis. It is our view that a reasonable, objective observer would, after reviewing both the voluminous record in this case, and our narrative, reach the inescapable conclusion that ID is an interesting theological argument, but that it is not science. . . .

CONCLUSION

The proper application of both the endorsement and *Lemon* tests to the facts of this case makes it abundantly clear that the Board's ID Policy violates the Establishment Clause. In making this determination, we have addressed the seminal question of whether ID is science. We have concluded that it is not, and moreover that ID cannot uncouple itself from its creationist, and thus religious, antecedents.

Both Defendants and many of the leading proponents of ID make a bedrock assumption which is utterly false. Their presupposition is that evolutionary theory is antithetical to a belief in the existence of a supreme being and to religion in general. Repeatedly in this trial, Plaintiffs' scientific experts testified that the theory of evolution represents good science, is overwhelmingly accepted by the scientific community, and that it in no way conflicts with, nor does it deny, the existence of a divine creator.

To be sure, Darwin's theory of evolution is imperfect. However, the fact that a scientific theory cannot yet render an explanation on every point should not be used as a pretext to thrust an untestable alternative hypothesis grounded in religion into the science classroom or to misrepresent well-established scientific propositions.

The citizens of the Dover area were poorly served by the members of the Board who voted for the ID Policy. It is ironic that several of these individuals, who so staunchly and proudly touted their religious convictions in public, would time and again lie to cover their tracks and disguise the real purpose behind the ID Policy.

With that said, we do not question that many of the leading advocates of ID have *bona fide* and deeply held beliefs which drive their scholarly endeavors. Nor do we controvert that ID should continue to be studied, debated, and discussed. As stated, our conclusion today is that it is unconstitutional to teach ID as an alternative to evolution in a public school science classroom.

Those who disagree with our holding will likely mark it as the product of an activist judge. If so, they will have erred as this is manifestly not an activist Court. Rather, this case came to us as the result of the activism of an ill-informed faction on a school board, aided by a national public interest law firm eager to find a constitutional test case on ID, who in combination drove the Board to adopt an imprudent and ultimately unconstitutional policy. The breathtaking inanity of the Board's decision is evident when considered against the factual backdrop which has now been fully revealed through this trial. The students, parents, and teachers of the Dover Area School District deserved better than to be dragged into this legal maelstrom, with its resulting utter waste of monetary and personal resources.

To preserve the separation of church and state mandated by the Establishment Clause of the First Amendment to the United States Constitution, and Art. I, § 3 of the Pennsylvania Constitution, we will enter an order permanently enjoining Defendants from maintaining the ID Policy in any school within the Dover Area School District, from requiring teachers to denigrate or disparage the scientific theory of evolution, and from requiring teachers to refer to a religious, alternative theory known as ID. We will also issue a declaratory judgment that Plaintiffs' rights under the Constitutions of the United States and the Commonwealth of Pennsylvania have been violated by Defendants' actions. Defendants' actions in violation of Plaintiffs' civil rights as guaranteed to them by the Constitution of the United States and 42 U.S.C. § 1983 subject Defendants to liability with respect to injunctive and declaratory relief, but also for nominal damages and the reasonable value of Plaintiffs' attorneys' services and costs incurred in vindicating Plaintiffs' constitutional rights.

SOURCE: Jones, John E. III. *Kitzmiller v. Dover* (2005).

TERMS

Apologetics—A term applied to those who provide formal arguments that support central tenets of religious sects to strengthen believers' faith and persuade the uncommitted. In the context of the creation/evolution controversy, an apologetic defends against entirely materialistic and naturalistic explanations of the origin of life and the diversity of living things, offering in its place a theistic or deistic account of creationism.

Intelligent Design—The belief that certain features of the universe and especially of living things are best explained by invoking the work of an intelligent agent rather than an undirected, naturalistic or materialistic process like evolution. Intelligent design is an updated form of natural theology, which sees in nature evidence of a benevolent and omnipotent Creator. Many of its proponents argue that it is in fact a secular movement. However, in 2005 the judge in *Kitzmiller v. Dover* decided that it was a form of scientific creationism, which has earlier been ruled inappropriate to teach in public school science classrooms because it violated the constitutional mandate for a separation between church and state.

Teach the Controversy—A campaign initiated by the Discovery Institute and promoted by adherents to intelligent design that seeks to undermine the teaching of evolution in public schools and ultimately belief in evolution by the general public. The approach asserts the claim that evolution is a theory in crisis and often cites, often inaccurately, statements from scientists about shortcomings in modern evolutionary theory. Opponents of the "teach the controversy" model assert that the only controversial aspects of evolution are issues involving religion and politics, not anything within science itself. They also argue that the approach is a disingenuous attack on evolution that seeks to inject religious teaching into public school classrooms in the guise of science.

RESOURCES: ELECTRONIC AND PRINT

WEB SITES

The information available online ranges from scholarly scientific and theological articles to the opinions and beliefs of Internet users around the world. There are thoughtful materials provided on education from various camps in this debate. The sampling below is intended to introduce readers to the most balanced and helpful sources.

Scientific Organizations

American Association for the Advancement of Science Web site:
    http://www.aaas.org/news/press_room/evolution/.
        The American Association for the Advancement of Science (AAAS) is a leading organization of scientists from all fields. They provide a range of resources online.

National Academy of Science Web site:
    http://nationalacademies.org/evolution/.
        The National Academy of Science (NAS) provides scientific expertise to a variety of public and governmental organizations. They offer publications on evolution and the teaching of evolution online.

Scientific American Web site:
    http://www.sciam.com/article.cfm?articleID=000D4FEC-7D5B-1D07-8E49809EC588EEDF&page Number=1&catID=2.
        The journal *Scientific American* published this article in 2002, providing a point-by-point refutation of the most common creationist critiques of evolution.

Educational Organizations

Evolution and the Nature of Science Institutes Web site:
    http://www.indiana.edu/~ensiweb/.
        Housed at Indiana University, the Evolution and the Nature of Science Institutes (ENSI) site contains activities, resources, and information for high school teachers of biology and evolution.

National Association of Biology Teachers Web site:
> http://nabt.org/sites/S1/index.php?p=65.
>> The National Association of Biology Teachers (NABT) provides support for biology instruction and articles on current topics of interest.

National Center for Science Education Web site:
> http://www.ncseweb.org/.
>> The National Center for Science Education (NCSE) maintains an up-to-date listing of news and information about the teaching of evolution and the challenges from creationism, creation science, intelligent design, and other antievolution modes.

National Science Teachers Association Web site:
> http://www.nsta.org/positionstatement&psid=10.
>> The National Science Teachers Association (NSTA) promotes science teaching and provides support for teachers.

At Large Evolution and Creation Sites

The Panda's Thumb Web site:
> http://www.pandasthumb.org/.
>> A "virtual pub" for discussion of evolution and critique of antievolution claims. The site contains abundant links to multimedia articles, videos, and recordings of debates and lectures.

TalkOrigins Web site:
> http://www.talkorigins.org.
>> An archive of postings, articles, questions, and answers exploring the connections between scientific theories, religious beliefs, and other origin accounts. The material is organized into useful categories.

Wikipedia Web sites:
> http://en.wikipedia.org/wiki/Young_Earth_Creationism
> http://en.wikipedia.org/wiki/Creation_science
> http://en.wikipedia.org/wiki/Intelligent_design
> http://en.wikipedia.org/wiki/Neo-Creationism
> http://en.wikipedia.org/wiki/Natural_selection
> http://en.wikipedia.org/wiki/Evolution
> http://en.wikipedia.org/wiki/Evidence_of_evolution
> http://en.wikipedia.org/wiki/Social_effect_of_evolutionary_theory
>> A favorite (and often unreliable) information source, Wikipedia offers a variety of perspectives, with links to related topics. The sites are easy to navigate and updated regularly with new information.

Yahoo! Directory Web site:
> http://dir.yahoo.com/Society_and_Culture/Religion_and_Spirituality/Science_and_Religion/Creation_vs_Evolution/.
>> Many of the relevant Internet sources are organized under this directory. The directory is more comprehensive and less analytical than most sources on this topic, but it provides a useful overview of what is out there.

Religious Perspectives

Access Research Network Web site:
> http://www.arn.org/.
>> The Access Research Network (ARN) provides the most recent and detailed information about developments in intelligent design publications.

Answers in Genesis Web site:

http://www.answersingenesis.org/

> Answers in Genesis (AiG) uses the biblical story of Genesis as the basis for understanding the basic tenets of Christianity as well as the origins of science. A museum is under construction in Cincinnati, Ohio.

Creation Research Society Web site:

http://www.creationresearch.org/.

> The Creation Research Society (CRS) was among the earliest organized groups established by scientists to explore the possibilities of biblical creation within a scientific framework, beginning in 1963. The Web site provides an overview of the organization.

Creation Science Evangelism Web site:

http://www.drdino.com/.

> Kent Hovind founded Creation Science Evangelism in 1989 and remains its primary spokesman. Referring to himself as "Dr. Dino," Hovind received a Ph.D. in Christian Education from Patriot University, a claim that his critics have challenged.

Institute for Creation Research Web site:

http://www.icr.org.

> The Institute for Creation Research (ICR) provides evidence to illustrate the accuracy of the Bible's account of creation. The Web site lists recent news headlines with links to articles written by ICR advocates to clarify or refute the claims of scientists or journalists.

Northwest Creation Network Web site:

http://www.nwcreation.net/.

> The Northwest Creation Network (NCN) provides a "megasite" for information on creation and biblical accounts of biological origins. The NCN features PowerPoint presentations available for download.

Unification Church Web site:

http://www.tparents.org/library/unification/books/evoltheo/0%2Dtoc.htm.

> The True Parents Organization of the Unification Church entered the antievolution camp primarily through the work of Jonathan Wells. Wells published the "Ten Icons of Evolution" but was not widely known to be associated with the Unification Church.

BIBLIOGRAPHY

Alters, B.J. and S.M. Alters. *Defending Evolution: A Guide to the Creation/Evolution Controversy.* Sudbury, MA: Jones and Bartlett, 2001.

Behe, Michael J. *Darwin's Black Box: The Biochemical Challenge to Evolution.* New York: Touchstone, 1996.

Benz, E. *Evolution and Christian Hope: Man's Concept of the Future, from the Early Fathers to Teilhard de Chardin.* Garden City, NY: Doubleday, 1966.

Caudill, Edward. *Darwinian Myths: The Legends and Misuses of a Theory.* Knoxville: University of Tennessee Press, 1997.

Dawkins, Richard. *The Blind Watchmaker: Why the Evidence of Evolution Reveals a Universe without Design.* New York: Norton, 1986.

———. *Unweaving the Rainbow: Science, Delusion, and the Appetite for Wonder.* London: Penguin, 1998.

———. *A Devil's Chaplain: Reflections on Hope, Lies, Science, and Love.* Boston: Houghton Mifflin, 2003.

———. *The Ancestor's Tale: A Pilgrimage to the Dawn of Evolution.* New York: Houghton Mifflin, 2004.

Degler, Carl N. *In Search of Human Nature: The Decline and Revival of Darwinism in American Social Thought.* New York: Oxford University Press, 1991.

Dembski, William A. *The Design Inference: Eliminating Chance through Small Probabilities.* Cambridge: Cambridge University Press, 1998.

———. *Intelligent Design: The Bridge between Science and Theology.* Downers Grove, IL: InterVarsity Press, 1999.

Dembski, William A. and Michael Ruse, eds. *Debating Design: Darwin to DNA.* Cambridge: Cambridge University Press, 2004.

Dembski, William A. and J.M. Kusiner, eds. *Signs of Intelligence: Understanding Intelligent Design.* Grand Rapids, MI: Brazos, 2002.

Dennett, Daniel C. *Darwin's Dangerous Idea: Evolution and the Meanings of Life.* New York: Simon and Schuster, 1995.

Farber, Paul L. *The Temptations of Evolutionary Ethics.* Berkeley: University of California Press, 1994.

Forrest, B. and P.R. Gross. *Creationism's Trojan Horse: The Wedge of Intelligent Design.* Oxford: Oxford University Press, 2004.

Futuyma, Douglas. *Science on Trial: The Case for Evolution.* Sunderland, MA: Sinauer Associates Inc., 1995.

Gilkey, Langdon. *Creationism on Trial: Evolution and God at Little Rock.* Minneapolis, MN: Winston Press, 1985.

Gillespie, Charles C. *Genesis and Geology.* Cambridge, MA: Harvard University Press, 1951.

Gish, Duane. *Creation, Evolution, and Public Education.* Institute for Creation Research.

Gould, Stephen J. *Rocks of Ages: Science and Religion in the Fullness of Life.* New York: Ballantine Publishing Group, 1999.

Green, John C. *The Death of Adam.* Ames: Iowa State University Press, 1959.

Hofstadter, Richard. *Social Darwinism in American Thought.* New York: Braziller, 1959.

Hull, David L. *Darwin and His Critics.* Cambridge, MA: Harvard University Press, 1973.

Johnson, Phillip. *Defeating Darwinism by Opening Minds.* Downers Grove, IL: InterVarsity Press, 1997.

———. *The Wedge of Truth: Splitting the Foundations of Naturalism.* Downers Grove, IL: InterVarsity Press, 2000.

Kitcher, Phillip. *Abusing Science: The Case against Creationism.* Cambridge, MA: Harvard University Press, 1982.

Larson, Edward J. *Trial and Error: The American Controversy over Creation and Evolution.* New York: Oxford University Press, 1985.

———. *Summer for the Gods: The Scopes Trial and America's Continuing Debate over Science and Religion.* New York: Basic Books, 1997.

———. *Evolution's Workshop: God and Science on the Galapagos Islands.* New York: Basic Books, 2001.

Lovejoy, Arthur O. *The Great Chain of Being.* Cambridge, MA: Harvard University Press, 1936.

Morris, Henry M. *The Long War against God: The History and Impact of the Creation/Evolution Conflict.* Grand Rapids, MI: Baker Book House, 1989.

Morris, Henry M. et al. *Scientific Creationism.* San Diego, CA: Creation-Life Publishers, 1974.

Numbers, Ronald L. *The Creationists: The Evolution of Scientific Creationism.* Berkeley: University of California, 1992.

———. *Darwinism Comes to America.* Cambridge, MA: Harvard University Press, 1998.

Numbers, Ronald L. and J. Stenhouse. *Disseminating Darwinism: The Role of Place, Race, Religion, and Gender.* Cambridge: Cambridge University Press, 1999.

Olshansky, S.J., B. A. Carnes, and R.N. Butler. "If Humans Were Built to Last." *Scientific American* 284 (2001): 50–55.

Ospovot, Dov. *The Development of Darwin's Theory: Natural History, Natural Theology, and Natural Selection, 1838–1859.* Cambridge: Cambridge University Press, reissue 1995.

Pennock, Robert. *Tower of Babel: The Evidence against the New Creationism.* Cambridge, MA: MIT Press, 1999.

Petto, Andrew J. and Laurie R. Godfrey, eds. *Scientists Confront Intelligent Design and Creationism.* New York: W.W. Norton and Company, 2007.

Pigliucci, Massimo. *Tales of the Rational: Skeptical Essays about Nature and Science.* Atlanta, GA: Freethought Press, 2000.

———. *Denying Evolution: Creationism, Scientism, and the Nature of Science.* Sunderland, MA: Sinauer Associates Inc., 2002.

Richards, Robert. *The Meaning of Evolution: The Morphological Construction and Ideological Reconstruction of Darwin's Theory*. Chicago: University of Chicago Press, 1992.

Ruse, Michael. *The Darwinian Revolution: Science Red in Tooth and Claw*. 2nd ed. Chicago: University of Chicago Press, 1999.

———. *Can a Darwinian Be a Christian? The Relationship between Science and Religion*. Cambridge: Cambridge University Press, 2001.

———. *Darwin and Design: Does Evolution Have a Purpose?* Cambridge, MA: Harvard University Press, 2003.

———. *The Evolution-Creation Struggle*. Cambridge, MA: Harvard University Press, 2005.

———, ed. *But Is It Science? The Philosophical Question in the Creation/Evolution Controversy*. Buffalo: Prometheus, 1988.

Scott, Eugenie C. *Evolution vs. Creationism: An Introduction*. Westport, CT: Greenwood Press, 2004.

Shermer, Michael. *How We Believe: The Search for God in an Age of Science*. New York: W.H. Freeman, 2000.

Sulloway, Frank J. "Darwin and His Finches: The Evolution of a Legend." *Journal of the History of Biology* 15 (1982): 1–53.

Webb, G.E. *The Evolution Controversy in America*. Lexington: University Press of Kentucky, 1994.

Wells, Jonathan. *Icons of Evolution: Science or Myth?: Why Much of What We Teach about Evolution Is Wrong*. Washington, DC: National Book Network, 2000.

Whitcomb, J.C. and H.M. Morris. *The Genesis Flood: The Biblical Record and Its Scientific Implications*. Philadelphia, PA: Presbyterian and Reformed Publishing Company, 1961.

Wilson, Edward O. *The Creation: An Appeal to Save Life on Earth*. New York: Norton, 2006.

INDEX

Index

About the Authors

CHRISTIAN C. YOUNG teaches introductory biology, evolution, environmental studies, and the relationship between science and society at Alverno College in Milwaukee, Wisconsin. He is assistant professor of biology and a member of the department for Developing a Global Perspective. His research focuses on social controversies about and within science, particularly over environmental topics.

MARK A. LARGENT is assistant professor of science policy in James Madison College at Michigan State University, where he teaches courses on the history of science and U.S. science policy. His research and writing focuses on the history of biology, in particular the evolution/creation debates and the history of the American eugenics movement. Trained as a historian of science and technology, his work explores the role of American biologists in various political and social movements as well as the impact of science on policy debates in the early twentieth century.